D1264691

An Introduction to the Theory of Real Functions

An Introduction to the Theory of Real Functions

Stanisław Łojasiewicz

Translated by
G. H. Lawden (University of Sussex)
Edited by
A. V. Ferreira (University of Bologna)

A Wiley–Interscience Publication

JOHN WILEY & SONS
Chichester · New York · Brisbane · Toronto · Singapore

Copyright © 1988 by John Wiley & Sons Ltd.
Originally published as Wstep do teorii funkcji rzeczywistych,
© copyright by Państwowe Wydawnictwo Naukowe, Warszawa

All rights reserved.

No part of this book may be reproduced by any means,
or transmitted, or translated into a machine language
without the written permission of the publisher

Library of Congress Cataloging-in-Publication Data:

Łojasiewicz, Stanisław.
[Wstęp do teorii funkcji rzeczywistych. English]
An introduction to the theory of real functions/Stanisław Łojasiewicz; translated by G. H. Lawden; edited by A. V. Ferreira.
p. cm.
Translation of: Wstep do teorii funckji rzeczywistych.
'A Wiley–Interscience publication.'
Bibliography: p.
Includes index.
ISBN 0 471 91414 2
1. Functions. I. Ferreira, A. V. (A. Vaz) II. Title.
QA331.L7913 1988 87–30432
515—dc19 CIP

British Library Cataloguing in Publication Data:

Łojasiewicz, Stanisław
An introduction to the theory of real functions.
1. Calculus 2. Functions of real variables
I. Title II. Ferreira, A. V.
515.8 QA331.5
ISBN 0 471 91414 2

Typeset by Thomson Press (India) Ltd
Printed and bound in Great Britain by Anchor Brendon Ltd, Tiptree, Essex

CONTENTS

Preface ix

Introduction **1**
0.1 Infima and Suprema 2
0.2 Limits 4
0.3 The 'Diagonal' Principle of Choice 8
Notes 9

Chapter 1 Functions of Bounded Variation **10**
1.1 Monotone Functions 10
1.2 Sequences of Monotone Functions 12
1.3 The Variation of a Function 14
The Relationship with Arc-length 17
1.4 Jordan Canonical Decomposition 18
1.5 The Riemann–Stieltjes Integral. Existence Criteria 20
1.6 Properties of the Integral 26
Taking the Limit under the Integral Sign 29
Notes 30

Chapter 2 Approximation of Continuous Functions by Polynomials **31**
2.1 Weierstrass' Theorem 31
2.2 Tonelli Polynomials 31
2.3 Bernstein Polynomials 33
Application to Absolutely Monotone Functions 35
2.4 The Stone–Weierstrass Theorem 39
Notes 43

Chapter 3 Functions on Metric Spaces **45**
3.1 Continuous Functions 45
Uniform and Continuous Convergence 47
3.2 Equicontinuous Families of Functions 49
3.3 Semicontinuous Functions 51
3.4 Maximum and Minimum at a Point 55
3.5 Functions of the First Class of Baire 57
Notes 64

Chapter 4 Algebras of Sets and Measurable Functions 65
4.1 Algebras of Sets 65
4.2 Cartesian Products 68
4.3 Borel Sets 69
Classification of Borel Sets 71
4.4 Measurable Functions 73
4.5 Baire Functions 78
Notes 80

Chapter 5 Measure and Measurable Functions 81
5.1 Measure 81
Sets of Measure Zero 82
5.2 Outer Measure 85
Outer Measure on a Countably Additive Ideal 91
Extension of a Finitely Additive Function to a Measure 91
5.3 Content of an Interval 92
5.4 Lebesgue Measure 96
A Characterization of the Modulus of a Determinant 104
Non-Measurable Sets 107
5.5 Measurable Functions 108
$\mathscr{L}_n$-measurable Functions 109
5.6 Sequences of Measurable Functions 111
Notes 115

Chapter 6 Integration 117
6.1 Integration of Non-negative Functions 117
The Banach–Vitali Theorem 123
6.2 Integration of Functions of Arbitrary Sign. Summability 124
Lebesgue Sums 130
Relationship of Lebesgue Measure and Integral with the Riemann Integral 130
6.3 Cartesian Product of Measures. Fubini's Theorem 132
The case of Lebesgue Measure and Integrals 136
6.4 Riesz's Theorem 139
Proof of Riesz's Theorem 140
Uniqueness of Measure in Riesz's Theorem 142
Notes 143

Chapter 7 Differentiation 145
7.1 Differentiability Almost Everywhere 145
Rademacher's Theorem 152
7.2 Interval Functions of Bounded Variation 156
7.3 Absolutely Continuous Interval Functions 160
7.4 The case of Functions of a Single Variable 163
Arc Length 170
Dini Derivatives 172
Non-differentiable Continuous Functions 176
7.5 Countably Additive Set Functions 178
Riesz's Theorem 183

Connection with Interval Functions 185
Lebesgue–Stieltjes Integral 190
7.6 The Radon–Nikodym Theorem 192
Change of Variable in a Lebesgue Integral 199
Notes 201

Appendix 1 Another Proof of Lebesgue's Theorem on the Differentiability of the Integral 204
Appendix 2 Another Proof of Fubini's Theorem for Lebesgue Integrals 205
Appendix 3 Stepanov's Theorem on Differentiability Almost Everywhere 208
Appendix 4 Proofs of the Sierpiński–Young and Denjoy–Young–Saks Theorems 210

Exercises (Compiled by M. Kosiek, W. Mlak and Z. Opial.) 211

Bibliography 226

Subject Index 227

PREFACE

This work is a slightly modified and enlarged[1] version of a text with the same title intended for mathematics students and containing material from the lectures on Real Functions which I gave for a number of years at Jagiello University (1st edition (1957), 2nd edition (1964), Jagiello University). It is supplemented by a set of exercises compiled by M. Kosiek, W. Mlak and Z. Opial.

It assumes a familiarity with the elements of set theory, topology and the differential and integral calculus. The topological context of the lectures was that of metric spaces which, from the viewpoint of the most natural level of generality, is a blemish. The reader himself will be able to determine what type of topological space can be substituted for a metric space in any particular argument or theorem and can carry out the necessary modifications.

The book contains classical material and aims to provide a concise account of the most typical theorems. In writing the text I made use of the handbooks and monographs of the authors named in the references given at the end of the book: G. Aumann, N. Bourbaki, H. Hahn and A. Rosenthal, P. Halmos, O. Haupt and G. Aumann, K. Kuratowski, L. P. Natanson, F. Riesz and B. Sz. Nagy, S. Saks, W. Sierpiński and Ch. J. de la Vallée Poussin.

The reader who is interested in studying the material further should read, besides the classical work of S. Saks, the wide-ranging monographs of P. Halmos, R. Sikorski and H. Federer as well as the handbooks of analysis by W. Rudin, I. E. Segal and R. A. Kunz, and K. Maurin. The special case of functions of a single variable can be found, in a concise form, in the book by S. Hartman and J. Mikusiński.

I wish to thank Professor W. Mlak and Professor A. Vaz-Ferreira most warmly for valuable discussions and observations made concerning the incorporated material.

Cracow, July 1986 S. Łojasiewicz

NOTE

1. Additional material includes, among other things, the remarks on the extension of finitely additive functions to a measure (Ch. 5, §2) and a particular construction of a continuous, non-differentiable function of a certain general type (Ch. 7, §4). The Banach–Vitali theorem (Ch. 6, §1) and Stepanov's theorem (Ch. 7, §1) are also new additions.

PREFACE

[illegible] is a slightly modified and enlarged version of a text with the same [illegible] for mathematics students and containing material from the lectures [illegible] which I gave for a number of years at Jagello University [illegible] edition (1973), 2nd ed. [illegible] It is supplemented [illegible] set of exercises [illegible] and Zajac [illegible].

[illegible] familiarity with the elements of [illegible] topology and the differential and integral calculus. [illegible] of the lectures was [illegible] space [illegible] from the viewpoint of [illegible] generality [illegible] will be able to determine [illegible] of [illegible] space [illegible] substituted [illegible] in any particular [illegible] and [illegible].

The book contains [illegible] material and [illegible] provide [illegible] the most typical [illegible] in writing the [illegible] the [illegible] given in the end of the book [illegible] Aronszajn [illegible] Rosenthal [illegible] Halmos [illegible] Riesz and [illegible] and [illegible].

[illegible] interested in studying the material further should read [illegible] the classic [illegible] Halmos, [illegible] as well as [illegible] by W. Rudin, [illegible] Segal and [illegible] The special case of functions of a single variable [illegible] by S. [illegible]

I wish to thank Professor W. Mlak and [illegible] for [illegible] and observations [illegible]

[illegible]

[illegible] 1980 [illegible]

NOTE

[illegible] material [illegible] the contents [illegible] the construction of a [illegible] type [illegible] The Banach [illegible] also new additions.

INTRODUCTION

We will be concerned with functions taking values in the set $\bar{\mathbb{R}}$ made up of the Real Numbers and the two additional elements ∞ and $-\infty$:

$$\bar{\mathbb{R}} = \mathbb{R} \cup \{-\infty\} \cup \{\infty\},$$

where $\mathbb{R}$ denotes the set of Real Numbers. In Chapters 1 and 2 we will have to do with finite functions only, that is, functions taking values in $\mathbb{R}$.

Operations and convergence[1] in $\bar{\mathbb{R}}$ are defined in the usual way, except that we make use of the additional rules:

$$-\infty + \infty = \infty - \infty = 0 \quad \text{and} \quad (\pm\infty)\cdot 0 = 0\cdot(\pm\infty) = 0.^2$$

We also define $|\pm\infty| = \infty$; hence the condition $|\eta| < \infty$ means that the number η is finite.

Convergence in $\bar{\mathbb{R}}$ is convergence relative to a metric, for example

$$\rho(x, y) = |\text{arc tan}\, x - \text{arc tan}\, y|,$$

where arc tan $\pm\infty = \pm\frac{1}{2}\pi$. This metric is the inverse image of the Euclidean metric on $[-\pi/2, +\pi/2]$ by the function $u \to x(u)$, where $x(u) = \tan u$ for $-\pi/2 < u < \pi/2$, and $x(\pm\pi/2) = \pm\infty$. The intervals $(a, \infty]$ are neighbourhoods of the number ∞. Similarly the intervals $[-\infty, a)$ are neighbourhoods of $-\infty$. The unbounded intervals[3] (a, ∞), $(a, \infty]$, $(-\infty, a)$, $[-\infty, a)$ are open but not closed. The intervals $[a, \infty]$, $[-\infty, a]$ are closed but not open. The interval $[-\infty, \infty] = \bar{\mathbb{R}}$ is both open and closed.

Let G be an open set in $\bar{\mathbb{R}}$. The set $(-\infty, \infty) \cap G$ is open in $\mathbb{R}$, and is therefore a union of at most a countable number of disjoint, open intervals. If $\infty \in G$ then G contains an interval of the form $(a, \infty]$ so that the above union contains a term of the form $(a', +\infty)$. Similarly, if $-\infty \in G$ then the union contains a term of the form $(-\infty, a')$. Thus, every open set in $\bar{\mathbb{R}}$ is the union of at most a countable number of disjoint, open intervals.

Note (about terminology): We will denote by f_E the restriction of the function f to the set E. The domain and range of a function are respectively the set of its arguments and the set of its values. The composition of the functions f and g is denoted by $f \circ g$.

0.1 INFIMA AND SUPREMA

For a set $Z \subset \bar{\mathbb{R}}$ we denote its greatest lower bound and its least upper bound by $\inf Z$ and $\sup Z$ respectively. Let φ be a real-valued function whose argument x varies in a certain set E, let $w(x)$ be a condition imposed on x and let A be the set of all $x \in E$ which satisfy the condition $w(x)$. We introduce the notation:

$$\inf_{w(x)} \varphi(x) = \inf_A \varphi = \inf \varphi(A)$$

and

$$\sup_{w(x)} \varphi(x) = \sup_A \varphi = \sup \varphi(A).$$

We now present the fundamental properties of inf and sup:

$$\inf_A \varphi \leqslant \varphi(x) \leqslant \sup_A \varphi \qquad \text{if } x \in A; \tag{0.1.1}$$

moreover there are sequences $x_n \in A$, $z_n \in A$ such that

$$\varphi(x_n) \to \inf_A \varphi \quad \text{and} \quad \varphi(z_n) \to \sup_A \varphi.$$

$$\left.\begin{aligned} \varphi(x) &\geqslant \alpha \quad \text{for } x \in A \Leftrightarrow \inf_A \varphi \geqslant \alpha, \\ \varphi(x) &\leqslant \beta \quad \text{for } x \in A \Leftrightarrow \sup_A \varphi \leqslant \beta. \end{aligned}\right\} \tag{0.1.2}$$

From this it follows that

$$\left.\begin{aligned} &\sup_A \varphi > \alpha \Leftrightarrow \varphi(x) > \alpha \quad \text{for some } x \in A, \\ &\inf_A \varphi < \alpha \Leftrightarrow \varphi(x) < \alpha \quad \text{for some } x \in A; \end{aligned}\right\} \tag{0.1.3}$$

$$\left.\begin{aligned} &\sup_A \varphi \geqslant \alpha \Leftrightarrow \quad \text{for all } \alpha' < \alpha, \text{ there exists} \\ &\qquad x \in A \text{ such that } \varphi(x) > \alpha', \\ &\inf_A \varphi \leqslant \alpha \Leftrightarrow \quad \text{for all } \alpha' > \alpha, \text{ there exists} \\ &\qquad x \in A \text{ such that } \varphi(x) < \alpha'.^4 \end{aligned}\right\} \tag{0.14}$$

Next we have the so-called principle of taking suprema and infima on inequalities:

$$\varphi(x) \leqslant \psi(x) \quad \text{for all} \quad x \in A \Rightarrow \sup_A \varphi \leqslant \sup_A \psi \quad \text{and} \quad \inf_A \varphi \leqslant \inf_A \psi. \tag{0.1.5}$$

Indeed, by (0.1.1), $\varphi(x) \leqslant \sup_A \psi$ for all $x \in A$, and so, by (0.1.2) $\sup_A \varphi \leqslant \sup_A \psi$.

$$A_1 \subset A_2 \Rightarrow \inf_{A_1} \varphi \geqslant \inf_{A_2} \varphi \quad \text{and} \quad \sup_{A_1} \varphi \leqslant \sup_{A_2} \varphi, \tag{0.1.6}$$

for in this case $\varphi(A_1) \subset \varphi(A_2)$.

Let λ be a finite or non-finite function which is monotonic[5] and continuous in the interval θ which is closed in $\bar{\mathbb{R}}$ (e.g. $\theta = [-\infty, \infty]$ and $\lambda(t) = \arctan t$). If λ is increasing then for non-void $Z \subset \theta$ we have $\lambda(\sup Z) = \sup \lambda(Z)$. Indeed, there exists a sequence $\xi_n \in Z$ such that $\xi_n \to \sup Z$, so $\sup Z \in \theta$ and

$\lambda(\xi_n)\to\lambda(\sup Z)$. But since $\lambda(\xi_n)\leqslant\sup\lambda(Z)$ (for $\lambda(\xi_n)\in\lambda(Z)$), we have $\lambda(\sup Z)\leqslant \sup\lambda(Z)$. The reverse inequality follows from the fact that $\lambda(\xi)\leqslant\lambda(\sup Z)$ for $\xi\in Z$.

Similarly, $\lambda(\inf Z)=\inf\lambda(Z)$, whereas, if λ is decreasing, $\lambda(\sup Z)=\inf\lambda(Z)$ and $\lambda(\inf Z)=\sup\lambda(Z)$. If, therefore, $\varphi(A)\subset\theta$, then

$$\left.\begin{aligned}&\lambda \text{ increasing}\Rightarrow\sup_A\lambda(\varphi(x))=\lambda\left(\sup_A\varphi(x)\right),\inf_A\lambda(\varphi(x))=\lambda\left(\inf_A\varphi(x)\right),\\&\lambda \text{ decreasing}\Rightarrow\sup_A\lambda(\varphi(x))=\lambda\left(\inf_A\varphi(x)\right),\inf_A\lambda(\varphi(x))=\lambda\left(\sup_A\varphi(x)\right).\end{aligned}\right\}\tag{0.1.7}$$

In particular, taking $\lambda(t)=\alpha t$ or $\lambda(x)=\beta+x$ ($|\alpha|,|\beta|<\infty$), we obtain

$$\left.\begin{aligned}&\sup_A\alpha\varphi=\alpha\sup_A\varphi,\quad \inf_A\alpha\varphi=\alpha\inf_A\varphi,\qquad \text{if } \alpha\geqslant 0,\\&\sup_A\alpha\varphi=\alpha\inf_A\varphi,\quad \inf_A\alpha\varphi=\alpha\sup_A\varphi,\qquad \text{if } \alpha\leqslant 0,\end{aligned}\right\}\tag{0.1.8}$$

$$\sup_A(\beta+\varphi)=\beta+\sup_A\varphi,\quad \inf_A(\beta+\varphi)=\beta+\inf_A\varphi.\tag{0.1.9}$$

It is easily checked that when $|\beta|=\infty$, the first of the formulae (0.1.9) remains true except in the case where the right-hand side becomes $-\infty+\infty$, while the second formula is true except when the right-hand side is $\infty-\infty$

Next, we have the inequalities:

$$\sup_A(\varphi+\psi)\leqslant\sup_A\varphi+\sup_A\psi,\quad \inf_A(\varphi+\psi)\geqslant\inf_A\varphi+\inf_A\psi.\tag{0.1.10}$$

Indeed, for $x\in A$ we have $\varphi(x)+\psi(x)\leqslant\sup\varphi+\sup\psi$, which yields the first inequality, using (0.1.2). Similarly, if $\varphi(x)\geqslant 0$ and $\psi(x)\geqslant 0$ in A, then

$$\sup_A(\varphi.\psi)\leqslant\left(\sup_A\varphi\right)\cdot\left(\sup_A\psi\right),\quad \inf_A(\varphi\cdot\psi)\geqslant\left(\inf_A\varphi\right)\cdot\left(\inf_A\psi\right).\tag{0.1.11}$$

We have further

$$\sup_{A\times B}\varphi(x,y)=\sup_{x\in A}\left[\sup_{y\in B}\varphi(x,y)\right],\quad \inf_{A\times B}\varphi(x,y)=\inf_{x\in A}\left[\inf_{y\in B}\varphi(x,y)\right].\tag{0.1.12}$$

Indeed, for $x\in A$, $y\in B$ we have $\varphi(x,y)\leqslant\sup_{\eta\in B}\varphi(x,\eta)\leqslant\sup_{\xi\in A}[\sup_{\eta\in B}\varphi(\xi,\eta)]$, so that

$$\sup_{A\times B}\varphi\leqslant\sup_{x\in A}\left[\sup_{y\in B}\varphi(x,y)\right];$$

also, if $x\in A$, $y\in B$ then $\varphi(x,y)\leqslant\sup_{A\times B}\varphi$, so that

$$\sup_{y\in B}\varphi(x,y)\leqslant\sup_{A\times B}\varphi$$

for $x\in A$, hence

$$\sup_{x\in A}\left[\sup_{y\in B}\varphi(x,y)\right]\leqslant\sup_{A\times B}\varphi.$$

It follows that

$$\left.\begin{aligned}\sup_{A\times B}[\varphi(x)+\psi(y)]&=\sup_A\varphi+\sup_B\psi,\\ \inf_{A\times B}[\varphi(x)+\psi(y)]&=\inf_A\varphi+\inf_B\psi\end{aligned}\right\}\tag{0.1.13}$$

provided the right-hand sides are not of the form $\mp\infty\pm\infty$. For example, in order to obtain the first of these equalities, it suffices to use the first of equations (0.1.9) twice, in the case $\sup_B\psi<\infty$, whereas, when $\sup_B\psi=\infty$ we can check that both sides are equal to $+\infty$ (again making use of the first of the formulae (0.1.9)).

In particular, we conclude that $\sup_{x,y\in A}[\varphi(x)-\varphi(y)]=\sup_A\varphi-\inf_A\varphi$ if the right member is not $\infty-\infty$ and $A\neq\varnothing$. But

$$\sup_{x,y\in A}[\varphi(x)-\varphi(y)]=\sup_{x,y\in A}|\varphi(x)-\varphi(y)|$$

(for the inequality $\leqslant$ is clear, while the inequality $\geqslant$ follows from $\sup_{\xi,\eta\in A}[\varphi(\xi)-\varphi(\eta)]\geqslant|\varphi(x)-\varphi(y)|$ with $x,y\in A$), therefore

$$\sup_{x,y\in A}|\varphi(y)-\varphi(x)|=\sup_A\varphi-\inf_A\varphi,\tag{0.1.14}$$

provided that the right-hand side is not $\infty-\infty$ and $A\neq\varnothing$. This is the so-called *oscillation of the function* φ on the set A.

We also note that when $g(C)=A$, then it follows from the definition that

$$\sup_A\varphi=\sup_C\varphi\circ g,\quad \inf_A\varphi=\inf_C\varphi\circ g.\tag{0.1.15}$$

0.2 LIMITS

Let φ, η be real-valued functions of a variable x which belongs to a certain set E, let α be a real number (finite or infinite), while $w(x)$ is a condition imposed on x. A *cluster value* is a number which is the limit of a sequence $\varphi(x_n)$, where $\{x_n\}$ is a sequence for which $w(x_n)$ holds for each n and $\eta(x_n)\to\alpha$. Let us assume that there exists at least one such sequence $\{x_n\}$. Then, some particular sequence $\varphi(x_{\alpha_n})$ will converge, that is, the set of cluster values will be non-empty. The supremum and infimum of this set will be denoted by

$$\limsup_{\eta(x)\to\alpha,\,w(x)}\varphi(x),\qquad \liminf_{\eta(x)\to\alpha,\,w(x)}\varphi(x)$$

respectively.[6] If both these limits are equal (i.e. if there exists exactly one cluster value), then we denote their common value β by $\lim_{\eta(x)\to\alpha,\,w(x)}\varphi(x)$ and we say that $\varphi(x)\to\beta$ as $\eta(x)\to\alpha$ while the condition $w(x)$ is satisfied. This occurs if and only if it is the case that for every sequence $\{x_n\}$ such that $w(x_n)$ holds and $\eta(x_n)\to\alpha$, we have $\varphi(x_n)\to\beta$.[7] If, for example, $\sum(P)$ denotes the approximating sum for the Riemann integral of the function f, constructed for the interval P with intermediate points: $a=t_0<\cdots<t_n=b$, $t_{i-1}\leqslant\tau_i\leqslant t_i$, and $\delta(P)=$

$\max_i |t_i - t_{i-1}|$, then the definition of the Riemann integral can be written

$$\int_a^b f(t)\,\mathrm{d}t = \lim_{\delta(P)\to 0} \sum(P).$$

In the above definitions we may write $\eta(x)\to\alpha+$ instead of $\eta(x)\to\alpha$ and $\eta(x)>\alpha$, and $\eta(x)\to\alpha-$ instead of $\eta(x)\to\alpha$ and $\eta(x)<\alpha$. If $|\alpha|<\infty$, then $\eta(x)\to\alpha$ is equivalent to $\eta(x)-\alpha\to 0$. But, when $\alpha=\infty$ or $-\infty$ then $\eta(x)\to\alpha$ is equivalent to $1/\eta(x)\to 0+$ or $1/\eta(x)\to 0-$, respectively. It therefore suffices to consider only the case $\alpha=0$. If, instead of the set E we take the subset $E_0=\{x: w(x)\}$, and if we replace the functions φ and η by their restrictions to E_0, we can omit the condition $w(x)$ beneath the limit sign.

Let

$$A_\delta = \{x: |\eta(x)| < \delta, w(x)\}.$$

We have, by assumption, that $A_\delta \neq \varnothing$ for $\delta > 0$. Also $0<\delta\leqslant\delta'$ implies that $A_\delta \subset A_{\delta'}$. We will prove that

$$\left.\begin{aligned} \limsup_{\eta(x)\to 0,\, w(x)} \varphi(x) &= \inf_{\delta>0}\left(\sup_{A_\delta}\varphi\right), \\ \liminf_{\eta(x)\to 0,\, w(x)} \varphi(x) &= \sup_{\delta>0}\left(\inf_{A_\delta}\varphi\right). \end{aligned}\right\} \tag{0.2.1}$$

Let us show, namely, that

$$\lambda_0 = \inf_{\delta>0}\left(\sup_{A_\delta}\varphi\right)$$

is the largest cluster value.

We have $\sup_{A_{\delta'_n}}\varphi\to\lambda_0$ for some sequence $\delta'_n>0$. Let $\delta_n=\min(1/n, \delta'_n)$. Then

$$\lambda_0 \leqslant \sup_{A_{\delta_n}}\varphi \leqslant \sup_{A_{\delta'_n}}\varphi \to \lambda_0.$$

Let λ_n be a sequence such that $\lambda_n\to\lambda_0$ and $\lambda_n<\lambda_0$ in the case $\lambda_0>-\infty$, and $\lambda_n=-\infty$ in the case $\lambda_0=-\infty$. Thus there exist $x_n\in A_{\delta_n}$ such that $\lambda_n\leqslant\varphi(x_n)\leqslant \sup_{A_{\delta_n}}\varphi$ and so $\varphi(x_n)\to\lambda_0$. But the condition $w(x_n)$ is satisfied and $|\eta(x_n)|<\delta_n\leqslant 1/n$ so that $\eta(x_n)\to 0$. Hence λ_0 is a cluster point. Now let λ be a cluster point; then $\varphi(x_n)\to\lambda$ for some sequence x_n such that $w(x_n)$ holds and $\eta(x_n)\to 0$. Thus, for every $\delta>0$ we have

$$x_n\in A_\delta \quad\text{and}\quad \varphi(x_n)\leqslant\sup_{A_\delta}\varphi$$

for sufficiently large n, whence

$$\lambda\leqslant\sup_{A_\delta}\varphi \quad\text{and}\quad \lambda\leqslant\inf_{\delta>0}\left(\sup_{A_\delta}\varphi\right)=\lambda_0.$$

Hence λ_0 is the largest cluster point.

Similarly, one can show that $\sup_{\delta>0}(\inf_{A_\delta}\varphi)$ is the smallest cluster point.

Let E be a subset of a metric space and let x_0 be a point in this space. Then we define

$$\begin{aligned}\limsup_{x\to x_0}\varphi(x) &= \limsup_{\rho(x,x_0)\to 0+}\varphi(x)\\ \liminf_{x\to x_0}\varphi(x) &= \liminf_{\rho(x,x_0)\to 0+}\varphi(x)\end{aligned}\tag{0.2.2}$$

(x_0 must be an accumulation point of the set E). In the case where $x_0 \notin E$, $\rho(x,x_0)\to 0+$ is equivalent to $\rho(x,x_0)\to 0$. In the case in which $x_0 \in E$, if we replace $\rho(x,x_0)\to 0+$ by $\rho(x,x_0)\to 0$ then the number $\varphi(x_0)$ will be included in the set of cluster points and we have

$$\begin{aligned}\limsup_{\rho(x,x_0)\to 0}\varphi(x) &= \max\left(\varphi(x_0), \limsup_{x\to x_0}\varphi(x)\right),\\ \liminf_{\rho(x,x_0)\to 0}\varphi(x) &= \min\left(\varphi(x_0), \liminf_{x\to x_0}\varphi(x)\right),\end{aligned}\tag{0.2.3}$$

Using the relations (0.2.1) we can obtain a series of properties of limits from the properties of suprema and infima given in §0.1. Thus, from (0.1.3) and (0.1.4) we have

$$\left.\begin{aligned}\limsup_{\eta(x)\to 0}\varphi(x) < \alpha \Leftrightarrow &\text{there exist } \delta>0 \text{ and } \bar{\alpha}<\alpha \text{ such that}\\ &\varphi(x)\leqslant\bar{\alpha} \text{ when } |\eta(x)|<\delta;\\ \liminf_{\eta(x)\to 0}\varphi(x) > \alpha \Leftrightarrow &\text{there exist } \delta>0 \text{ and } \bar{\alpha}>\alpha \text{ such that}\\ &\varphi(x)\geqslant\bar{\alpha} \text{ when } |\eta(x)|<\delta.\end{aligned}\right\}\tag{0.2.4}$$

$$\left.\begin{aligned}\limsup_{\eta(x)\to 0}\varphi(x) \leqslant \alpha \Leftrightarrow &\text{for every } \alpha'>\alpha \text{ there exists } \delta>0 \text{ such}\\ &\text{that } \varphi(x)\leqslant\alpha' \text{ if } |\eta(x)|<\delta;\\ \liminf_{\eta(x)\to 0}\varphi(x) \geqslant \alpha \Leftrightarrow &\text{for every } \alpha'<\alpha \text{ there exists } \delta>0 \text{ such}\\ &\text{that } \varphi(x)\geqslant\alpha' \text{ if } |\eta(x)|<\delta.^8\end{aligned}\right\}\tag{0.2.5}$$

Since the condition $\lim_{\eta(x)\to 0}\varphi(x)=\alpha$ is equivalent to $\limsup_{\eta\to 0}\varphi\leqslant\alpha\leqslant\liminf_{\eta\to 0}\varphi$, it follows from (0.2.5) that when $|\alpha|<\infty$ we obtain Cauchy's definition of a limit: given $\varepsilon>0$ there exists $\delta>0$ such that $|\varphi(x)-\alpha|\leqslant\varepsilon$ if $|\eta(x)|<\delta$.

Next, from the relations (0.1.5)–(0.1.11) we obtain the following properties:

$$\left.\begin{gathered}\varphi(x)\leqslant\psi(x) \quad \text{on } E \text{ implies that}\\ \limsup_{\eta(x)\to 0}\varphi(x)\leqslant\limsup_{\eta(x)\to 0}\psi(x) \quad\text{and}\quad \liminf_{\eta(x)\to 0}\varphi(x)\leqslant\liminf_{\eta(x)\to 0}\psi(x)\end{gathered}\right\}\tag{0.2.6}$$

If the condition $w(x)$ implies the condition $v(x)$, then

$$\limsup_{\eta(x)\to 0,\, w(x)}\varphi(x)\leqslant\limsup_{\eta(x)\to 0,\, v(x)}\varphi(x) \quad\text{and}\quad \liminf_{\eta(x)\to 0,\, w(x)}\varphi(x)\geqslant\liminf_{\eta(x)\to 0,\, v(x)}\varphi(x).\tag{0.2.7}$$

Let λ be a function as in (0.1.7) and let $\varphi(E) \subset \theta$. Then

$$\left.\begin{aligned} \limsup_{\eta\to 0} \lambda(\varphi) &= \lambda\left(\limsup_{\eta\to 0} \varphi\right), \\ \liminf_{\eta\to 0} \lambda(\varphi) &= \lambda\left(\liminf_{\eta\to 0} \varphi\right), \end{aligned}\right\} \text{if } \lambda \text{ is increasing,}$$
$$\left.\begin{aligned} \limsup_{\eta\to 0} \lambda(\varphi) &= \lambda\left(\liminf_{\eta\to 0} \varphi\right), \\ \liminf_{\eta\to 0} \lambda(\varphi) &= \lambda\left(\limsup_{\eta\to 0} \varphi\right), \end{aligned}\right\} \text{if } \lambda \text{ is decreasing} \tag{0.2.8}$$

$$\begin{aligned} &\limsup \alpha\varphi = \alpha \limsup \varphi \quad \text{and} \quad \liminf \alpha\varphi = \alpha \liminf \varphi, \qquad \text{if } \alpha \geqslant 0, \\ &\limsup \alpha\varphi = \alpha \liminf \varphi \quad \text{and} \quad \liminf \alpha\varphi = \alpha \limsup \varphi, \qquad \text{if } \alpha \leqslant 0. \end{aligned} \tag{0.2.9}$$

We have also

$$\begin{aligned} \limsup(\varphi + \psi) &\leqslant \limsup \varphi + \limsup \psi, \\ \liminf(\varphi + \psi) &\geqslant \liminf \varphi + \liminf \psi, \end{aligned} \tag{0.2.10}$$

provided the right-hand sides are not of the form $\mp\infty \pm \infty$. This follows from (0.2.1) and the inequalities (0.1.10) by taking the limit as $\delta \to 0$. Similarly, if $\varphi \geqslant 0$, $\psi \geqslant 0$ then

$$\begin{aligned} \limsup(\varphi\cdot\psi) &\leqslant (\limsup \varphi)\cdot(\limsup \psi), \\ \liminf(\varphi\cdot\psi) &\geqslant (\liminf \varphi)\cdot(\liminf \psi). \end{aligned} \tag{0.2.11}$$

provided the right-hand sides are not of the form $0\cdot\infty$. If $\lim \psi$ exists, then in (0.2.10) and (0.2.11) (under their conditions of validity) we have equality:

$$\begin{aligned} \limsup(\varphi + \psi) &= \limsup \varphi + \lim \psi, \\ \liminf(\varphi + \psi) &= \liminf \varphi + \lim \psi, \end{aligned} \tag{0.2.12}$$

$$\begin{aligned} \limsup(\varphi\cdot\psi) &= \limsup \varphi\cdot\lim \psi, \\ \liminf(\varphi\cdot\psi) &= \liminf \varphi\cdot\lim \psi. \end{aligned} \tag{0.2.13}$$

Indeed, taking x_n such that

$$\eta(x_n) \to 0 \quad \text{and} \quad \varphi(x_n) \to \limsup \varphi$$

we have also $\psi(x_n) \to \lim \psi$ from which it follows that $\limsup \varphi + \lim \psi$ is a cluster value for $\varphi + \psi$, that is we have the reverse inequality $\geqslant$.

By (0.2.1) and (0.1.6), (0.1.14) if φ is a finite function then the existence of a finite limit $\lim \varphi(x)$ is equivalent to:

$$\inf_{\delta > 0}\left(\sup_{x,y \in A_\delta} |\varphi(y) - \varphi(x)|\right) = 0.$$

From this we obtain Cauchy's necessary and sufficient condition:

$$\begin{gathered}\text{given } \varepsilon > 0 \text{ there exists } \delta > 0 \text{ such that} \\ |\varphi(y) - \varphi(x)| \leqslant \varepsilon \quad \text{if} \quad |\eta(x)| < \delta \quad \text{and} \quad |\eta(y)| < \delta.\end{gathered} \tag{0.2.14}$$

0.3 THE 'DIAGONAL' PRINCIPLE OF CHOICE

LEMMA. *Let there be given a sequence of strictly increasing sequences of natural numbers*

$$\{\alpha_\nu^{(1)}\}, \{\alpha_\nu^{(2)}\}, \ldots$$

with the property that each sequence $\{\alpha_\nu^{(k+1)}\}$ is chosen from the preceding sequence $\{\alpha_\nu^{(k)}\}$. Then the sequence $\{\alpha_\nu^{(\nu)}\}$ is strictly increasing and, with the exception of the first $k-1$ terms, it is a subsequence of the sequence $\{\alpha_\nu^{(k)}\}$ $(k = 1, 2\ldots)$.

Proof. Fix k. Since $\{\alpha_\nu^{(k+1)}\}$ is a sequence chosen from the strictly increasing sequence of natural numbers $\{\alpha_\nu^{(k)}\}$, we have $\alpha_\nu^{(k)} \leqslant \alpha_\nu^{(k+1)}$ (proof by induction on ν). Therefore

$$\alpha_\nu^{(\nu)} \leqslant \alpha_\nu^{(\nu+1)} < \alpha_{\nu+1}^{(\nu+1)},$$

so that the sequence $\{\alpha_\nu^{(\nu)}\}$ is strictly increasing. Since $\alpha_s^{(s)}$ is a term of each of the sequences $\{\alpha_\nu^{(1)}\}, \ldots, \{\alpha_\nu^{(s)}\}$, therefore, for $\nu \geqslant k$, $\alpha_\nu^{(\nu)}$ is a term of the sequence $\{\alpha_\nu^{(k)}\}$. ■

THE PRINCIPLE OF CHOICE. *Let there be given a sequence of sequences of numbers*

$$\{\xi_n^{(1)}\}, \{\xi_n^{(2)}\}, \ldots \tag{0.3.1}$$

There exists an increasing sequence of indices $\{n_\nu\}$ such that all the sequences $\{\xi_{n_\nu}^{(1)}\}, \{\xi_{n_\nu}^{(2)}\}, \ldots$ are convergent.

Proof. From the sequence $\{\xi_n^{(1)}\}$ we select a convergent subsequence $\{\xi_{\alpha_\nu^{(1)}}^{(1)}\}$, from the sequence $\{\xi_{\alpha_\nu^{(1)}}^{(2)}\}$ we select a convergent subsequence $\{\xi_{\alpha_\nu^{(2)}}^{(2)}\}$ etc. in such a way that, given the convergent subsequence $\{\xi_{\alpha_\nu^{(k)}}^{(k)}\}$ we select from the sequence $\{\xi_{\alpha_\nu^{(k)}}^{(k+1)}\}$ a convergent subsequence $\{\xi_{\alpha_\nu^{(k+1)}}^{(k+1)}\}$. In this way we obtain a sequence of sequences of indices $\{\alpha_\nu^{(1)}\}, \{\alpha_\nu^{(2)}\}, \ldots$ satisfying the conditions of the lemma. Thus if we put $n_\nu = \alpha_\nu^{(\nu)}$, the sequences

$$\{\xi_{n_\nu}^{(1)}\}, \{\xi_{n_\nu}^{(2)}\}, \{\xi_{n_\nu}^{(3)}\}, \ldots$$

are (after excluding a finite number of terms) subsequences of the sequences

$$\{\xi_{\alpha_\nu^{(1)}}^{(1)}\}, \{\xi_{\alpha_\nu^{(2)}}^{(2)}\}, \ldots,$$

and are therefore convergent. ■

Let f_n be a sequence of functions with numerical values defined on a set Z

which is at most countable. Then this set can be ordered as a sequence $a_1, a_2, \ldots$ and we can take $\xi_n^{(k)} = f_n(a_k)$. We thus have a second form of the 'diagonal' principle of choice:

If a sequence of numerical valued functions f_n is defined on a set Z which is at most countable, then it is possible to select a subsequence f_{n_ν} which is convergent on Z.

NOTES

1. $x_\nu \to \infty$ ($x_\nu \to -\infty$) if for any finite a, $x_\nu > a$ ($x_\nu < a$) for all values of the index ν greater than a certain value: $x_\nu \to g$ (g finite), if, for all except a finite number of indices, the x_ν are finite and $x_\nu \to g$ in the usual sense.
2. The product xy is a discontinuous function at the points $x = 0$, $y = \pm\infty$ and $x = \pm\infty$, $y = 0$! A similar remark applies to sums (for $x = -\infty$, $y = +\infty$ and for $x = +\infty$, $y = -\infty$). Note also that addition and multiplication so defined are commutative; multiplication is also associative. On the other hand addition is not associative (for this reason it is not permissible to take an infinite quantity from one side of an equation to the other after changing its sign) and multiplication is not distributive with respect to addition! Both these rules, i.e. associativity of addition and distributivity of multiplication with respect to addition, hold for the non-negative numbers on their own (whether finite or not).
3. We assume that $|a| < \infty$.
4. Here the strict inequalities $\varphi(x) > \alpha'$ and $\varphi(x) < \alpha'$ can be replaced by non-strict inequalities.
5. See the definitions in §1.1.
6. Clearly, the set of cluster values is identical with the set of all numbers of the form $\limsup_{n\to\infty} \varphi(x_n)$ (and equally with the set of all numbers of the form $\liminf_{n\to\infty} \varphi(x_n)$), where $\{x_n\}$ is a sequence such that $w(x_n)$ holds and $\eta(x_n) \to \alpha$.
7. For $\lim \varphi$ to exist it suffices that $\varphi(x_n)$ should be convergent whenever $w(x_n)$ holds and $\eta(x_n) \to \alpha$; indeed, if the sequences x_n, x_n' satisfy these conditions then so does the sequence $x_1, x_1', x_2, x_2', \ldots$ so that the sequence $\varphi(x_1), \varphi(x_1'), \ldots$ is convergent. Hence $\varphi(x_n)$ and $\varphi(x_n')$ must have the same limit.
8. The non-strict inequalities $\varphi(x) \leqslant \alpha'$ and $\varphi(x) \geqslant \alpha'$ can be replaced here by strict inequalities.

CHAPTER 1

FUNCTIONS OF BOUNDED VARIATION

1.1 MONOTONE FUNCTIONS

A *real function* f of a single real variable, defined on a set Z, is called *increasing* (*decreasing*) on that set, if for $x_1, x_2 \in Z$ such that $x_1 < x_2$ we have $f(x_1) \leqslant f(x_2)$ ($f(x_1) \geqslant f(x_2)$). A function is called *monotone* on a set Z if it is increasing or decreasing on that set.

We use the notation

$$f(a+0) = \lim_{x \to a+} f(x), \quad f(a-0) = \lim_{x \to a-} f(x)$$

for a function f which has a right-hand or left-hand limit at a.

THEOREM 1.1.1. *A function f which is defined and monotone on a set Z has a right-hand (left-hand) limit at every right-hand (left-hand) accumulation point of the set Z. If the function is increasing we have for $x < y$ the inequality*

$$f(x-0) \leqslant f(x) \leqslant f(x+0) \leqslant f(y-0) \leqslant f(y) \leqslant f(y+0) \tag{1.1.1}$$

(*where expressions which are not defined are to be omitted*); *there is a similar inequality for decreasing functions.*

Proof. Let f be an increasing function on Z and let x be a right-hand accumulation point of this set. Put

$$k = \inf\{f(\xi): x < \xi, \xi \in Z\}$$

Now let $\xi_\nu \to x$, $\xi_\nu > x$, $\xi_\nu \in Z$; then $k \leqslant f(\xi_\nu)$. From the definition of infimum, for any $l > k$ there exists $\xi' > x$, $\xi' \in Z$, such that $f(\xi') < l$. For sufficiently large ν we have $\xi_\nu < \xi'$, so that $f(\xi_\nu) \leqslant f(\xi')$, or $k \leqslant f(\xi_\nu) < l$. Hence $f(\xi_\nu) \to k$. We have shown that $f(x+0) = \lim_{\xi \to x+} f(\xi)$ exists and equals k:

$$f(x+0) = \inf\{f(\xi): x < \xi, \xi \in Z\}. \tag{1.1.2}$$

Similarly, when x is a left-hand accumulation point

$$f(x-0) = \sup\{f(\xi): \xi < x, \xi \in Z\}. \tag{1.1.3}$$

Since f is increasing the relations (1.1.2) and (1.1.3) imply that $f(x-0) \leqslant f(x) \leqslant f(x+0)$ and moreover that $f(x+0) \leqslant f(\xi)$ for $\xi > x$, $\xi \in Z$ and $f(\xi) \leqslant f(x-0)$ for $\xi < x$, $\xi \in Z$. Hence, if x is a right-hand accumulation point and y a left-hand accumulation point of Z and $x < y$ then taking $z \in Z$ such that $x < z < y$ we have

$$f(x+0) \leqslant f(z) \leqslant f(y-0). \quad \blacksquare$$

It follows from this theorem that if, at points of discontinuity of f in an open interval, we replace the value $f(x)$ by an arbitrary number in the interval $[f(x-0), f(x+0)]$, then the function is still monotone. Clearly the condition

$$f(x-0) = f(x+0)$$

is a necessary and sufficient condition for f to be continuous at x.

THEOREM 1.1.2. *A monotone function has at most a countable number of points of discontinuity.*

Proof. For definiteness assume that the function f is increasing in the interval (a, b) and let D be the set of points of discontinuity of f. With each $x \in D$ we associate a rational number w_x such that

$$f(x-0) < w_x < f(x+0).$$

From the inequality (1.1.1) it follows that if x, $y \in D$ and $x < y$ then $w_x < w_y$. Hence the mapping $x \to w_x$ is a bijection. Thus, the set D has the same cardinality as a subset of the rational numbers. Hence D is at most countable. $\blacksquare$

COROLLARY 1.1.1. *If f is a monotone function in the interval (a, b) and if $\{x_\nu\}$ is a finite or infinite sequence of distinct points in this interval, then*

$$\sum_\nu |f(x_\nu+0) - f(x_\nu-0)| \leqslant |f(b-0) - f(a+0)| \tag{1.1.4}$$

In particular $\{x_\nu\}$ may be a sequence of all the points of discontinuity of the function $f(x)$.

THEOREM 1.1.3. *A function f increasing (decreasing) in the interval $[a, b]$ may be expressed as a sum*

$$f(x) = g(x) + s(x)$$

of a continuous increasing (decreasing) function g and a saltus-function $s(x) = \sum_\nu u_\nu(x)$, where $u_\nu(x) = 0$ for $x < x_\nu$, $u_\nu(x_\nu) = f(x_\nu) - f(x_\nu - 0)$, $u_\nu(x) = f(x_\nu+0) - f(x_\nu-0)$ for $x > x_\nu$, and $\{x_\nu\}$ is a sequence of all the points of discontinuity of the function f, (where we put $f(a-0) = f(a)$, $f(b+0) = f(b)$).

Proof. The inequality

$$|u_\nu(x)| \leqslant |f(x_\nu+0) - f(x_\nu-0)|$$

together with the inequality (1.1.4) imply that the sum $s(x)=\sum_\nu u_\nu(x)$ is a uniformly convergent series, in the case where the number of terms is infinite. We write

$$g(x)=f(x)-s(x).$$

If $\bar{x}$ is a point of continuity of the function f, then $\bar{x}\neq x_\nu$ for all ν. Hence all u_ν are continuous at $\bar{x}$ and so s is continuous at $\bar{x}$ and hence g is also.

Let $\bar{x}=x_{\nu_0}$ for some ν_0. We have

$$g(x)=f(x)-u_{\nu_0}(x)-\sum_{\nu\neq\nu_0}u_\nu(x).$$

We know that $\sum_{\nu\neq\nu_0}u_\nu(x)$ is continuous at $\bar{x}$. The function $h(x)=f(x)-u_{\nu_0}(x)$ is also continuous at $\bar{x}$, for it is easily seen that

$$h(\bar{x}-0)=h(\bar{x})=h(\bar{x}+0)=f(x_{\nu_0}-0).$$

Hence g is continuous at $\bar{x}$. We have therefore proved that the function g is continuous in the interval $[a,b]$ and it remains to show that it is increasing, if f is increasing.

Let $a\leqslant x<y\leqslant b$. The difference $u_\nu(y)-u_\nu(x)$ takes values in the cases $x_\nu<x$, $x_\nu=x$, $x<x_\nu<y$, $x_\nu=y$ or $x_\nu>y$ which are respectively equal to 0, $f(x+0)-f(x)$, $f(x_\nu+0)-f(x_\nu-0)$, $f(y)-f(y-0)$ or 0. Hence, using inequality (1.1.4),

$$\begin{aligned}s(y)-s(x)&=\sum_\nu(u_\nu(y)-u_\nu(x))\\&\leqslant f(x+0)-f(x)+f(y-0)-f(x+0)+f(y)-f(y-0)\\&=f(y)-f(x).\end{aligned}$$

Therefore $g(y)-g(x)\geqslant 0$, as required. Similarly if f is decreasing. ■

EXTENSION LEMMA. *If the function f is increasing (decreasing) on the set Z and $a=\inf Z$, $b=\sup Z$, then there exists a function $\bar{f}$ increasing (decreasing) on (a,b) and such that $\bar{f}(x)=f(x)$ for $x\in Z$.*

Proof. It suffices to define

$$\bar{f}(x)=\sup_{(a,x]\cap Z}f\left(\text{respectively } \bar{f}(x)=\inf_{(a,x]\cap Z}f\right). \quad ■$$

1.2 SEQUENCES OF MONOTONE FUNCTIONS

The limit of a convergent sequence of functions which are monotone on Z is a monotone function on Z (for non-strict inequalities are preserved in the limit).

THEOREM 1.2.1 (Helly's First Theorem). *Given a sequence of increasing (decreasing) functions on an interval Δ which is bounded at each point of the interval, there exists a subsequence which is convergent on the whole interval Δ.*

Proof. For the sake of definiteness assume that the functions f_n are increasing in the interval $\Delta=(a,b)$ and that for each $\xi\in(a,b)$ the sequence $f_n(\xi)$ is bounded.

Let Z be a countable set which is dense in (a,b); for example, it could be the set of rational numbers in this interval. Since the sequence $\{f_n(x)\}$ is bounded at each point of the set Z, we can use the diagonalization principle to choose a subsequence $\{f_{\alpha_\nu}(x)\}$ which is convergent for all $x\in Z$. The limit function $\lim_{\nu\to\infty} f_{\alpha_\nu}(x)$ is an increasing function in Z. By the Extension Lemma in §1.1 there is a function g increasing in (a,b) and such that

$$g(x)=\lim_{\nu\to\infty} f_{\alpha_\nu}(x) \quad \text{in } Z.$$

Let x be any point in (a,b). There exist sequences $\{u_n\}$, $\{v_n\}$ of points in Z such that

$$u_n<x<v_n, \quad u_n\to x,\ v_n\to x.$$

Fix n. Then, since f_{α_ν} are increasing functions, we have

$$f_{\alpha_\nu}(u_n)\leqslant f_{\alpha_\nu}(x)\leqslant f_{\alpha_\nu}(v_n),$$

and so, proceeding to the limit as $\nu\to\infty$,

$$g(u_n)\leqslant \liminf_{\nu\to\infty} f_{\alpha_\nu}(x)\leqslant \limsup_{\nu\to\infty} f_{\alpha_\nu}(x)\leqslant g(v_n).$$

Now, proceeding to the limit as $n\to\infty$ we obtain

$$g(x-0)\leqslant \liminf_{\nu\to\infty} f_{\alpha_\nu}(x)\leqslant \limsup_{\nu\to\infty} f_{\alpha_\nu}(x)\leqslant g(x+0).$$

From these inequalities it follows that when x is a point of continuity of g then $g(x)=\lim_{\nu\to\infty} f_{\alpha_\nu}(x)$. Hence, the sequence $\{f_{\alpha_\nu}(x)\}$ is convergent in (a,b) outside the set D of points of discontinuity of the function g. But, by Theorem 1.1.2 the set D is at most countable. By hypothesis, the sequence $\{f_{\alpha_\nu}\}$ is bounded at each point of D. Hence, we may use the diagonalization principle to choose a subsequence $\{f_{\beta_\nu}\}$ which converges at each point of D. Since this sequence is taken from the sequence $\{f_{\alpha_\nu}\}$ it must also be convergent at every point in $(a,b)\backslash D$ and hence in the whole interval (a,b). ■

THEOREM 1.2.2. *A sequence of increasing (decreasing) functions f_n defined on $[a,b]$ which converges on a dense subset Z of $[a,b]$ containing a and b to a function f continuous on $[a,b]$ is uniformly convergent on $[a,b]$.*

Proof. Let $\varepsilon>0$. The function f which is continuous on the closed interval $[a,b]$ must be uniformly continuous on $[a,b]$. Hence we can choose a subdivision $a=x_0<x_1<\cdots<x_p=b$, $x_k\in Z$, sufficiently fine so that

$$|f(x_k)-f(x_{k-1})|\leqslant \tfrac{1}{2}\varepsilon \qquad (k=1,\ldots,p).$$

Since $\lim_{\nu\to\infty} f_\nu(x_k)=f(x_k)$ there is an N such that for $\nu\geqslant N$ and $k=0,\,1,\ldots,p$

we have

$$|f_v(x_k) - f(x_k)| < \tfrac{1}{2}\varepsilon.$$

Suppose the functions f_n are increasing.

Now let $\xi \in [a, b]$ be chosen arbitrarily. We must have $x_{k-1} \leqslant \xi \leqslant x_k$ for some k. Since the functions f_v and the function f are increasing, we have, for $v \geqslant N$, the inequalities

$$f(x_{k-1}) - \tfrac{1}{2}\varepsilon \leqslant f_v(x_{k-1}) \leqslant f_v(\xi) \leqslant f_v(x_k) \leqslant f(x_k) + \tfrac{1}{2}\varepsilon, \quad f(x_{k-1}) \leqslant f(\xi) \leqslant f(x_k)$$

which together imply

$$|f_v(\xi) - f(\xi)| \leqslant f(x_k) - f(x_{k-1}) + \tfrac{1}{2}\varepsilon \leqslant \varepsilon,$$

as required. Similarly if f in decreasing ∎

1.3 THE VARIATION OF A FUNCTION

Let f be a function defined on the interval $[a, b]$. The number

$$W_a^b(f) = \sup_{a = x_0 < \cdots < x_n = b} \sum_1^n |f(x_i) - f(x_{i-1})| \tag{1.3.1}$$

(finite or infinite) is called the *variation of the function* f on the interval $[a, b]$. We say that f is a *function of bounded variation* on $[a, b]$, if $W_a^b(f) < \infty$.

It follows from the definition of variation that a change of variable through a continuous and strictly increasing function[1]

$$x = \varphi(t), \quad \alpha \leqslant t \leqslant \beta, \quad a = x(\alpha), \quad b = x(\beta)$$

does not affect the value of the variation: $W_a^b(f) = W_\alpha^\beta\,(f \circ \varphi)$. For we have a bijection between the set of subdivisions $\alpha = t_0 < \cdots < t_n = \beta$ and the set of subdivisions $a = x_0 < \cdots < x_n = b$ defined by $x_i = \varphi(t_i)$. Hence (taking account of (0.1.15)) the suprema of the sums

$$\sum_1^n |f(x_i) - f(x_{i-1})| \quad \text{and} \quad \sum_1^n |f(\varphi(t_i)) - f(\varphi(t_{i-1}))|$$

are equal. The function f on $[a, b]$ is of bounded variation if and only if the function $f \circ \varphi$ on $[\alpha, \beta]$ is of bounded variation.

A monotone function on $[a, b]$ is a function of bounded variation on $[a, b]$ for in this case the sums on the right-hand side of (1.3.1) all have the same value $|f(b) - f(a)|$ that is $W_a^b(f) = |f(b) - f(a)|$.

Thus a function of bounded variation can be discontinuous. On the other hand, a continuous function, and even a function differentiable on the whole interval $[a, b]$, can have a variation equal to $+\infty$. For example, this is the case for the function f defined on the interval $[0, 1]$ by the formulae

$$f(0) = 0, \quad f(x) = x^2 \cos(\pi/x^2).$$

But, if f has a continuous derivative, or more generally, if f satisfies a *Lipschitz condition* on $[a, b]$, that is, there is a constant M such that

$$|f(x')-f(x)| \leqslant M|x'-x| \qquad \text{for } x,\ x' \in [a, b],$$

then f is a function of bounded variation on $[a, b]$. For in this case

$$\sum_{i=1}^{n} |f(x_i)-f(x_{i-1})| \leqslant \sum_{i=1}^{n} M|x_i - x_{i-1}| = M(b-a),$$

so that taking the supremum gives

$$W_a^b(f) \leqslant M(b-a).$$

THEOREM 1.3.1. *The variation is an additive function of interval:*

$$W_a^c(f) = W_a^b(f) + W_b^c(f), \quad a<b<c. \tag{1.3.2}$$

Proof. If $a = x_0 < \cdots < x_m = b = y_0 < \cdots < y_n = c$, then

$$\sum_{i=1}^{m} |f(x_i)-f(x_{i-1})| + \sum_{j=1}^{n} |f(y_j)-f(y_{j-1})| \leqslant W_a^c(f).$$

In this inequality take the supremum first over the first sum and then over the second sum. This gives

$$W_a^b(f) + W_b^c(f) \leqslant W_a^c(f).$$

Thus it suffices to prove the reverse inequality. If $a = x_0 < \cdots < x_n = c$ then $x_{k-1} \leqslant b \leqslant x_k$ for some k. We then have

$$\begin{aligned}\sum_{i=1}^{n} |f(x_i)-f(x_{i-1})| &\leqslant |f(x_1)-f(x_0)| + \cdots + |f(b)-f(x_{k-1})| + |f(x_k)-f(b)| \\ &\quad + \cdots + |f(x_n)-f(x_{n-1})| \\ &\leqslant W_a^b(f) + W_b^c(f).\end{aligned}$$

Taking the supremum of the left-hand side we have

$$W_a^c(f) \leqslant W_a^b(f) + W_b^c(f). \quad \blacksquare$$

COROLLARY 1.3.1. *The variation is a monotonic function of interval:*

$$W_c^d(f) \leqslant W_a^b(f) \qquad \text{if } a \leqslant c < d \leqslant b. \tag{1.3.3}$$

COROLLARY 1.3.2. *A function of bounded variation on the intervals $[a, b]$ and $[b, c]$ is of bounded variation on their union $[a, c]$. A function of bounded variation on the interval $[a, b]$ is of bounded variation on every sub-interval of $[a, b]$.*

It follows from the definition of variation (1.3.1) that $|f(d)-f(c)| \leqslant W_c^d(f)$, and hence, from (1.3.3),

$$|f(b)-f(a)| \leqslant \operatorname*{osc}_{[a,b]} f \leqslant W_a^b(f), \tag{1.3.4}$$

where $\operatorname{osc}_{[a,b]} f = \sup_{a \leqslant x \leqslant y \leqslant b} |f(y)-f(x)|$.

Further

$$|f(x)| \leqslant |f(a)| + W_a^b(f) \qquad \text{for } a \leqslant x \leqslant b \tag{1.3.5}$$

and we have

THEOREM 1.3.2. *A function of bounded variation on $[a,b]$ is bounded on $[a,b]$.*

Let the functions f, g defined on the interval $[a,b]$ take values in the sets Δ_1, Δ_2 and let F be a function satisfying a *Lipschitz condition* on $\Delta_1 \times \Delta_2$

$$|F(\bar{u},\bar{v}) - F(u,v)| \leqslant M(|\bar{u}-u| + |\bar{v}-v|).$$

Let $\varphi(x) = F(f(x), g(x))$. Then

$$\sum_{i=1}^{n} |\varphi(x_i) - \varphi(x_{i-1})| \leqslant M\left(\sum_{i=1}^{n} |f(x_i) - f(x_{i-1})| + \sum_{i=1}^{n} |g(x_i) - g(x_{i-1})|\right)$$
$$\leqslant M(W_a^b(f) + W_a^b(g)).$$

Taking the supremum on the left-hand side we have

$$W_a^b(F(f,g)) \leqslant M(W_a^b(f) + W_a^b(g)). \tag{1.3.6}$$

We can take $F(u,v) = \alpha u + \beta v$ and obtain in a similar way

$$W_a^b(\alpha f + \beta g) \leqslant |\alpha|\, W_a^b(f) + |\beta|\, W_a^b(g). \tag{1.3.7}$$

Analogously, by Theorem 1.3.2 one can take:

(1) $F(u,v) = uv$, where $M = \max(\sup|f|, \sup|g|) < \infty$ and $\Delta_1 = \Delta_2 = [-M, M]$;

(2) $F(u,v) = u/v$, if $|g(x)| \geqslant \varepsilon > 0$ on $[a,b]$, where $\Delta_1 = [-\sup|f|, \sup|f|]$, $\Delta_2 = (-\infty, -\varepsilon] \cup [\varepsilon, \infty)$, $M = \max(1/\varepsilon, (1/\varepsilon^2)\sup|f|)$.

We then have

THEOREM 1.3.3. *If f and g are functions of bounded variation on $[a,b]$ then the function $F(f,g)$ is of bounded variation on $[a,b]$. In particular, a linear combination and the product of functions of bounded variation are functions of bounded variation. The quotient of functions of bounded variation is a function of bounded variation if the modulus of the divisor is bounded below by a positive number.*

THEOREM 1.3.4. *If f is a function of bounded variation on $[a,b]$ and is continuous on the right (on the left) at x_0 then the variation*

$$x \to W_a^x(f)$$

is a function which is continuous on the right (on the left) at x_0.

Proof. By Theorem 1.3.1 it suffices to show that

$$\lim_{x \to x_0+} W_{x_0}^x = 0$$

Let $\varepsilon > 0$. There exists $\xi > x_0$ such that when $x_0 < x < \xi$ then

$$|f(x) - f(x_0)| < \tfrac{1}{2}\varepsilon.$$

There is a subdivision $x_0 < x_1 < \cdots < x_n = b$, such that $W_{x_0}^b \leqslant |f(x_1) - f(x_0)| + \cdots + |f(x_n) - f(x_{n-1})| + \tfrac{1}{2}\varepsilon$.

Let $x_0 < x < \min(\xi, x_1)$. Then

$$W_{x_0}^b \leqslant |f(x) - f(x_0)| + |f(x_1) - f(x)| + \cdots + |f(x_n) - f(x_{n-1})| + \tfrac{1}{2}\varepsilon \leqslant \varepsilon + W_x^b.$$

Hence (Th. 1.3.1) $W_{x_0}^x \leqslant \varepsilon$. It follows that

$$\lim_{x \to x_0+} W_{x_0}^x = 0.$$

Similarly for the continuity on the left. ■

Note: Conversely, it is clear that, by inequality (1.3.4), the continuity of the variation $x \to W_a^x(f)$ implies that the function $f(x)$ is continuous.

THEOREM 1.3.5. *If $f(x) = \lim_{n\to\infty} f_n(x)$ in $[a, b]$, then*

$$W_a^b(f) \leqslant \liminf_{n\to\infty} W_a^b(f_n). \tag{1.3.8}$$

Proof. If the right-hand side is not finite then there is nothing to prove. Assume, therefore, that it is finite and choose $L > \liminf_{n\to\infty} W_a^b(f_n)$. It suffices to prove that

$$W_a^b(f) \leqslant L.$$

Take a subdivision $a = x_0 < \cdots < x_n = b$. There must be an index sequence such that $W_a^b(f_{\alpha_\nu}) < L$ and hence such that

$$\sum_{i=1}^{n} |f_{\alpha_\nu}(x_i) - f_{\alpha_\nu}(x_{i-1})| < L.$$

Now take the limit as $\nu \to \infty$ and we have

$$\sum_{i=1}^{n} |f(x_i) - f(x_{i-1})| \leqslant L.$$

Hence, taking the supremum of the left-hand side

$$W_a^b(f) \leqslant L. \quad ■$$

COROLLARY 1.3.3. *The limit of a convergent sequence of functions of uniformly bounded variation is a function of bounded variation.*

The relationship with arc-length

The *arc-length* of a curve with equations $x = x(t)$, $y = y(t)$, where $\alpha \leqslant t \leqslant \beta$, is defined by

$$L = \sup_{\alpha = t_0 < \cdots < t_n = \beta} \sum_{1}^{n} \sqrt{\{[x(t_i) - x(t_{i-1})]^2 + [y(t_i) - y(t_{i-1})]^2\}}$$

Taking the suprema in the inequalities

$$\left.\begin{array}{l}\sum_{i=1}^{n}|x(t_i)-x(t_{i-1})| \\ \sum_{i=1}^{n}|y(t_i)-y(t_{i-1})|\end{array}\right\} \leqslant \sum_{i=1}^{n}\sqrt{\{[x(t_i)-x(t_{i-1})]^2+[y(t_i)-y(t_{i-1})]^2\}}$$

$$\leqslant \sum_{i=1}^{n}|x(t_i)-x(t_{i-1})|+\sum_{i=1}^{n}|y(t_i)-y(t_{i-1})|$$

we obtain the inequality

$$\left.\begin{array}{l}W_\alpha^\beta(x) \\ W_\alpha^\beta(y)\end{array}\right\} \leqslant L \leqslant W_\alpha^\beta(x)+W_\alpha^\beta(y). \tag{1.3.9}$$

This gives

THEOREM 1.3.6. *The curve*

$$x=x(t),\ y=y(t) \qquad (\alpha \leqslant t \leqslant \beta)$$

is rectifiable (i.e. has finite length), if and only if x and y are functions of bounded variation on $[\alpha, \beta]$.

1.4 JORDAN CANONICAL DECOMPOSITION

It follows from Th. 1.3.3 that the difference of two increasing functions is a function of bounded variation. We will show conversely:

THEOREM 1.4.1 (Jordan). *Every function f of bounded variation in the interval* $[a, b]$ *is the difference of two increasing functions:*

$$f=\varphi-\psi \tag{1.4.1}$$

where $\varphi(x)=\frac{1}{2}[W_a^x(f)+f(x)]$ *and* $\psi(x)=\frac{1}{2}[W_a^x(f)-f(x)]$. *If also f is continuous on the right (on the left) at* x_0, *then* φ *and* ψ *are continuous on the right (on the left) at* x_0.

Formula (1.4.1) is called the *Jordan canonical decomposition.*

Proof. Let $x<y$. We have

$$\varphi(y)-\varphi(x)=\tfrac{1}{2}[W_x^y(f)+f(y)-f(x)], \quad \psi(y)-\psi(x)=\tfrac{1}{2}[W_x^y(f)-(f(y)-f(x))]$$

By inequality (1.3.4):

$$-(f(y)-f(x)) \leqslant W_x^y(f) \quad \text{and} \quad f(y)-f(x) \leqslant W_x^y(f),$$

hence

$$\varphi(y)-\varphi(x) \geqslant 0 \quad \text{and} \quad \psi(y)-\psi(x) \geqslant 0.$$

Thus the functions φ and ψ are increasing. From Th. 1.3.4, it follows that these functions are continuous on the right (on the left) at points at which f is continuous on the right (on the left). ■

Remark 1. In the Jordan canonical decomposition the functions φ and ψ increase at the least possible rate in the sense that if $f=\bar{\varphi}-\bar{\psi}$, where $\bar{\varphi}$, $\bar{\psi}$ are increasing functions, then for $x<y$ we must have

$$\varphi(y)-\varphi(x)\leqslant\bar{\varphi}(y)-\bar{\varphi}(x) \quad\text{and}\quad \psi(y)-\psi(x)\leqslant\bar{\psi}(y)-\bar{\psi}(x).$$

Indeed, by (1.3.7) we have

$$W_x^y(f)\leqslant W_x^y(\bar{\varphi})+W_x^y(\bar{\psi})=\bar{\varphi}(y)-\bar{\varphi}(x)+\bar{\psi}(y)-\bar{\psi}(x),$$

whence

$$\varphi(y)-\varphi(x)=\tfrac{1}{2}[W_x^y(f)+f(y)-f(x)]\leqslant\bar{\varphi}(y)-\bar{\varphi}(x)$$

and

$$\psi(y)-\psi(x)=\tfrac{1}{2}[W_x^y(f)-(f(y)-f(x))]\leqslant\bar{\psi}(y)-\bar{\psi}(x).$$

Remark 2. Let $a\leqslant x<y\leqslant b$. Since $W_x^y(f)=\varphi(y)-\varphi(x)+\psi(y)-\psi(x)$, therefore

$$W_x^y(f)=W_x^y(\varphi)+W_x^y(\psi). \tag{1.4.2}$$

Using Jordan's theorem we may extend the theorems of §1.1 and §1.2 on monotone functions to the case of functions of bounded variation. We thus have

THEOREM 1.4.2. *A function f of bounded variation in the interval $[a,b]$ has, at each interior point of the interval, a right-hand limit $f(x+0)$ and a left-hand limit $f(x-0)$. Moreover, $f(a+0)$ and $f(b-0)$ exist.*

THEOREM 1.4.3. *A function of bounded variation has at most a countable number of points of discontinuity.*

Let f be a function of bounded variation on $[a,b]$ and let $\{x_\nu\}$ be a sequence, with distinct terms, of all its points of discontinuity. By Jordan's theorem

$$f(x)=f_1(x)-f_2(x),$$

where f_1, f_2 are increasing functions on $[a,b]$ and where all the points of discontinuity of the functions f_1 and f_2 are contained in the sequence $\{x_\nu\}$. By Th. 1.1.3,

$$f_i(x)=g_i(x)+s_i(x) \qquad (i=1,2),$$

where g_i is a continuous, increasing function in $[a,b]$, and $s_i(x)=\sum u_{i\nu}(x)$, where $u_{i\nu}(x)=0$ for $x<x_\nu$, $u_{i\nu}(x_\nu)=f_i(x_\nu)-f_i(x_\nu-0)$ and $u_{i\nu}(x)=f_i(x_\nu+0)-f_i(x_\nu-0)$ for $x>x_\nu$. (If f is continuous at x_ν then $u_{i\nu}\equiv 0$.) Hence $f(x)=g(x)+s(x)$ where $g=g_1-g_2$, $s=\sum u_\nu$, $u_\nu=u_{1\nu}-u_{2\nu}$. We now have

THEOREM 1.4.4. *Every function f of bounded variation on $[a,b]$ is the sum of a continuous function g of bounded variation and of a saltus-function $s(x)=\sum u_\nu(x)$, where $u_\nu(x)=0$ for $x<x_\nu$, $u_\nu(x_\nu)=f(x_\nu)-f(x_\nu-0)$, $u_\nu(x)=f(x_\nu+0)-f(x_\nu-0)$ for $x>x_\nu$ and where $\{x_\nu\}$ is a sequence of all the points of discontinuity of the function f.*

THEOREM 1.4.5 (Helly's First Theorem). *Let $\{f_n\}$ be a sequence of functions bounded at a, $(|f_n(a)| \leqslant M)$, and of uniformly bounded variation on $[a, b]$, $(W_a^b(f_n) \leqslant M)$. Then, there is a subsequence of functions which converges on $[a, b]$ to a function of bounded variation.*

Proof. By inequality (1.3.5) we have $|f_n(x)| \leqslant 2M$. By Jordan's theorem $f_n = \varphi_n - \psi_n$, where φ_n and ψ_n are increasing and jointly bounded:

$$|\varphi_n(x)| = \tfrac{1}{2}|W_a^x(f_n) + f_n(x)| \leqslant \tfrac{3}{2}M,$$
$$|\psi_n(x)| = \tfrac{1}{2}|W_a^x(f_n) - f_n(x)| \leqslant \tfrac{3}{2}M.$$

Now make two applications of Th. 1.2.1, first to the sequence $\{\varphi_n\}$ to obtain a subsequence φ_{β_n} convergent in $[a, b]$, and then to the sequence $\{\psi_{\beta_n}\}$. In this way we obtain two subsequences convergent in $[a, b]$: $\varphi_{\alpha_n}(x) \to \varphi(x)$, $\psi_{\alpha_n}(x) \to \psi(x)$. Hence

$$f_{\alpha_n}(x) \to f(x) = \varphi(x) - \psi(x). \quad \blacksquare$$

To apply Th. 1.2.2 in the case of a sequence of functions of bounded variation $\{f_n\}$, it is further necessary to suppose that $W_a^b(f_n) \to W_a^b(f)$.

1.5 THE RIEMANN–STIELTJES INTEGRAL. EXISTENCE CRITERIA

Let f and g be functions defined (and finite) on the interval $[a, b]$. Given a subdivision $a = x_0 < \cdots < x_k = b$ and intermediate points ξ_i, where $x_{i-1} \leqslant \xi_i \leqslant x_i$, we form the approximating sum

$$S = \sum_{i=1}^{k} f(\xi_i)[g(x_i) - g(x_{i-1})]. \tag{1.5.1}$$

If for every sequence of subdivisions $a = \overset{n}{x}_0 < \cdots < \overset{n}{x}_{k_n} = b$ satisfying the conditions

$$\lim_{n \to \infty} \max_i |\overset{n}{x}_i - \overset{n}{x}_{i-1}| = 0, \tag{1.5.2}$$

with arbitrary intermediate points $\overset{n}{x}_{i-1} \leqslant \overset{n}{\xi}_i \leqslant \overset{n}{x}_i$, the sequence of approximating sums

$$S_n = \sum_{i=1}^{k_n} f(\overset{n}{\xi}_i)[g(\overset{n}{x}_i) - g(\overset{n}{x}_{i-1})]$$

always tends to the same finite limit, then we call this limit the *Riemann–Stieltjes Integral* over the interval $[a, b]$ of the function f relative to the function g and we denote it by the symbol

$$\int_a^b f(x)\, dg(x).$$

We also say that the function f is *Riemann–Stieltjes integrable* on the interval $[a, b]$ relatively to the function g. In other words (cf. §0.2), the Riemann–Stieltjes integral is defined to be the finite limit

$$\int_a^b f(x)\,dg(x) = \lim_{\max_i |x_i - x_{i-1}| \to 0} \sum_i f(\xi_i)[g(x_i) - g(x_{i-1})] \tag{1.5.3}$$

where the limit is taken over all subdivisions and intermediate points.

Clearly, in this definition it suffices to require the convergence of each sequence of approximating sums when condition (1.5.2) is fulfilled (see Note 0.7).

It follows from the definition that a continuous and strictly increasing substitution[1] $x = x(t)$, $\alpha \leqslant t \leqslant \beta$, $a = x(\alpha)$, $b = x(\beta)$ does not affect the property of integrability or the value of the integral:

$$\int_a^b f(x)\,dg(x) = \int_\alpha^\beta f(x(t))\,dg(x(t)). \tag{1.5.4}$$

For we have a bijection between subdivisions with intermediate points

$$a = x_0 < \cdots < x_k = b, \quad x_{i-1} \leqslant \xi_i \leqslant x_i$$

and

$$\alpha = t_0 < \cdots < t_k = \beta, \quad t_{i-1} \leqslant \tau_i \leqslant t_i,$$

given by $x_i = x(t_i)$, $\xi_i = x(\tau_i)$. Under this bijection the sum (1.5.1) is the same as the sum

$$\sum_{i=1}^k f(x(\tau_i))[g(x(t_i)) - g(x(t_{i-1}))]$$

and condition (1.5.2) is equivalent to the condition

$$\lim_{n\to\infty} \max_i |\overset{n}{t}_i - \overset{n}{t}_{i-1}| = 0$$

($\overset{n}{x}_i = x(\overset{n}{t}_i)$), by the uniform continuity of the function $x(t)$ on $[a, b]$ and of its inverse function on $[\alpha, \beta]$.

It is easy to verify that $\int_a^b dg(x) = g(b) - g(a)$. Also if $g(x) = \text{const.}$, then $\int_a^b f(x)\,dg(x) = 0$. If $a < c < b$, $f(x)$ is continuous at c, $g(x) = 0$ for $x \leqslant c$, $g(x) = 1$ for $x > c$, then $\int_a^b f(x)\,dg(x) = f(c)$. If f is continuous and g is of class $\mathscr{C}^1$, then writing the approximating sum (1.5.1) in the form

$$\sum_{i=1}^k f(\xi_i) g'(\tilde{\xi}_i)(x_i - x_{i-1})$$

using the mean-value theorem, we see that

$$\int_a^b f(x)\,dg(x) = \int_a^b f(x) g'(x)\,dx. \tag{1.5.5}$$

We say that $a = x'_0 < \cdots < x'_l = b$ is a *sub-subdivision* of the subdivision $a = x_0 < \cdots < x_k = b$, if $x_i = x'_{\alpha_i}$ where $0 = \alpha_0 < \cdots < \alpha_k = l$. Clearly, any two subdivisions have a common sub-subdivision. This leads to

LEMMA 1.5.1. *For the integral*

$$\int_a^b f(x)\,\mathrm{d}g(x)$$

to exist it is necessary and sufficient that for all $\varepsilon > 0$ there exists $\delta > 0$ such that if S is an approximating sum for the subdivision P, where $\delta(P) = \max_i |x_i - x_{i-1}| < \delta$, and if S' is an approximating sum for a sub-subdivision of P, then

$$|S - S'| < \varepsilon.$$

This condition is equivalent to Cauchy's criterion (see (0.2.14)) for the convergence of the right-hand side of (1.5.3): given $\varepsilon > 0$ there exists $\delta > 0$ such that if S_1, S_2 are approximating sums for the subdivisions P_1, P_2 with $\delta(P_1)$, $\delta(P_2) < \delta$ then $|S_1 - S_2| < \varepsilon$.

LEMMA 1.5.2. *If S is an approximating sum for the subdivision $a = x_0 < \cdots < x_k = b$ and S' is an approximating sum for a sub-subdivision, then*

$$|S' - S| \leqslant \sum_{i=1}^{k} \left(\operatorname*{osc}_{[x_{i-1}, x_i]} f \right) W_{x_{i-1}}^{x_i}(g). \tag{1.5.6}$$

Proof. If

$$S' = \sum_{\nu=1}^{l} f(\xi'_\nu)[g(x'_\nu) - g(x'_{\nu-1})] = \sum_{i=1}^{k} \sum_{\nu=\alpha_{i-1}+1}^{\alpha_i} f(\xi'_\nu)[g(x'_\nu) - g(x'_{\nu-1})],$$

where $x'_{\nu-1} \leqslant \xi'_\nu \leqslant x'_\nu$, and

$$S = \sum_{i=1}^{k} f(\xi_i)[g(x_i) - g(x_{i-1})] = \sum_{i=1}^{k} f(\xi_i) \sum_{\nu=\alpha_{i-1}+1}^{\alpha_i} [g(x'_\nu) - g(x'_{\nu-1})],$$

where $x_{i-1} \leqslant \xi_i \leqslant x_i$, then

$$\begin{aligned} |S' - S| &= \left| \sum_{i=1}^{k} \sum_{\nu=\alpha_{i-1}+1}^{\alpha_i} [f(\xi'_\nu) - f(\xi_i)][g(x'_\nu) - g(x'_{\nu-1})] \right| \\ &\leqslant \sum_{i=1}^{k} \left(\operatorname*{osc}_{[x_{i-1}, x_i]} f \right) \sum_{\nu=\alpha_{i-1}+1}^{\alpha_i} |g(x'_\nu) - g(x'_{\nu-1})| \\ &\leqslant \sum_{i=1}^{k} \left(\operatorname*{osc}_{[x_{i-1}, x_i]} f \right) W_{x_{i-1}}^{x_i}(g). \quad \blacksquare \end{aligned}$$

If the integral $\int_a^b f\,\mathrm{d}g$ exists, then by taking a sequence of sums S'_n for sub-subdivisions of the subdivision $a = x_0 < \cdots < x_n = b$ converging to the integral, we can proceed to the limit in the inequality (1.5.6). This gives

LEMMA 1.5.3. *If the integral $\int_a^b f\,\mathrm{d}g$ exists and if S is an approximating sum for the subdivision $a = x_0 < \cdots < x_n = b$ and $\operatorname{osc}_{[x_{i-1}, x_i]} f < \varepsilon$ for $i = 1, \ldots, n$ then*

$$\left| S - \int_a^b f\,\mathrm{d}g \right| < \varepsilon W_a^b(g).$$

Suppose that the integral $\int_a^b f dg$ exists. Let $\delta > 0$ be a number chosen to correspond with $\varepsilon > 0$ as in Lemma 1.5.1. Consider a subdivision for which $\max_i |x_i - x_{i-1}| < \delta$. Then

$$\sum_{i=1}^{n} [f(\xi_i'') - f(\xi_i')][g(x_i) - g(x_{i-1})] \leqslant \varepsilon.$$

Taking the supremum in each of the terms in this sum, we get

$$\sum_{i=1}^{n} \left(\operatorname*{osc}_{[x_{i-1}, x_i]} f \right) |g(x_i) - g(x_{i-1})| \leqslant \varepsilon. \tag{1.5.7}$$

We will show that this last condition (for every $\varepsilon > 0$ there exists $\delta > 0$ such that the inequality (1.5.7) holds for all subdivisions for which $\max_i |x_i - x_{i-1}| < \delta$) implies that at least one of the functions f, g is continuous at any point of the interval $[a, b]$.

Indeed, suppose that g is not continuous at the point $x_0 \in [a, b]$. Then

$$g(x_0 + 0) \neq g(x_0) \quad \text{or} \quad g(x_0 + 0) \neq g(x_0 - 0);$$

let, respectively, $x_h = x_0$ or $x_h = x_0 - h$. If $0 < h < \min(\delta, b - x_0)$, then taking a subdivision (with $\max_i |x_i - x_{i-1}| < \delta$) which contains x_h, $x_0 + h$ as two successive points, we obtain, by (1.5.7),

$$|f(x_0 + h) - f(x_0)| \cdot |g(x_0 + h) - g(x_h)| < \varepsilon.$$

Hence, for $h \to 0+$ we have

$$|f(x_0 + h) - f(x_0)| \cdot |g(x_0 + h) - g(x_h)| \to 0,$$

whence $f(x_0 + h) \to f(x_0)$. Thus f is continuous on the right. Similarly we can show that f is continuous on the left. This gives

THEOREM 1.5.1. *If the integral $\int_a^b f dg$ exists, then at every point of the interval $[a, b]$ at least one of the functions f, g must be continuous.*

THEOREM 1.5.2. *Let f be a bounded function and let g be a function of bounded variation on $[a, b]$. A necessary and sufficient condition for the existence of the integral $\int_a^b f dg$ is given by each of the following conditions:*

(A) *For each $\varepsilon > 0$ there exists $\delta > 0$ such that for every subdivision with $\max_i |x_i - x_{i-1}| < \delta$,*

$$\sum_{i=1}^{n} \left(\operatorname*{osc}_{[x_{i-1}, x_i]} f \right) |g(x_i) - g(x_{i-1})| \leqslant \varepsilon. \tag{1.5.8}$$

(B) *For each $\varepsilon > 0$ there exists $\delta > 0$ such that for every subdivision with $\max_i |x_i - x_{i-1}| < \delta$,*

$$\sum_{i=1}^{n} \left(\operatorname*{osc}_{[x_{i-1}, x_i]} f \right) W_{x_{i-1}}^{x_i}(g) \leqslant \varepsilon. \tag{1.5.9}$$

Proof. We have already shown that the existence of the integral $\int_a^b f dg$ implies condition (A). From Lemmas 1.5.1 and 1.5.2 it follows that (B) is a sufficient condition for the integral to exist. It therefore suffices to show that condition (B) follows from condition (A). Suppose, therefore, that condition (A) holds.

Let $\varepsilon > 0$ and let $\delta_0 > 0$ be chosen to correspond to $\frac{1}{3}\varepsilon > 0$ in condition (A). Let $M = \operatorname{osc}_{[a,b]} f$ and let $a = z_0 < \cdots < z_N = b$ be a subdivision such that

$$W_a^b(g) - \frac{\varepsilon}{3M} \leqslant \sum_{i=1}^{N} |g(z_i) - g(z_{i-1})|.$$

This inequality will also hold for every sub-subdivision. We have already shown that condition (A) implies that at any point of the interval $[a,b]$ one of the functions f, g is continuous. Hence, by Th. 1.3.4, at each of the points $z_0, \ldots, z_N$ one of the functions f, $x \to W_a^x(g)$ must be continuous. Since both of these functions are bounded, there exists $\delta_1 > 0$ such that

$$\left(\operatorname*{osc}_{[x',x]} f\right) \cdot W_{x'}^x(g) < \frac{\varepsilon}{3M} \tag{1.5.10}$$

when $x' < z_k < x$, $|x - x'| < \delta_1$ $(k = 0, \ldots, N)$. Let $a = x_0 < \cdots < x_n = b$ be a subdivision such that

$$\max_i |x_i - x_{i-1}| < \delta = \min(\delta_0, \delta_1).$$

Then, for this subdivision the inequality (1.5.8) holds (with $\frac{1}{3}\varepsilon$ in place of ε). By adding the points z_k (distinct from the x_i) we obtain a new subdivision $a = x'_0 < \cdots < x'_m = b$, for which

$$W_a^b(g) - \frac{\varepsilon}{3N} \leqslant \sum_{i=1}^{m} |g(x'_i) - g(x'_{i-1})|,$$

that is

$$\sum_{i=1}^{m} [W_{x'_{i-1}}^{x'_i}(g) - |g(x'_i) - g(x'_{i-1})|] \leqslant \frac{\varepsilon}{3M}.$$

From this we have

$$\begin{aligned} S_0 &= \sum_{i=1}^{m} \left(\operatorname*{osc}_{[x'_{i-1},x'_i]} f\right) \cdot W_{x'_{i-1}}^{x'_i}(g) \\ &\leqslant \sum_{i=1}^{m} \left(\operatorname*{osc}_{[x'_{i-1},x'_i]} f\right) [W_{x'_{i-1}}^{x'_i}(g) - |g(x'_i) - g(x'_{i-1})|] \\ &\quad + \sum_{i=1}^{m} \left(\operatorname*{osc}_{[x'_{i-1},x'_i]} f\right) |g(x'_i) - g(x'_{i-1})| \\ &\leqslant \tfrac{2}{3}\varepsilon. \end{aligned}$$

Now

$$S = \sum_{i=1}^{n} \left(\operatorname*{osc}_{[x_{i-1},x_i]} f\right) W_{x_{i-1}}^{x_i}(g) = S' + S''$$

where S'' is the sum of all those terms for which $x_{i-1} < z_k < x_i$ for some k, and S' is a partial sum of the sum S_0. Thus $S' \leqslant S_0 \leqslant \frac{2}{3}\varepsilon$ and by (1.5.10) we have $S'' \leqslant \frac{1}{3}\varepsilon$. Hence $S \leqslant \varepsilon$ and the proof is complete. ■

If $g = \varphi - \psi$ is the Jordan canonical decomposition, then (see (1.4.2)):

$$W_{x_{i-1}}^{x_i}(g) = W_{x_{i-1}}^{x_i}(\varphi) + W_{x_{i-1}}^{x_i}(\psi).$$

Hence given the function f, condition (B) holds for a function g if and only if it also holds for both functions φ and ψ. We therefore have

THEOREM 1.5.3. *If $g = \varphi - \psi$ is the Jordan canonical decomposition of the function g of bounded variation on the interval $[a, b]$ and if f is a bounded function, then the integral $\int_a^b f\,\mathrm{d}g$ exists if and only if the integrals $\int_a^b f\,\mathrm{d}\varphi$ and $\int_a^b f\,\mathrm{d}\psi$ exist.*

If $[c, d] \subset [a, b]$ then every subdivision of the interval $[c, d]$ which satisfies the condition $\max_i |x_i - x_{i-1}| < \delta$ can be extended to a subdivision of the interval $[a, b]$ for which the condition is still satisfied. From Lemma 1.5.1 we obtain.

THEOREM 1.5.4. *If f is integrable relatively to g on the interval $[a, b]$ then it is integrable relatively to g on each subinterval $[c, d] \subset [a, b]$.*[2]

THEOREM 1.5.5. *If f is continuous and if g is of bounded variation on the interval $[a, b]$ then the integral $\int_a^b f(x)\,\mathrm{d}g(x)$ exists.*

Proof. Let $\varepsilon > 0$. By the uniform continuity of $f(x)$ on $[a, b]$ there exists $\delta > 0$ such that for every subdivision $a = x_0 < \cdots < x_n = b$ with $\max_i |x_i - x_{i-1}| < \delta$ we have

$$\operatorname*{osc}_{[x_{i-1}, x_i]} f < \frac{\varepsilon}{W_a^b(g)}.$$

This yields the inequality (1.5.9). ■

THEOREM 1.5.6. *Let f, g be functions of bounded variation on $[a, b]$. Then the integral $\int_a^b f(x)\,\mathrm{d}g(x)$ exists if and only if f and g have no common point of discontinuity.*

Proof. Necessity follows from Th. 1.5.1. Sufficiency. Let $\varepsilon > 0$. We define

$$\varepsilon_1 = \frac{\varepsilon}{W_a^b(f) + W_a^b(g)}$$

We assert that there exists $\delta > 0$ such that if $|x'' - x'| < \delta$ then either $\operatorname{osc}_{[x', x'']} f < \varepsilon$, or $|g(x'') - g(x')| < \varepsilon_1$. For otherwise there would exist x_n', x_n'' such that

$$|x_n'' - x_n'| < \frac{1}{n}, \quad |g(x_n'') - g(x_n')| \geqslant \varepsilon_1 \quad \text{and} \quad \operatorname*{osc}_{[x_n', x_n'']} f \geqslant \varepsilon_1,$$

that is

$$|f(\xi_n'') - f(\xi_n')| \geqslant \tfrac{1}{2}\varepsilon_1 \quad \text{for some } \xi_n'', \xi_n' \in [x_n', x_n''].$$

Selecting a convergent subsequence $x'_{\alpha_n} \to x_0$ we would have

$$x''_{\alpha_n} \to x_0, \quad \xi'_{\alpha_n} \to x_0, \quad \xi''_{\alpha_n} \to x_0$$

and so both functions would be discontinuous at x_0.

Let $a = x_0 < \cdots < x_n = b$ be a subdivision such that $\max_i |x_i - x_{i-1}| < \delta$. Then for each i we have either

$$\operatorname*{osc}_{[x_{i-1}, x_i]} f < \varepsilon_1 \quad \text{or} \quad |g(x_i) - g(x_{i-1})| < \varepsilon_1.$$

Using (1.3.4) this gives

$$\left(\operatorname*{osc}_{[x_{i-1}, x_i]} f\right) \cdot |g(x_i) - g(x_{i-1})| < \varepsilon_1 [W_{x_{i-1}}^{x_i}(f) + W_{x_{i-1}}^{x_i}(g)]$$

which implies inequality (1.5.8). ■

1.6 PROPERTIES OF THE INTEGRAL

If $|f(x)| \leqslant M$ in $[a, b]$ then the approximating sums satisfy the inequality

$$\left|\sum_{i=1}^{k} f(\xi_i)[g(x_i) - g(x_{i-1})]\right| \leqslant M \sum_{i=1}^{k} |g(x_i) - g(x_{i-1})| \leqslant M W_a^b(g).$$

We therefore have

THEOREM 1.6.1. *If f is integrable on $[a, b]$ relatively to g and $|f(x)| \leqslant M$ on $[a, b]$ then*

$$\left|\int_a^b f(x)\,\mathrm{d}g(x)\right| \leqslant M W_a^b(g). \tag{1.6.1}$$

THEOREM 1.6.2. *If f_1 and f_2 are integrable on $[a, b]$ relatively to an increasing function g, and if $f_1(x) \leqslant f_2(x)$ in $[a, b]$ then*

$$\int_a^b f_1(x)\,\mathrm{d}g(x) \leqslant \int_a^b f_2(x)\,\mathrm{d}g(x). \tag{1.6.2}$$

Indeed, in this case we have the corresponding inequality between the approximating sums.

THEOREM 1.6.3. *If f is integrable on $[a, c]$ relatively to g, then*

$$\int_a^c f\mathrm{d}g = \int_a^b f\mathrm{d}g + \int_b^c f\mathrm{d}g. \tag{1.6.3}$$

Indeed, for the subdivision $a=x_0<\cdots<x_k=b=y_0<\cdots<y_l=c$,

$$\sum_{i=1}^{k} f(\xi_i)[g(x_i)-g(x_{i-1})]+\sum_{i=1}^{l} f(\eta_i)[g(y_i)-g(y_{i-1})]$$

is both an approximating sum for the integral $\int_a^c$ and a sum of approximating sums for the integrals $\int_a^b$ and $\int_b^c$ which exist by Th. 1.5.4. Hence, passing to the limit we obtain equality (1.6.3).

THEOREM 1.6.4. *If f_1, f_2 are integrable on $[a,b]$ relatively to g then so is $c_1 f_1 + c_2 f_2$ and*

$$\int_a^b (c_1 f_1 + c_2 f_2)\,\mathrm{d}g = c_1 \int_a^b f_1\,\mathrm{d}g + c_2 \int_a^b f_2\,\mathrm{d}g. \tag{1.6.4}$$

For suppose that the sequence of subdivisions $a=\overset{n}{x}_0<\cdots<\overset{n}{x}_{k_n}=b$ satisfy condition (1.5.2), then for $n\to\infty$ we have

$$\begin{aligned} S_n &= \sum_{i=1}^{k_n} [c_1 f_1(\overset{n}{\xi}_i)+c_2 f_2(\overset{n}{\xi}_i)]\cdot[g(\overset{n}{x}_i)-g(\overset{n}{x}_{i-1})] \\ &= c_1 \sum_{i=1}^{k_n} f_1(\overset{n}{\xi}_i)[g(\overset{n}{x}_i)-g(\overset{n}{x}_{i-1})] + c_2 \sum_{i=1}^{k_n} f_2(\overset{n}{\xi}_i)[g(\overset{n}{x}_i)-g(\overset{n}{x}_{i-1})] \\ &\to c_1 \int_a^b f_1\,\mathrm{d}g + c_2 \int_a^b f_2\,\mathrm{d}g. \end{aligned}$$

In the same way we can establish

THEOREM 1.6.5. *If f is integrable on $[a,b]$ relatively to both g_1 and g_2, then it is integrable relatively to $c_1 g_1 + c_2 g_2$ and*

$$\int_a^b f\,\mathrm{d}(c_1 g_1 + c_2 g_2) = c_1 \int_a^b f\,\mathrm{d}g_1 + c_2 \int_a^b f\,\mathrm{d}g_2. \tag{1.6.5}$$

THEOREM 1.6.6 (Mean Value). *If f is integrable on $[a,b]$ relatively to a monotone function g then*

$$\int_a^b f(x)\,\mathrm{d}g(x) = \mu[g(b)-g(a)] \tag{1.6.6}$$

where $\inf_{[a,b]} f(x) \leqslant \mu \leqslant \sup_{[a,b]} f(x)$.

Proof. We can assume that g is increasing, for otherwise we can replace g by $-g$ and use the formula $\int_a^b f\mathrm{d}(-g) = -\int_a^b f\mathrm{d}g$ by Th. 1.6.5. If $g(a)=g(b)$ then $\int_a^b f\mathrm{d}g = 0$. So let $g(a)<g(b)$ and let

$$\mu = \frac{\int_a^b f(x)\,\mathrm{d}g(x)}{g(b)-g(a)},$$

$m = \inf_{[a,b]} f(x)$, $M = \sup_{[a,b]} f(x)$. By Th. 1.6.2

$$m[g(b)-g(a)] = \int_a^b m\,\mathrm{d}g \leqslant \int_a^b f\mathrm{d}g \leqslant \int_a^b M\,\mathrm{d}g = M[g(b)-g(a)].$$

Now, dividing by $g(b)-g(a)$ we obtain $m \leqslant \mu \leqslant M$.

THEOREM 1.6.7 (Integration by parts). *If f, g are functions of bounded variation on the interval $[a,b]$ which do not possess any common points of discontinuity (see Th. 1.5.6), then*

$$\int_a^b f\mathrm{d}g + \int_a^b g\,\mathrm{d}f = f(b)g(b) - f(a)g(a). \tag{1.6.7}$$

Proof. To obtain equation (1.6.7) it suffices to proceed to the limit in the following identity for approximating sums

$$\sum_{i=1}^{k} f(x_i)[g(x_i)-g(x_{i-1})] + \sum_{i=1}^{k} g(x_{i-1})[f(x_i)-f(x_{i-1})] = f(b)g(b)-f(a)g(a). \quad \blacksquare$$

THEOREM 1.6.8 (Change of integrating function). *If the bounded function φ is integrable relatively to the function g of bounded variation on $[a,b]$, then $G(x) = \int_a^x \varphi(t)\,\mathrm{d}g(t)$ is a function of bounded variation on the interval $[a,b]$. Also, for any function f bounded on $[a,b]$*

$$\int_a^b f(x)\,\mathrm{d}G(x) = \int_a^b f(x)\varphi(x)\,\mathrm{d}g(x) \tag{1.6.8}$$

provided one of these integrals exists.

Proof. Let $M_0 = \sup_{a\leqslant x\leqslant b} |\varphi(x)|$. Then, for $a = x_0 \leqslant \cdots \leqslant x_k = b$ (by Ths 1.6.1 and 1.6.3),

$$\sum_1^k |G(x_i) - G(x_{i-1})| = \sum_1^k \left| \int_{x_{i-1}}^{x_i} \varphi\,\mathrm{d}g \right| \leqslant \sum_1^k M_0 W_{x_{i-1}}^{x_i}(g) = M_0 W_a^b(g).$$

Thus G is of bounded variation on $[a,b]$.

Similarly, with $x_{i-1} \leqslant \xi_i \leqslant x_i$, we have the inequality

$$\begin{aligned} &\left| \sum_1^k f(\xi_i)[G(x_i) - G(x_{i-1})] - \sum_1^k f(\xi_i)\varphi(\xi_i)[g(x_i) - g(x_{i-1})] \right| \\ &\quad = \left| \sum_1^k f(\xi_i) \int_{x_{i-1}}^{x_i} (\varphi(x) - \varphi(\xi_i))\,\mathrm{d}g(x) \right| \\ &\quad \leqslant M_1 \sum_1^k \left(\operatorname*{osc}_{[x_{i-1},x_i]} \varphi \right) W_{x_{i-1}}^{x_i}(g) \end{aligned}$$

where $M_1 = \sup|f(x)|$. But, by Th. 1.5.2, the right-hand side tends to zero as $\max_i |x_i - x_{i-1}| \to 0$. This implies (1.6.8). $\blacksquare$

Taking the limit under the integral sign

If g is of bounded variation and $f_n(x)\to f(x)$ uniformly on $[a,b]$ then

$$\varepsilon_n=\sup_{[a,b]}|f_n(x)-f(x)|\to 0.$$

Thus, by Ths 1.6.4 and 1.6.1

$$\left|\int f_n\,\mathrm{d}g-\int f\mathrm{d}g\right|=\left|\int (f_n-f)\,\mathrm{d}g\right|\leqslant \varepsilon_n W_a^b(g)\to 0$$

provided the integrals exist. So we have

THEOREM 1.6.9. *If f_n and f are integrable*[3] *on $[a,b]$ relatively to g and $f_n(x)\to f(x)$ uniformly on $[a,b]$, then*

$$\int_a^b f_n\,\mathrm{d}g\to\int_a^b f\mathrm{d}g. \tag{1.6.9}$$

THEOREM 1.6.10 (Helly's Second Theorem). *Let f be continuous and let g be of bounded variation on $[a,b]$. If $g_n(x)\to g(x)$ on a set Z dense in $[a,b]$, with a and b in Z, and if the sequence $\{W_a^b(g_n)\}$ is bounded, then*

$$\int_a^b f\mathrm{d}g_n\to\int_a^b f\mathrm{d}g.$$

Proof. It suffices to show that

$$\int_a^b f\mathrm{d}\varphi_n\to 0 \qquad \text{where } \varphi_n=g_n-g.$$

By (1.3.7) we have $W_a^b(\varphi_n)\leqslant W_a^b(g)+W_a^b(g_n)$. Thus, there is a constant M such that

$$|W_a^b(\varphi_n)|\leqslant M.$$

Let $\varepsilon>0$. Since f is uniformly continuous we can select a subdivision $a=x_0<\cdots<x_N=b$, $x_0,\ldots,x_N\in Z$, sufficiently fine so that

$$\operatorname*{osc}_{[x_{i-1},x_i]} f\leqslant\frac{\varepsilon}{2M} \qquad (i=1,\ldots,N)$$

Then, by Lemma 1.5.3, we have

$$\left|\int_a^b f\mathrm{d}\varphi_n-\sum_{i=1}^N f(\xi_i)[\varphi_n(x_i)-\varphi_n(x_{i-1})]\right|\leqslant\frac{\varepsilon}{2M}W_a^b(\varphi_n)\leqslant\frac{\varepsilon}{2}.$$

Since $\lim_{n\to\infty}\varphi_n(x_i)=0$ $(i=0,\ldots,N)$, therefore

$$\left|\sum_{i=1}^N f(\xi_i)[\varphi_n(x_i)-\varphi_n(x_{i-1})]\right|\leqslant\frac{\varepsilon}{2}$$

for all n greater than some index n_0. Then

$$\left|\int_a^b f\,\mathrm{d}\varphi_n\right| \leqslant \varepsilon.$$

Hence, we have shown that $\int_a^b f\,\mathrm{d}\varphi_n \to 0$. ■

NOTES

1. It suffices to assume that the function is continuous and increasing.
2. It follows from Th. 1.5.1 that the function f may be integrable relative to g on the intervals $[a, c]$ and $[c, b]$ but not be integrable on $[a, b]$.
3. The integrability of f follows from the remaining hypothesis.

CHAPTER 2

APPROXIMATION OF CONTINUOUS FUNCTIONS BY POLYNOMIALS

2.1 WEIERSTRASS' THEOREM

Let E be a compact subset of the space $\mathbb{R}^k$. We have the following

THEOREM 2.1.1. *Every function f which is continuous on the set E is the limit of a uniformly convergent sequence of polynomials on E.*

It suffices to prove this theorem for a single special *k-dimensional interval* $E = [c_1, d_1] \times \cdots \times [c_k, d_k]$. For, by Tietze's theorem, a function f which is continuous on a compact set E has a continuous extension $\bar{f}$ over the whole space. Since E is bounded, it is contained in some interval $[a_1, b_1] \times \cdots \times [a_k, b_k]$. If $\{W_n\}$ is a sequence of polynomials which converge to the function

$$u \to \bar{f}\left(a_1 + \frac{b_1 - a_1}{d_1 - c_1}(u_1 - c_1), \ldots, a_k + \frac{b_k - a_k}{d_k - c_k}(u_k - c_k)\right)^1$$

uniformly in $[c_1, d_1] \times \cdots \times [c_k, d_k]$ then the sequence of polynomials

$$x \to W_n\left(c_1 + \frac{d_1 - c_1}{b_1 - a_1}(x_1 - a_1), \ldots, c_k + \frac{d_k - c_k}{b_k - a_k}(x_k - a_k)\right)$$

converges to the function f uniformly on $[a_1, b_1] \times \cdots \times [a_k, b_k]$ and hence *a fortiori* on E.

We will give two constructions of a sequence of polynomials converging to a given function on an interval. These are the Tonelli and the Bernstein polynomials.

2.2 TONELLI POLYNOMIALS

Consider the sequence of polynomials in a single variable defined by

$$t_n(x) = \frac{(1 - x^2)^n}{\int_{-1}^{1}(1 - u^2)^n \, du} \qquad (n = 1, 2, \ldots). \tag{2.2.1}$$

Let $0 < \delta < 1$. Since

$$\int_{-1}^{1} (1-u^2)^n \, du \geqslant 2\int_{0}^{1} (1-u)^n \, du = \frac{2}{n+1}$$

hence

$$|t_n(x)| \leqslant \frac{n+1}{2}(1-\delta^2)^n, \qquad \text{for } \delta \leqslant |x| \leqslant 1. \tag{2.2.2}$$

Thus

$$\left(\int_{-1}^{-\delta} + \int_{\delta}^{1}\right) t_n(u)\, du \leqslant (n+1)(1-\delta^2)^n,$$

and since $\int_{-1}^{1} t_n(u)\, du = 1$, therefore

$$\left|\int_{-\delta}^{\delta} t_n(u)\, du - 1\right| \leqslant (n+1)(1-\delta^2)^n. \tag{2.2.3}$$

Now suppose that we have given a function f continuous on an interval $[a, b]$ such that $0 < b - a < 1$. We can find a continuous extension of it to an interval $[\alpha, \beta]$ such that $\alpha < a < b < \beta$, $\beta - \alpha < 1$. The polynomial

$$T_n(x) = \int_{\alpha}^{\beta} f(u) t_n(u - x)\, du \tag{2.2.4}$$

is called the nth *Tonelli polynomial* for the function f.

THEOREM 2.2.1. *$T_n(x) \to f(x)$ uniformly on $[a, b]$.*

Proof. Since f is continuous, $|f(x)| \leqslant M$ in $[\alpha, \beta]$ for some constant M. Let $\varepsilon > 0$. The function f is uniformly continuous on $[\alpha, \beta]$, so there exists $\delta < 0$ such that $|f(x'') - f(x')| < \frac{1}{2}\varepsilon$ if $\alpha \leqslant x' \leqslant x'' \leqslant \beta$ and $|x'' - x'| < \delta$. We can also require that $\delta < \min(a - \alpha, \beta - b)$. Then for $x \in [a, b]$ we must have

$$[x - \delta, x + \delta] \subset [\alpha, \beta] \subset [x - 1, x + 1].$$

Thus, if $x \in [a, b]$ then

$$T_n(x) - f(x) = \int_{\alpha}^{\beta} f(u) t_n(u - x)\, du - f(x) = \int_{\alpha}^{x-\delta} + \int_{x+\delta}^{\beta} + \int_{x-\delta}^{x+\delta} - f(x)$$

$$= \int_{\alpha}^{x-\delta} + \int_{x+\delta}^{\beta} + \int_{-\delta}^{\delta} [f(x+u) - f(x)] t_n(u)\, du$$

$$+ f(x)\left\{\int_{-\delta}^{\delta} t_n(u)\, du - 1\right\}.$$

Hence, using (2.2.2) and (2.2.3) we have

$$|T_n(x) - f(x)| \leqslant \tfrac{1}{2} M(n+1)(1-\delta^2)^n + \tfrac{1}{2}\varepsilon + M(n+1)(1-\delta^2)^n.$$

Choosing N so large that $\frac{3}{2} M(n+1)(1-\delta^2)^n \leqslant \frac{1}{2}\varepsilon$ for $n \geqslant N$ we have $|T_n(x) - f(x)| \leqslant \varepsilon$ when $x \in [a, b]$ and $n \geqslant N$. ■

Suppose that the function f is of class $\mathscr{C}^p$ (i.e. it has a continuous pth derivative in $[a,b]$). Extend the function over the interval $[\alpha,\beta]$ while preserving the class $\mathscr{C}^p$ and so that $f^{(i)}(\alpha)=f^{(i)}(\beta)=0$ for $i=0, 1,\ldots,p-1$.[2] Then

$$T_n'(x) = -\int_\alpha^\beta f(u)t_n'(u-x)\,\mathrm{d}u = \int_\alpha^\beta f'(u)t_n(u-x)\,\mathrm{d}u$$

is the nth Tonelli polynomial for f' and so, by Th. 2.2.1, $T_n'(x)\to f'(x)$ uniformly on $[a,b]$. Continuing in this way for the second and higher derivatives we obtain

THEOREM 2.2.2. *If the function f is of class $\mathscr{C}^p$ on $[a,b]$ and its extension (of class $\mathscr{C}^p$) to $[\alpha,\beta]$ satisfies the condition*

$$f^{(i)}(\alpha)=f^{(i)}(\beta)=0 \qquad (i=0,1,\ldots,p-1),$$

then $T_n^{(i)}$ is the nth Tonelli polynomial for $f^{(i)}$ and $T_n^{(i)}(x)\to f^{(i)}(x)$ uniformly on $[a,b]$ $(i=0,1,\ldots,p)$

For a function f of several variables which is continuous in the interval $P=[a_1,b_1]\times\cdots\times[a_k,b_k]$, where $0<b_i-a_i<1$, the nth Tonelli polynomial takes the form

$$T_n(x_1,\ldots,x_k)=\int\cdots\int_Q f(u_1,\ldots,u_k)t_n(u_1-x_1)\cdots t_n(u_k-x_k)\,\mathrm{d}u_1\ldots\mathrm{d}u_k \qquad (2.2.5)$$

where $Q=[\alpha_1,\beta_1]\times\cdots\times[\alpha_k,\beta_k]$, $\alpha_i<a_i<b_i<\beta_i$, $\beta_i-\alpha_i<1$. In this case $f(x)$ must be extended continuously to Q. Theorems analogous to Theorems 2.2.1 and 2.2.2 can be proved in a similar way for this case also.[3]

2.3 BERNSTEIN POLYNOMIALS

Let f be a continuous function on $[0,1]$. For $\delta>0$ define

$$\omega(\delta)=\sup_{|x''-x'|<\delta}|f(x'')-f(x')|$$

This is called the *modulus of continuity* of the function f in the interval $[0,1]$. As the function f is uniformly continuous on $[0,1]$, then $\omega(\delta)\to 0$ as $\delta\to 0$.[4]

The polynomial

$$B_n(x)=\sum_{i=0}^{n}\binom{n}{i}f\left(\frac{i}{n}\right)x^i(1-x)^{n-i} \qquad (2.3.1)$$

is called the nth *Bernstein polynomial* of the function f.

THEOREM 2.3.1. $B_n(x)\to f(x)$ *uniformly in* $[0,1]$. *An error estimate is given by*

$$|B_n(x)-f(x)|\leqslant \tfrac{3}{2}\omega\left(\frac{1}{\sqrt{n}}\right). \qquad (2.3.2)$$

Proof. It is clearly sufficient to prove the inequality (2.3.2) since this implies the uniform convergence of B_n to f. We have the inequality

$$|f(x'')-f(x')| \leqslant (|x''-x'|\sqrt{(n)}+1)\omega\left(\frac{1}{\sqrt{n}}\right). \tag{2.3.3}$$

To obtain this it suffices to take a subdivision $x'=x_0<\cdots<x_N=x''$ of $[x',x'']$ into N equal parts, where $(x''-x')\sqrt{n}\leqslant N<(x''-x')\sqrt{(n)}+1$ and to make use of the inequality

$$|f(x_i)-f(x_{i-1})| \leqslant \omega\left(\frac{1}{\sqrt{n}}\right).$$

Next we have

$$\sum_{i=0}^{n}\binom{n}{i}x^i(1-x)^{n-i}=1. \tag{2.3.4}$$

Differentiating the identity

$$(e^y+(1-x))^n=\sum_{i=0}^{n}\binom{n}{i}e^{iy}(1-x)^{n-i}$$

twice with respect to y and substituting $e^y=x$, we obtain

$$nx=\sum_{i=0}^{n}\binom{n}{i}ix^i(1-x)^{n-i} \quad \text{and} \quad nx+n(n-1)x^2=\sum_{i=0}^{n}\binom{n}{i}i^2x^i(1-x)^{n-i}.$$

Hence

$$\sum_{i=0}^{n}\binom{n}{i}(i-nx)^2x^i(1-x)^{n-i}=nx+n(n-1)x^2-2nxnx+n^2x^2$$
$$=nx(1-x).$$

Since $x(1-x)\leqslant\frac{1}{4}$ on $[0,1]$, therefore

$$\sum_{i=0}^{n}\binom{n}{i}\left(\frac{i}{n}-x\right)^2x^i(1-x)^{n-i}\leqslant\frac{1}{4n} \quad \text{on } [0,1].$$

Now substitute

$$a_i=\left|\frac{i}{n}-x\right|\sqrt{\left[\binom{n}{i}x^i(1-x)^{n-i}\right]},\quad b_i=\sqrt{\left[\binom{n}{i}x^i(1-x)^{n-i}\right]}$$

into Schwarz's inequality $\sum a_ib_i\leqslant\sqrt{(\sum a_i^2)}\sqrt{(\sum b_i^2)}$ and use (2.3.4) to obtain the bound

$$\sum_{i=0}^{n}\binom{n}{i}\left|\frac{i}{n}-x\right|x^i(1-x)^{n-i}\leqslant\frac{1}{2\sqrt{n}}. \tag{2.3.5}$$

From (2.3.4), (2.3.3) we have

$$|B_n(x)-f(x)|=\left|\sum_{i=0}^{n}\binom{n}{i}\left[f\left(\frac{i}{n}\right)-f(x)\right]x^i(1-x)^{n-i}\right|$$
$$\leqslant\omega\left(\frac{1}{\sqrt{n}}\right)\left[1+\sqrt{n}\sum_{i=0}^{n}\binom{n}{i}\left|\frac{i}{n}-x\right|x^i(1-x)^{n-i}\right]$$

which, using (2.3.5), yields inequality (2.3.2). ■

Application to absolutely monotone functions

The *p*th-*order difference* $\Delta_h^p f(x)$ $(h>0)$ of a function f is defined inductively as follows:

$$\Delta_h^0 f(x)=f(x),\quad \Delta_h^1 f(x)=f(x+h)-f(x),$$
$$\Delta_h^p f(x)=\Delta_h^{p-1}f(x+h)-\Delta_h^{p-1}f(x).$$

Then

$$\Delta_h^p f(x)=\sum_{i=0}^{p}(-1)^{p-i}\binom{p}{i}f(x+ih) \tag{2.3.6}$$

If f is of class $\mathscr{C}^p$ in $[x,x+ph]$ then we have the generalized mean-value theorem

$$\frac{1}{h^p}\Delta_h^p f(x)=f^{(p)}(\xi)\qquad \text{for some } \xi\in(x,x+ph). \tag{2.3.7}$$

We say that the function f is *absolutely monotone* in the interval $[a,b)$ if

$$\Delta_h^p f(x)\geqslant 0\quad \text{for}\quad a\leqslant x\leqslant x+ph<b,\quad p=0,1,\ldots$$

We then have

THEOREM 2.3.2 (Bernstein). *Every absolutely monotone function f on $[0,a)$ is analytic*:

$$f(x)=\sum_{n=0}^{\infty}a_nx^n\qquad \textit{in } [0,a) \textit{ with } a_n\geqslant 0,\ n=0,1,\ldots \tag{2.3.8}$$

Proof. Choose α arbitrarily such that $0<\alpha<a$. It suffices to obtain the expansion (2.3.8) in $[0,\alpha)$. We can also assume that $\alpha=1$ by making the substitution $x=\alpha u$ if necessary.

Since $\Delta_h' f(x)=f(x+h)-f(x)\geqslant 0$ the function f is increasing. Taking the limit $(h\to 0+)$ in the inequalities

$$\Delta_h^2 f(0)=f(0)-2f(0+h)+f(0+2h)\geqslant 0$$

and

$$\Delta_h^2 f(x-\tfrac{1}{2}h)=f(x-\tfrac{1}{2}h)-2f(x+\tfrac{1}{2}h)+f(x+\tfrac{3}{2}h)\geqslant 0$$

we obtain $f(0)-f(0+0)\geqslant 0$ and $f(x-0)-f(x+0)\geqslant 0$ for $0<x<a$. This shows that the function f is continuous in $[0,a)$ and hence in $[0,1]$.

By Th. 2.3.1 the sequence of Bernstein polynomials B_n converges uniformly to f in $[0,1]$. We can write

$$\begin{aligned} B_n(x) &= \sum_{i=0}^{n}\binom{n}{i} f\left(\frac{i}{n}\right) x^i(1-x)^{n-i} \\ &= \sum_{i=0}^{n}\sum_{\nu=0}^{n-i}\binom{n}{i}\binom{n-i}{\nu} f\left(\frac{i}{n}\right)(-1)^\nu x^{i+\nu} \\ &= \sum_{i=0}^{n}\sum_{\nu=i}^{n}\binom{n}{i}\binom{n-i}{\nu-i} f\left(\frac{i}{n}\right)(-1)^{\nu-i} x^\nu. \end{aligned}$$

Changing the order of summation and using the identity $\binom{n}{i}\binom{n-i}{\nu-i}=\binom{n}{\nu}\binom{\nu}{i}$ and the formula (2.3.6) we obtain

$$\begin{aligned} B_n(x) &= \sum_{\nu=0}^{n}\sum_{i=0}^{\nu}\binom{n}{\nu}\binom{\nu}{i} f\left(\frac{i}{n}\right)(-1)^{\nu-i} x^\nu \\ &= \sum_{\nu=0}^{n}\binom{n}{\nu}\Delta_{1/n}^{\nu} f(0) x^\nu = \sum_{\nu=0}^{n} a_\nu^{(n)} x^\nu \end{aligned}$$

where $a_\nu^{(n)}\geqslant 0$.[5] Since the sequence $B_n(1)=\sum_{\nu=0}^{n} a_\nu^{(n)}$ converges it is also bounded: $\sum_{\nu=0}^{n} a_\nu^{(n)}\leqslant M$ for some constant M, so that $0\leqslant a_\nu^{(n)}\leqslant M$. Applying the diagonalization principle to the sequences $\{a_1^{(n)}\}, \{a_2^{(n)}\},\ldots$ we obtain a sequence of indices $\alpha_n\to\infty$ such that the corresponding subsequences are convergent:

$$a_1^{(\alpha_n)}\to a_1,\quad a_2^{(\alpha_n)}\to a_2,\ldots$$ [6]

Let $s_n(x)=\sum_{\nu=0}^{n} a_\nu x^\nu$. We have $0\leqslant a_\nu\leqslant M$, so it suffices to show that

$$s_{\alpha_n}(x)\to f(x) \qquad \text{in } [0,1].$$

Now let $0\leqslant x<1$ and let $\varepsilon>0$. Choose k so large that $2Mx^k(1-x)^{-1}<\varepsilon/3$ and then choose N so large that for $n\geqslant N$

$$|B_{\alpha_n}(x)-f(x)|<\tfrac{1}{3}\varepsilon \quad\text{and}\quad \sum_{\nu=0}^{k-1}|a_\nu^{(\alpha_n)}-a_\nu|<\tfrac{1}{3}\varepsilon.$$

Then

$$\begin{aligned} |s_{\alpha_n}(x)-f(x)| &\leqslant |s_{\alpha_n}(x)-B_{\alpha_n}(x)|+|B_{\alpha_n}(x)-f(x)| \\ &\leqslant \sum_{\nu=0}^{\alpha_n}|a_\nu^{(\alpha_n)}-a_\nu|x^\nu+\tfrac{1}{3}\varepsilon \\ &\leqslant \sum_{\nu=0}^{k-1}|a_\nu^{(\alpha_n)}-a_\nu|+2Mx^k(1-x)^{-1}+\tfrac{1}{3}\varepsilon\leqslant\varepsilon. \end{aligned}$$

Hence $s_{\alpha_n}(x)\to f(x)$ in $[0,1)$. ■

Remark. From Bernstein's theorem and equation (2.3.7) it follows that both condition (2.3.8) and the condition $f^{(n)}(x) \geqslant 0$ on $[0, a)$ $(n = 0, 1, \ldots)$ are equivalent to the absolute monotonicity of f on $[0, a)$.

THEOREM 2.3.3. *If f is of class $\mathscr{C}^p$ on $[0, 1]$, then*

$$B_n^{(i)}(x) \to f^{(i)}(x)$$

uniformly on $[0, 1]$, $(i = 1, \ldots, p)$.

Proof. Differentiating the polynomial

$$W_n(x) = \sum_{\nu=0}^{n} \binom{n}{\nu} a_\nu x^\nu (1-x)^{n-\nu}$$

we have

$$W_n'(x) = \sum_{\nu=1}^{n} \binom{n}{\nu} a_\nu \nu x^{\nu-1}(1-x)^{n-\nu} - \sum_{\nu=0}^{n-1} \binom{n}{\nu} a_\nu (n-\nu) x^\nu (1-x)^{n-\nu-1}$$

Hence, using the identity

$$\binom{n}{\nu+1}(\nu+1) = \binom{n}{\nu}(n-\nu) = n\binom{n-1}{\nu}$$

we obtain

$$W_n'(x) = \sum_{\nu=0}^{n-1} \binom{n-1}{\nu} n(a_{\nu+1} - a_\nu) x^\nu (1-x)^{n-1-\nu}$$

It follows that successive differentiation of Bernstein polynomials gives:

$$B_n'(x) = \sum_{\nu=0}^{n-1} \binom{n-1}{\nu} n \Delta_{1/n}^1 f\left(\frac{\nu}{n}\right) x^\nu (1-x)^{n-1-\nu},$$

$$\cdots\cdots\cdots\cdots\cdots\cdots\cdots\cdots$$

$$B_n^{(i)}(x) = \sum_{\nu=0}^{n-i} \binom{n-i}{\nu} n(n-1)\ldots(n-i+1) \Delta_{1/n}^i f\left(\frac{\nu}{n}\right) x^\nu (1-x)^{n-i-\nu},$$

$$\cdots\cdots\cdots\cdots\cdots\cdots\cdots\cdots$$

Fix i, $0 < i \leqslant p$. Let $\varepsilon > 0$ and let $\tilde{B}_k$ be the kth Bernstein polynomial for $f^{(i)}$. There exists $\delta > 0$ such that

$$|f^{(i)}(x'') - f^{(i)}(x')| \leqslant \tfrac{1}{3}\varepsilon \quad \text{for} \quad |x'' - x'| < \delta.$$

We have also $|f^{(i)}(x)| \leqslant M$ for some constant M. Choose N so large that for $n \geqslant N$

$$\left|1 - \left(1 - \frac{1}{n}\right)\cdots\left(1 - \frac{i-1}{n}\right)\right| \leqslant \frac{\varepsilon}{3M}, \quad \frac{i}{n} < \delta$$

and

$$|\tilde{B}_{n-i}(x) - f^{(i)}(x)| < \tfrac{1}{3}\varepsilon \qquad \text{in } [0, 1].$$

Then for $n \geqslant N$ and $x \leqslant \bar{x} \leqslant x+(i/n)$ we obtain, using (2.3.7),

$$\begin{aligned}
&|n(n-1)\cdots(n-i+1)\Delta_{1/n}^{i}f(x)-f^{(i)}(\bar{x})| \\
&=\left|\left(1-\frac{1}{n}\right)\cdots\left(1-\frac{i-1}{n}\right)f^{(i)}(\xi)-f^{(i)}(\bar{x})\right| \\
&\leqslant\left|1-\left(1-\frac{1}{n}\right)\cdots\left(1-\frac{i-1}{n}\right)\right|M+|f^{(i)}(\xi)-f^{(i)}(\bar{x})|\leqslant\tfrac{2}{3}\varepsilon,
\end{aligned}$$

for some ξ, $x<\xi<x+(i/n)$ and hence $|\xi-\bar{x}|<\delta$. It follows from this that

$$\begin{aligned}
&|B_n^{(i)}(x)-\tilde{B}_{n-i}(x)| \\
&\leqslant\sum_{\nu=0}^{n-i}\binom{n-i}{\nu}\left|n(n-1)\cdots(n-i+1)\Delta_{1/n}^{i}f\left(\frac{\nu}{n}\right)-f^{(i)}\left(\frac{\nu}{n-i}\right)\right|x^{\nu}(1-x)^{n-i-\nu} \\
&\leqslant\tfrac{2}{3}\varepsilon.
\end{aligned}$$

Hence

$$|B_n^{(i)}(x)-f^{(i)}(x)|\leqslant|B_n^{(i)}(x)-\tilde{B}_{n-i}(x)|+|\tilde{B}_{n-i}(x)-f^{(i)}(x)|\leqslant\varepsilon.$$

So $B_n^{(i)}(x)\rightarrow f^{(i)}(x)$ uniformly in $[0,1]$. ∎

We now turn to the case of several variables. Let f be a continuous function on $[0,1]\times[0,1]$, and let

$$\omega(\delta)=\sup_{\substack{|x'-x''|<\delta\\|y'-y''|<\delta}}|f(x'',y'')-f(x',y')|$$

be the modulus of continuity. The polynomial

$$B_n(x,y)=\sum_{i=0}^{n}\sum_{j=0}^{n}\binom{n}{i}\binom{n}{j}f\left(\frac{i}{n},\frac{j}{n}\right)x^i(1-x)^{n-i}y^j(1-y)^{n-j} \tag{2.3.9}$$

is called the nth *Bernstein polynomial* for f. We show that

$$|B_n(x,y)-f(x,y)|\leqslant 3\omega\left(\frac{1}{\sqrt{n}}\right) \tag{2.3.10}$$

from which it follows that $B_n(x,y)\rightarrow f(x,y)$ uniformly on $[0,1]\times[0,1]$.

Define $\varphi_y(x)=f(x,y)$ for y fixed and $\psi_x(y)=f(x,y)$ for x fixed. Let $\bar{\omega}_y(\delta)$, $\bar{\omega}_x(\delta)$ be the moduli of continuity for the functions φ_y, ψ_x. Then $\bar{\omega}_y(\delta)\leqslant\omega(\delta)$ and $\bar{\omega}_x(\delta)\leqslant\omega(\delta)$. The functions

$$\begin{aligned}
B_n^{(1)}(x,y)&=\sum_{i=0}^{n}\binom{n}{i}f\left(\frac{i}{n},y\right)x^i(1-x)^{n-i}\\
B_n^{(2)}(x,y)&=\sum_{j=0}^{n}\binom{n}{j}f\left(x,\frac{j}{n}\right)y^j(1-y)^{n-j}
\end{aligned} \tag{2.3.11}$$

are the nth Bernstein polynomials for φ_y and ψ_x (with y and x respectively fixed).

Then, by Th. 2.3.1 we have

$$|B_n^{(1)}(x,y)-f(x,y)|\leqslant\tfrac{3}{2}\bar{\omega}_y\left(\frac{1}{\sqrt{n}}\right)\leqslant\tfrac{3}{2}\omega\left(\frac{1}{\sqrt{n}}\right),$$

$$|B_n^{(2)}(x,y)-f(x,y)|\leqslant\tfrac{3}{2}\bar{\omega}_x\left(\frac{1}{\sqrt{n}}\right)\leqslant\tfrac{3}{2}\omega\left(\frac{1}{\sqrt{n}}\right).$$

Comparing (2.3.9) and (2.3.11) we see that

$$B_n(x,y)=\sum_{i=0}^{n}\binom{n}{i}B_n^{(2)}\left(\frac{i}{n},y\right)x^i(1-x)^{n-i},$$

so that, by (2.3.11)

$$\begin{aligned}|B_n(x,y)-B_n^{(1)}(x,y)|&\leqslant\sum_{i=0}^{n}\binom{n}{i}\left|B_n^{(2)}\left(\frac{i}{n},y\right)-f\left(\frac{i}{n},y\right)\right|x^i(1-x)^{n-i}\\&\leqslant\tfrac{3}{2}\omega\left(\frac{1}{\sqrt{n}}\right).\end{aligned}$$

Inequality (2.3.10) now follows.

Similarly, for a function f of k variables which is continuous in $[0,1]^k$ we can define the nth Bernstein polynomial

$$B_n(x_1,\ldots,x_k)=\sum_{i_1=0}^{n}\cdots\sum_{i_k=0}^{n}\binom{n}{i_1}\cdots\binom{n}{i_k}f\left(\frac{i_1}{n},\ldots,\frac{i_k}{n}\right)x_1^{i_1}(1-x_1)^{n-i_1}\cdots x_k^{i_k}(1-x_k)^{n-i_k}\qquad(2.3.12)$$

and we can show inductively, by the above argument, that

$$|B_n(x_1,\ldots,x_k)-f(x_1,\ldots,x_k)|\leqslant\frac{3k}{2}\omega\left(\frac{1}{\sqrt{n}}\right),\qquad(2.3.13)$$

where

$$\omega(\delta)=\sup_{|x_1''-x_1'|<\delta,\ldots,|x_k''-x_k'|<\delta}|f(x_1'',\ldots,x_k'')-f(x_1',\ldots,x_k')|$$

is the modulus of continuity of the function f.

Also, we can show that when the function f is of class $\mathscr{C}^p$, then the partial derivatives of order $\leqslant p$ of the Bernstein polynomials coverage uniformly to the corresponding partial derivatives of the function f.

2.4 THE STONE–WEIERSTRASS THEOREM

Let $\mathscr{H}$ be a class of functions defined on a given set E. We say that the function f, defined on E, can be *approximated uniformly* (on E) by functions of the family $\mathscr{H}$ if there exists a sequence $h_n\in\mathscr{H}$ which converges to f uniformly on E or, equivalently, if for any $\varepsilon>0$ there is a function $h\in\mathscr{H}$ such that

$$|h(x)-f(x)|\leqslant\varepsilon\qquad\text{for } x\in E.$$ [7]

Let $\mathscr{H}_1$ be a class of functions defined on E such that every function in this class can be uniformly approximated by functions in the class $\mathscr{H}$. Then, if a function f defined on E can be approximated uniformly by functions in the family $\mathscr{H}_1$, then they can also be approximated uniformly by functions in the family $\mathscr{H}$.[8] Indeed, in this case, given $\varepsilon>0$ there exists $h_1 \in \mathscr{H}_1$ such that $|h_1(x)-f(x)| \leqslant \frac{1}{2}\varepsilon$ on E, and $h \in \mathscr{H}$ such that $|h(x)-h_1(x)| \leqslant \frac{1}{2}\varepsilon$ on E, whence $|h(x)-f(x)| \leqslant \varepsilon$ on E.

Weierstrass' theorem (for functions of one variable) states that *every function continuous in a closed interval* $[a,b]$ *can be uniformly approximated by polynomials.*

Every continuous function f on $[a,b]$ can be uniformly approximated by polygonal functions, that is, functions whose graphs are connected line segments. For, let $\varepsilon>0$ and take $\delta>0$ such that

$$|f(x'')-f(x')| \leqslant \varepsilon \qquad \text{for } |x''-x'| \leqslant \delta.$$

Let the subdivision $a=x_0<\cdots<x_n=b$ be such that $\max_i |x_i-x_{i-1}|<\delta$ and let φ be the polygonal function with vertices $(x_i, f(x_i))$, then we will have

$$|\varphi(x)-f(x)| \leqslant \varepsilon \qquad \text{in } [a,b].$$

So, to prove Weierstrass' theorem it suffices to show that every polygonal function can be uniformly approximated on $[a,b]$ by polynomials. But, since every polygonal function can be written in the form

$$\sum_{i=1}^{n} c_i|x-a_i| + cx + d \quad \text{in } [a,b],$$

it suffices, in turn, to prove the lemma that the function $x \to |x|$ can be uniformly approximated by polynomials in any bounded interval.

This lemma (whose proof follows below) will enable us also to prove a much more general theorem which contains Weierstrass' theorem as a special case.

LEMMA. *For any $M>0$ there exists a sequence of polynomials which converges uniformly in the interval $[-M,M]$ to the function $x \to |x|$.*

Proof. We can put $M=1$, for if p_n is a sequence of polynomials converging to $|x|$ uniformly on $[-1,1]$ then the sequence of polynomials $Mp_n(x/M)$ converges to $M|x/M|=|x|$ uniformly on $[-M,M]$. We have

$$\sqrt{(1-u)} = 1 - c_1u - c_2u^2 - \cdots \qquad \text{for } 0 \leqslant u < 1,$$

where

$$c_n = -(-1)^n \binom{\frac{1}{2}}{n} = \frac{1.3\cdots(2n-3)}{2.4\cdots(2n-2)2n} \geqslant 0.$$

Now $c_n = a_{n-1} - a_n$, where

$$a_0 = 1,\ a_n = \frac{1.3\cdots(2n-1)}{2.4\cdots 2n} \geqslant 0,$$

hence $c_1 + \cdots + c_n = a_0 - a_n \leqslant 1$, and so

$$\sum_{n=1}^{\infty} c_n < \infty.$$

Hence, the series $\sum_{n=1}^{\infty} c_n u^n$ is uniformly convergent for $0 \leqslant u \leqslant 1$. Thus

$$1 - c_1 u - \cdots - c_n u^n \to \sqrt{(1-u)} \text{ uniformly for } 0 \leqslant u \leqslant 1$$

whence

$$1 - c_1(1-x^2) - \cdots - c_n(1-x^2)^n \to \sqrt{[1-(1-x^2)]} = |x| \text{ uniformly in } [-1, 1].$$

A sequence of polynomials $p_n(x)$ converging uniformly to $|x|$ on $[-1, 1]$ can also be obtained by the following method. We define inductively:

$$p_0 = 0,\ p_{n+1} = p_n + \tfrac{1}{2}(x^2 - p_n^2).$$

From this we obtain the relations

$$|x| - p_{n+1} = (|x| - p_n)[1 - \tfrac{1}{2}(|x| + p_n)],$$
$$|x| + p_{n+1} = (|x| + p_n)[1 + \tfrac{1}{2}(|x| - p_n)]$$

from which we can prove by induction that $|p_n| \leqslant |x|$ on $[-1, 1]$. It follows from the definition of the sequence p_n that this sequence is increasing and bounded and so converges to a bounded function $p(x) \geqslant 0$ which must satisfy the equation $p = p + \frac{1}{2}(x^2 - p^2)$, that is

$$p(x) = |x|.$$

It follows from Dini's theorem (Th. 3.1.7) that the convergence is uniform on $[-1, 1]$.

THEOREM 2.4.1. *Let $\mathscr{H}$ be a family of continuous functions on a compact metric[9] space X. Suppose that the family $\mathscr{H}$ has the following property:*

$$g \in \mathscr{H}, h \in \mathscr{H} \quad \textit{implies that} \quad \min(g, h) \in \mathscr{H}, \quad \max(g, h) \in \mathscr{H}. \qquad (2.4.1)$$

If a function f continuous on X satisfies the condition: for each $\varepsilon > 0$ and for each pair of points a, $b \in X$ there is a function $h_{ab} \in \mathscr{H}$ such that

$$|f(a) - h_{ab}(a)| < \varepsilon \quad \textit{and} \quad |f(b) - h_{ab}(b)| < \varepsilon.$$

then the function f can be approximated uniformly by functions from the family $\mathscr{H}$.

Proof. Let $\varepsilon > 0$. Fix $b \in X$. For each $a \in X$ we have $f(a) - \varepsilon < h_{ab}(a)$ and so by the continuity of f and h_{ab}

$$f(x) - \varepsilon < h_{ab}(x)$$

in some neighbourhood U_a of the point a. It follows from the Borel–Lebesgue theorem that $X \subset U_{a_1} \cup \cdots \cup U_{a_k}$ for some $a_1, \ldots, a_k \in X$. So if we define $h_b(x) = \max(h_{a_1 b}(x), \ldots, h_{a_k b}(x))$ then, by (2.4.1) $h_b \in \mathscr{H}$ and $f(x) - \varepsilon < h_b(x)$ on X and also

$h_b(b) < f(b) + \varepsilon$. From the last inequality it follows that

$$h_b(x) < f(x) + \varepsilon$$

in some neighbourhood V_b of the point b. Again, the Borel–Lebesgue theorem implies that $X \subset V_{b_1} \cup \cdots \cup V_{b_l}$ for some $b_1, \ldots, b_l \in X$. So, if we take $h(x) = \min(h_{b_1}(x), \ldots, h_{b_l}(x))$ then, by (2.4.1), $h \in \mathscr{H}$ and also $h(x) < f(x) + \varepsilon$ on X and $f(x) - \varepsilon < h(x)$ on X, that is

$$|h(x) - f(x)| < \varepsilon \quad \text{on } X. \quad \blacksquare$$

COROLLARY. *If the family $\mathscr{H}$, in addition to the properties specified in theorem 2.4.1, satisfies the condition: for each pair of points $a, b \in X$, $a \neq b$ and for each pair of numbers α, β there exists a function $h \in \mathscr{H}$ such that $h(a) = \alpha$, $h(b) = \beta$, then every continuous function on X can be uniformly approximated by a function of the family $\mathscr{H}$.*

Let $\mathscr{H}$ be a family of functions defined on a set E. We say that *the family $\mathscr{H}$ separates the points of the set E* if for every pair of distinct points $x, y \in E$ there exists a function $h \in \mathscr{H}$ such that $h(x) \neq h(y)$.

THEOREM 2.4.2 (Stone). *Let $\mathscr{H}$ be a family of continuous functions defined on a compact metric space X. Suppose that this family has the following properties:*

(i) *it is linear: $f, g \in \mathscr{H}$ implies that $\alpha f + \beta g \in \mathscr{H}$,*
(ii) *$f \in \mathscr{H}$ implies that $|f| \in \mathscr{H}$,*
(iii) *it contains all constant non-zero functions,*[10]
(iv) *it separates the points of X.*

Then, every continuous function on X can be approximated uniformly by functions of the family $\mathscr{H}$.

Proof. Since

$$\max(f, g) = \frac{f+g}{2} + \frac{|f-g|}{2} \quad \text{and} \quad \min(f, g) = \frac{f+g}{2} - \frac{|f-g|}{2}$$

therefore, by properties (i) and (ii), the family $\mathscr{H}$ satisfies condition (2.4.1) of theorem 2.4.1. It therefore suffices to show that H also satisfies the supplementary condition of the corollary to Th. 2.4.1. So, let $a, b \in X$, $a \neq b$, and let α, β be real numbers. By (iv) there is a function $f \in \mathscr{H}$ such that $f(a) \neq f(b)$. By (i) and (iii) the function

$$g(x) = \alpha + \frac{\beta - \alpha}{f(b) - f(a)} [f(x) - f(a)]$$

belongs to the family $\mathscr{H}$. But $g(a) = \alpha$ and $g(b) = \beta$. $\blacksquare$

THEOREM 2.4.3 (Stone–Weierstrass). *Let $\mathscr{H}$ be a family of functions which* *are continuous on a compact metric space X and which separate the points of this space. Then, every continuous function on X may be uniformly approximated by polynomials in the functions of this family, that is, by functions of the form*

$$x \to w(h_1(x), \ldots, h_l(x))^{11} \tag{2.4.2}$$

where w is a polynomial and $h_1, \ldots, h_l \in \mathscr{H}$.

Proof. We denote by $\mathscr{W}$ the family of all functions of the form (2.4.2). This family satisfies conditions (i), (iii), (iv) of Th. 2.4.2. Let $\bar{\mathscr{W}}$ denote the family of all functions which can be uniformly approximated by functions of the family $\mathscr{W}$. It therefore suffices to show that every continuous function can be uniformly approximated by functions of the family $\bar{\mathscr{W}}$.

Since $\mathscr{W} \subset \bar{\mathscr{W}}$, therefore $\bar{\mathscr{W}}$ satisfies the conditions (iii) and (iv). $\bar{\mathscr{W}}$ also satisfies condition (i). For if $f, g \in \bar{\mathscr{W}}$ then $f_n \to f$, $g_n \to g$ uniformly on X for some $f_n, g_n \in \mathscr{W}$. Hence $\alpha f_n + \beta g_n \in \mathscr{W}$ and $\alpha f_n + \beta g_n \to \alpha f + \beta g$ uniformly on X and so $\alpha f + \beta g \in \bar{\mathscr{W}}$. By Th. 2.4.2 it remains to show that $\bar{\mathscr{W}}$ satisfies condition (ii).

Let $f \in \bar{\mathscr{W}}$. Then $f_n \to f$ uniformly on X for some $f_n \in \mathscr{W}$. Since f is continuous on the compact space X, therefore $|f(x)| \leqslant M$ for $x \in X$ for some constant M. By the lemma there exists a sequence of polynomials q_n which converges uniformly to the function $t \to |t|$ on $[-M-1, M+1]$.

Since $x \to q_n(f_n(x))$ are functions of the form (2.4.2) it suffices to show that

$$q_n(f_n(x)) \to |f(x)| \qquad \text{uniformly on } X,$$

for this means that $|f| \in \bar{\mathscr{W}}$.

Let $0 < \varepsilon < 2$. Choose N so large that for $n \geqslant N$

$$|f_n(x) - f(x)| \leqslant \tfrac{1}{2}\varepsilon \qquad \text{on } X$$

and

$$|q_n(t) - |t|| \leqslant \tfrac{1}{2}\varepsilon \qquad \text{on } [-M-1, M+1].$$

Then (for $n \geqslant N$) $|f_n(x)| \leqslant M+1$ and so $|q_n(f_n(x)) - |f_n(x)|| \leqslant \tfrac{1}{2}\varepsilon$, whence

$$|q_n(f_n(x)) - |f(x)|| \leqslant |q_n(f_n(x)) - |f_n(x)|| + ||f_n(x)| - |f(x)|| \leqslant \varepsilon \text{ on } X.$$

This concludes the proof. ■

In particular, let X be a compact subset of the space $\mathbb{R}^k$. We see that the family consisting of the functions $x \to x_1, \ldots, x \to x_k$ separates the points of the set X. Hence, every continuous function on X can be uniformly approximated by polynomials $x \to w(x_1, \ldots, x_k)$. Thus, Th. 2.4.3 contains Weierstrass' theorem as a special case.

NOTES

1. We put $u = (u_1, \ldots, u_k)$, $x = (x_1, \ldots, x_k)$, etc.
2. For example, one can take $f(x) = T(x)\varphi(x)$ on $[\alpha, a]$, where T is the pth Taylor polynomial of f at the point a, and $\varphi(x) = [1 - (x - a/\alpha - a)^p]^p$.

3. The formula for the extension of f from P to Q while preserving the class $\mathscr{C}^p$ is much more complicated.
4. This condition is just equivalent to uniform continuity.
5. It is clear from this that it suffices to assume the continuity of the function f and $\Delta^{\nu}_{1/n,}f(0) \geqslant 0$ $(\nu = 0, 1, \ldots, n; n = 1, 2, \ldots)$
6. By induction we would demonstrate the convergence of the sequence $\{a_1^{(n)}\}, \{a_2^{(n)}\}, \ldots$, themselves.
7. This means that $f \in \bar{\mathscr{H}}$ in the space of all functions using the metric of uniform convergence (in the space of bounded functions $\rho'(g, h) = \sup|g - h|$ is a suitable metric; in the space of all functions it is necessary to take $\rho = \lambda \circ \rho'$, where $\lambda(t) = t/(1 + t)$ for $0 \leqslant t < \infty$ and $\lambda(\infty) = 1$).
8. In other words, $\mathscr{H}_1 \subset \bar{\mathscr{H}}$ implies $\bar{\mathscr{H}}_1 \subset \bar{\mathscr{H}}$.
9. In the remainder of this section the existence of a metric is not essential; the arguments carry over without change to the case of any (compact) topological space.
10. By (i) it suffices to assume that the family contains just one such constant function.
11. l is not fixed!

CHAPTER 3

FUNCTIONS ON METRIC SPACES

3.1 CONTINUOUS FUNCTIONS

Let f be a real function defined on a metric space E. We say that *f is continuous at x_0* if $f(x_n)\to f(x_0)$ whenever $x_n\to x_0$. (This definition is due to Heine.) We say that *f is continuous on a set $H\subset E$* if it is continuous at each point of H. A function f is continuous at every isolated point of a space E. When x_0 is an accumulation point of E then the continuity of f at x_0 is equivalent to the condition

$$\lim_{x\to x_0} f(x)=f(x_0).$$

In the case where $|f(x_0)|<\infty$ the above definition is equivalent to the following definition due to Cauchy: given $\varepsilon>0$ there exists $\delta>0$ such that

$$|f(x)-f(x_0)|<\varepsilon \quad \text{when} \quad \rho(x,x_0)<\delta.$$[1]

If the functions $f_1,\ldots,f_k$ are continuous at x_0 and if the function φ is defined on a subset of the space $\bar{\mathbb{R}}^k$ and is continuous at $(f_1(x_0),\ldots,f_k(x_0))$, then the function $\varphi(f_1,\ldots,f_k)$ is continuous at x_0. Hence a linear combination, product, quotient, maximum or minimum of continuous functions at x_0 is a continuous function at x_0 if it is defined at x_0 and it is not of the form $\mp\infty\pm\infty$ or $0\cdot(\mp\infty)$ at x_0 in the case of linear combinations and products respectively).

The continuity of f is equivalent to the continuity of the bounded function

$$x\to \arctan f(x) \qquad (\arctan(\pm\infty)=\pm\tfrac{1}{2}\pi).$$

Let $x_0\in A\subset E$. If $f(x)$ is continuous at x_0 then the restriction f_A is also continuous at x_0. If $x_0\in \operatorname{Int} A$ and if f_A is continuous at x_0 then f is continuous at x_0.

THEOREM 3.1.1. *Let f be a function defined on the union of closed sets $F_1,\ldots,F_m$. If the restriction f_{F_i} is continuous on F_i $(i=1,\ldots,n)$ then f is continuous on $F_1\cup\cdots\cup F_m$.*

Proof. Let $x_\nu\to x_0$ where x_ν, $x_0\in F_1\cup\cdots\cup F_m$. Consider the sequence $\{f(x_n)\}$.

Let $\{f(x_{\alpha_n})\}$ be an arbitrary convergent subsequence. Then, for some k the set F_k contains infinitely many terms of the sequence $\{x_{\alpha_n}\}$; that is, it contains a subsequence $\{x_{\beta_n}\}$ and also $x_0 \in F_k$. Hence

$$f(x_{\beta_n}) = f_{F_k}(x_{\beta_n}) \to f_{F_k}(x_0) = f(x_0).$$

Hence

$$f(x_{\alpha_n}) \to f(x_0) \quad \text{and so} \quad f(x_n) \to f(x_0).^2 \quad \blacksquare$$

THEOREM 3.1.2. *The function f is continuous on the space E if and only if for every finite number α the sets*

$$\{x: f(x) < \alpha\} \quad \textit{and} \quad \{x: f(x) > \alpha\}$$

are open.

Proof. Necessity. Let $x_0 \in \{x: f(x) < \alpha\}$, that is $f(x_0) < \alpha$. If $x_\nu \to x_0$ then $f(x_\nu) \to f(x_0)$, so that $f(x_\nu) < \alpha$, that is,

$$x_\nu \in \{x: f(x) < \alpha\} \qquad \text{for } \nu \text{ sufficiently large.}$$

Hence $\{x: f(x) < \alpha\}$ is open. Similarly $\{x: f(x) > \alpha\}$ will be open.

Sufficiency. Let $x_\nu \to x_0$. If $|f(x_0)| < \infty$ then for any $\varepsilon > 0$ the open set

$$\{x: f(x) > f(x_0) - \varepsilon\} \cap \{x: f(x) < f(x_0) + \varepsilon\}$$

contains x_0, and so for ν sufficiently large it contains x_ν also. Then $f(x_0) - \varepsilon < f(x_\nu) < f(x_0) + \varepsilon$. Hence

$$f(x_\nu) \to f(x_0).$$

If $f(x_0) = \infty$ then for $\alpha < \infty$, $x_0 \in \{x: f(x) > \alpha\}$ and so $x_\nu \in \{x: f(x) > \alpha\}$, that is $f(x_\nu) > \alpha$, for ν sufficiently large. Thus $f(x_\nu) \to \infty$. The argument is similar when $f(x_0) = -\infty$. $\blacksquare$

COROLLARY 3.1.1. *The function f is continuous on the space E if and only if for every finite number α the sets*

$$\{x: f(x) \leqslant \alpha\} \quad \textit{and} \quad \{x: f(x) \geqslant \alpha\}$$

are closed.

Remark. In the proof of necessity one could take in place of the set

$$\{x: f(x) < \alpha\} = f^{-1}([-\infty, \alpha))$$

any inverse image $f^{-1}(G)$ of an arbitrary set G open in $\bar{\mathbb{R}}$. Hence, for a function f to be continuous on a space E it is necessary and sufficient that the inverse image of any open set in $\bar{\mathbb{R}}$ is an open set in E (or, equivalently, that the inverse image of any closed set in $\bar{\mathbb{R}}$ is a closed set in E).

THEOREM 3.1.3 (Bolzano). *If the function f is continuous on a connected space E, then for each $\xi \in [f(a), f(b)]$ there exists $x \in E$ such that $f(x) = \xi$.*

Proof. Let $f(a) < \xi < f(b)$. Because

$$E = \{x{:}f(x) < \xi\} \cup \{x{:}f(x) = \xi\} \cup \{x{:}f(x) > \xi\}$$

it follows that if $\{x{:}f(x) = \xi\} = \varnothing$ then E would be the union of two non-empty, disjoint open sets. This would contradict the connectedness of E. ■

THEOREM 3.1.4 (Weierstrass). *A continuous function on a compact set attains its maximum and its minimum.*

THEOREM 3.1.5. *A finite, continuous function on a compact set is uniformly continuous, that is, given* $\varepsilon > 0$ *there exists* $\delta > 0$ *such that*

$$|f(x') - f(x)| \leqslant \varepsilon \quad \text{for} \quad \rho(x', x) \leqslant \delta.$$

Uniform and continuous convergence

In the remainder of this section f will be a finite function on E and f_n a sequence of functions defined on E. We say that

$$f_n(x) \to f(x) \quad \text{or} \quad f_n \to f$$

uniformly on E if, given $\varepsilon > 0$ there exists N such that

$$|f_n(x) - f(x)| \leqslant \varepsilon \qquad \text{for } x \in E \text{ and } n \geqslant N.$$

This means that

$$\varepsilon_n = \sup_E |f_n(x) - f(x)| \to 0.$$

Thus, in the space of bounded functions uniform convergence is equivalent to convergence relative to the metric $\rho(f, g) = \sup_E |f(x) - g(x)|$; see Note 2.7.

THEOREM 3.1.6. *If* $f_n(x) \to f(x)$ *uniformly on* E *and if the* f_n *are continuous at* x_0, *then* f *is continuous at* x_0.

THEOREM 3.1.7 (Dini). *If* E *is a compact space,* f *and* f_n *are continuous on* E *and the sequence* $f_n(x)$ *is monotone and convergent to* $f(x)$ *on* E, *then the convergence is uniform,*

Proof. The functions

$$\varphi_n(x) = |f_n(x) - f(x)|$$

are continuous and form a decreasing sequence converging to zero on E. It suffices to show that this sequence converges to zero uniformly on E. Let $\varepsilon > 0$. The sets $E_n = \{x{:}\varphi_n(x) \geqslant \varepsilon\}$ are compact and form a decreasing sequence. If $E_n \neq \varnothing$ for all n then, by Cantor's theorem, the sets E_n would have a common point x_0, so that $\varphi_n(x_0) \geqslant \varepsilon$ for all n. This contradicts the assumption $\varphi_n(x_0) \to 0$. Hence $E_N = \varnothing$ for some N, whence $\varphi_N(x) < \varepsilon$ on E and $\varphi_n(x) < \varepsilon$ for $n \geqslant N$, $x \in E$. ■

We say that f_n converges to f *continuously at* x_0 if

$$f_n(x_n) \to f(x_0) \qquad \text{whenever } x_n \to x_0.^3$$

It follows that every subsequence f_{α_n} also converges to f continuously at x_0. Indeed, if $x_n \to x_0$ then

$$x_n = x'_{\alpha_n} \qquad \text{for some } x'_n \to x_0$$

(for example, put $x'_{\alpha_n} = x_n$ and $x'_\nu = x_0$ for the remaining ν). Since $f_n(x'_n) \to f(x_0)$, therefore

$$f_{\alpha_n}(x_n) = f_{\alpha_n}(x'_{\alpha_n}) \to f(x_0).$$

THEOREM 3.1.8. *Let $f_n(x) \to f(x)$ on E and let the convergence be continuous at x_0, then f is continuous at x_0.*

Proof. Let $x_n \to x_0$. Since $f(x_n) = \lim_{\nu \to \infty} f_\nu(x_n)$ therefore

$$|f_{\alpha_n}(x_n) - f(x_n)| < \frac{1}{n}$$

for some indices α_n, and we can also require that the sequence $\{\alpha_n\}$ be strictly increasing. Since the subsequence f_{α_n} converges to f continuously at x_0, therefore $f_{\alpha_n}(x_n) \to f(x_0)$. Hence $f(x_n) \to f(x_0)$. ■

THEOREM 3.1.9. *If the space E is compact, then uniform convergence to a continuous function on E is equivalent to continuous convergence at every point.*

Proof. (i). Suppose that f_n tends uniformly to a continuous function f on E. Let $x_0 \in E$ and let $x_n \to x_0$. Then $f(x_n) \to f(x_0)$. Let $\varepsilon > 0$ and choose N so large that for $n \geqslant N$ we have

$$|f(x_n) - f(x_0)| \leqslant \tfrac{1}{2}\varepsilon \quad \text{and} \quad |f_n(x) - f(x)| \leqslant \tfrac{1}{2}\varepsilon \text{ on } E.$$

Then (for $n \geqslant N$) $|f_n(x_n) - f(x_n)| \leqslant \frac{1}{2}\varepsilon$ so that

$$|f_n(x_n) - f(x_0)| \leqslant \varepsilon$$

Hence $f_n(x_n) \to f(x_0)$; thus we have continuous convergence at x_0.

(ii). Suppose that $f_n \to f$ continuously at x_0 for every $x_0 \in E$. By Th. 3.1.8, the function f is continuous on E. Suppose that the convergence is not uniform on E. Then, there exist $\varepsilon > 0$, $\alpha_n \geqslant n$ and $x_n \in E$ such that

$$|f_{\alpha_n}(x_n) - f(x_n)| \geqslant \varepsilon > 0 \tag{3.1.1}$$

and we can require that the sequence α_n be strictly increasing. From the sequence $\{x_n\}$ select a convergent subsequence $x_{n_\nu} \to x_0$. Then $f(x_{n_\nu}) \to f(x_0)$. But $f_{\alpha_n} \to f$ continuously at x_0, so that $f_{\alpha_{n_\nu}}(x_{n_\nu}) \to f(x_0)$. Hence

$$f_{\alpha_{n_\nu}}(x_{n_\nu}) - f(x_{n_\nu}) \to 0$$

which contradicts (3.1.1). ■

3.2 EQUICONTINUOUS FAMILIES OF FUNCTIONS

Let $\mathscr{R}$ be a family of finite functions defined on a metric space E. We say that $\mathscr{R}$ *is equicontinuous at* x_0 if for all $\varepsilon > 0$ there exists $\delta > 0$ such that

$$|f(x) - f(x_0)| \leqslant \varepsilon \qquad \text{when } f \in \mathscr{R} \quad \text{and} \quad \rho(x, x_0) \leqslant \delta.$$

Every subfamily of a family equicontinuous at x_0 is an equicontinuous family at x_0. If all the functions of the family $\mathscr{R}$ satisfy a *Lipschitz condition* with a common constant M:

$$|f(x') - f(x)| \leqslant M\rho(x, x')$$

then the family $\mathscr{R}$ is equicontinuous at each point.

We say that the family $\mathscr{R}$ is *bounded* (*above, below*) at x_0 if there exists a constant M such that

$$|f(x_0)| \leqslant M \qquad (f(x_0) \leqslant M, f(x_0) \geqslant M) \qquad \text{for all } f \in \mathscr{R}.$$

The family $\mathscr{R}$ is called *uniformly bounded* (*above, below*) on the set $A \subset E$ if there exists a constant M such that

$$|f(x)| \leqslant M \qquad (f(x) \leqslant M, f(x) \geqslant M) \qquad \text{for } x \in A \quad \text{and} \quad f \in \mathscr{R}.$$

The function

$$\varphi(x) = \sup_{f \in \mathscr{R}} f(x)$$

is called the *upper envelope* of the family $\mathscr{R}$, while

$$\psi(x) = \inf_{f \in \mathscr{R}} f(x)$$

is the *lower envelope* of $\mathscr{R}$. Suppose that the family $\mathscr{R}$ is equicontinuous at x_0 and let $\delta > 0$ correspond to the number $\varepsilon > 0$ in the definition of equicontinuity. Thus, if $\rho(x, x_0) \leqslant \delta$ then

$$f(x_0) - \varepsilon \leqslant f(x) \leqslant f(x_0) + \varepsilon \qquad \text{for } f \in \mathscr{R},$$

from which follows taking the suprema

$$\varphi(x_0) - \varepsilon \leqslant \varphi(x) \leqslant \varphi(x_0) + \varepsilon.$$

So we have

THEOREM 3.2.1. *Suppose that the family* $\mathscr{R}$ *is equicontinuous at* x_0 *and let* φ *be the upper envelope of this family. If* $\varphi(x_0) < \infty$ *then* φ *is continuous at* x_0. *If* $\varphi(x_0) = \infty$ *then* $\varphi(x) = \infty$ *in some neighbourhood of* x_0. *Similarly for the lower envelope.*

In the case $\varphi(x_0) < \infty$, the value of δ corresponding to ε in establishing the continuity of φ was identical to that given in the condition of equicontinuity of the family $\mathscr{R}$. We therefore have

THEOREM 3.2.2. *If the family $\mathscr{R}$ is equicontinuous at x_0, then if we add to the family all finite upper and lower envelopes of subfamilies of the family $\mathscr{R}$ we obtain an equicontinuous family (with the same δ, ε correspondence) at x_0.*

THEOREM 3.2.3. *Let E be a compact space and let the family $\mathscr{R}$ be equicontinuous at each point of this space. If $\mathscr{R}$ is bounded (above, below) at every point of a set Z dense in E, then $\mathscr{R}$ is uniformly bounded (above, below) on E.*

Proof. Suppose that $\mathscr{R}$ is bounded above at each $x \in Z$, that is $\varphi(x) < \infty$ in Z, where φ is the upper envelope. But then $\varphi(x) < \infty$ in E, for if $\varphi(x_0) = \infty$ then by Th. 3.2.1 we would have $\varphi(x) = \infty$ in a neighbourhood of x_0, which is impossible, since Z is dense in E. Thus $\varphi(x)$ is finite and continuous in E, hence $\varphi(x) \leqslant M$ in E for some constant $M < \infty$, that is

$$f(x) \leqslant M \qquad \text{on } E \text{ for } f \in \mathscr{R}. \quad \blacksquare$$

THEOREM 3.2.4. *Suppose that E is a compact space and that $\{f_n\}$ is a sequence equicontinuous at each point of the space. If the sequence $\{f_n(x)\}$ is convergent to a finite limit at each point x of a set Z dense in E, then the sequence $\{f_n\}$ is uniformly convergent in E.*

Proof. The sequence $f_n(x)$ is bounded at each point of Z, so by Th. 3.2.3 it is (uniformly) bounded on E. By Th. 3.2.2 it follows that the sequences

$$\varphi_n(x) = \inf_{\nu \geqslant n} f_\nu(x), \quad \psi_n(x) = \sup_{\nu \geqslant n} f_\nu(x)$$

are equicontinuous and their envelopes

$$\varphi(x) = \sup_n \varphi_n(x), \quad \psi(x) = \inf_n \psi_n(x)$$

are continuous in E. The sequences $\{\varphi_n\}$ and $\{\psi_n\}$ are monotone and $\varphi_n(x) \to \varphi(x)$, $\psi_n(x) \to \psi(x)$, so, by Th. 3.1.7, the convergence is uniform. Since

$$\varphi(x) = \liminf_{\nu \to \infty} f_\nu(x) \quad \text{and} \quad \psi(x) = \limsup_{\nu \to \infty} f_\nu(x)$$

therefore $\varphi(x) = \psi(x)$ on E (for $\varphi(x)$ and $\psi(x)$ are continuous and Z is dense in E). Hence, from the inequalities $\varphi_n(x) \leqslant f_n(x) \leqslant \psi_n(x)$ on E, it follows that

$$f_n(x) \to \varphi(x) = \psi(x)$$

uniformly on E. $\blacksquare$

THEOREM 3.2.5 (Ascoli). *Suppose that E is a compact space and $\mathscr{R}$ is a family of functions finite and continuous on E. Then, for the family $\mathscr{R}$ to be relatively compact, that is, for every sequence of functions from $\mathscr{R}$ to have a uniformly convergent subsequence on E, it is necessary and sufficient that $\mathscr{R}$ be equicontinuous and bounded at each point of E.*

Proof. Sufficiency. Let $\{f_\nu\}$ be any sequence from the family $\mathscr{R}$. Since E is compact, it contains a set Z which is dense in E and at most countable. By the diagonalization principle, we can select a subsequence $\{f_{\alpha_\nu}\}$ which converges to a finite limit at every point in Z. Then, by Th. 3.2.4 it is uniformly convergent on E.

Necessity. If the family $\mathscr{R}$ is not bounded at some point x_0 then there exists a sequence f_n from $\mathscr{R}$ such that $|f_n(x_0)| \to \infty$. From this sequence it is not possible to select a subsequence which is uniformly convergent. Hence the family $\mathscr{R}$ must be bounded at every point. Next, suppose that $\mathscr{R}$ is not equicontinuous at some point x_0. Thus, there exist $\varepsilon > 0, f_n \in \mathscr{R}$ and x_n such that $\rho(x_n, x_0) < 1/n$ and

$$|f_n(x_n) - f_n(x_0)| \geqslant \varepsilon > 0. \tag{3.2.1}$$

From $\{f_n\}$ select a subsequence $\{f_{\alpha_n}\}$ which converges uniformly on E to some function f continuous on E. By Th. 3.1.9, it is continuously convergent at x_0, so that since $x_{\alpha_n} \to x_0$ we have $f_{\alpha_n}(x_{\alpha_n}) \to f(x_0)$. But, since $f_{\alpha_n}(x_0) \to f(x_0)$, therefore

$$f_{\alpha_n}(x_{\alpha_n}) - f_{\alpha_n}(x_0) \to 0$$

contradicting (3.2.1). Thus the family $\mathscr{R}$ must be equicontinuous at each point of E. ■

3.3 SEMICONTINUOUS FUNCTIONS

Let f be a real function defined on a space E. We say that the function f is *lower semicontinuous* at x_0, or *upper semicontinuous* at x_0 if

$$\liminf_{\rho(x, x_0) \to 0} f(x) \geqslant f(x_0) \quad \text{or} \quad \limsup_{\rho(x, x_0) \to 0} f(x) \leqslant f(x_0)^4 \tag{3.3.1}$$

respectively. Hence, if f is both lower and upper semicontinuous at x_0 then this is equivalent to the condition

$$\lim_{\rho(x, x_0) \to 0} f(x) = f(x_0),$$

that is, to the continuity of f at x_0.

From the properties of limits (see Eq. (0.2.9)) it follows that the function f is upper semicontinuous at x_0 if and only if the function $-f$ is lower semicontinuous at x_0. For the sake of definiteness we will formulate the theorems for the case of lower semicontinuity in what follows.

We say that the function f is *lower semicontinuous* on the set $H \subset E$ if it is lower semicontinuous at every point of this set.

Each of the following conditions, (3.3.2)–(3.3.5), is equivalent to the lower semicontinuity of the function f at x_0. In accordance with the definition of lim inf (see §0.2):

$$x_n \to x_0 \quad \text{and} \quad f(x_n) \to \lambda \quad \text{implies} \quad \lambda \geqslant f(x_0); \tag{3.3.2}$$

or alternatively (see Note 0.6),

$$x_n \to x_0 \quad \text{implies} \quad \liminf_{n \to \infty} f(x_n) \geqslant f(x_0); \tag{3.3.3}$$

next, in the case where x_0 is an accumulation point of E (see (0.2.3))

$$\liminf_{x\to x_0} f(x) \geqslant f(x_0); \tag{3.3.4}$$

finally (see (0.2.5)), we have a Cauchy type definition:

$$\begin{aligned} &\text{for all } a < f(x_0) \text{ there exists } \delta > 0 \text{ such that} \\ &a < f(x) \text{ when } \rho(x, x_0) < \delta. \end{aligned} \tag{3.3.5}$$

Let $x_0 \in A \subset E$. From the definitions, (3.3.2) or (3.3.3) it follows that if f is lower semicontinuous at x_0 then the restriction f_A is also lower semicontinuous at x_0. If, in addition, $x_0 \in \operatorname{Int} A$, then the lower semicontinuity at x_0 of the restriction f_A implies that f is also lower semicontinuous at x_0.

Inequality (3.3.1) is true for the function f if and only if it is also true for the function $\arctan f$ (see (0.2.8)). Thus, lower semicontinuity of f is equivalent to the lower semicontinuity of the finite bounded function $\arctan f$.

If f and g are lower semicontinuous at x_0 then so is the linear combination $\alpha f + \beta g$ with non-negative coefficients (provided it never takes the form $\mp\infty \pm \infty$). Indeed (see (0.2.9), (0.2.10)),

$$\liminf_{x\to x_0} (\alpha f + \beta g) \geqslant \alpha \liminf_{x\to x_0} f + \beta \liminf_{x\to x_0} g \geqslant \alpha f(x_0) + \beta g(x_0),$$

provided the middle term in the inequality is not of the form $\mp\infty \pm \infty$. But in this case we would have, for example, $\beta > 0$ and $\liminf_{x\to x_0} g = -\infty$, so that $g(x_0) = -\infty$. Thus the inequality $\geqslant$ does hold between the left-hand and right-hand terms.

Similarly, the product of non-negative functions lower semicontinuous at x_0 is a lower semicontinuous function at x_0 (provided that the product never takes the form $0 \cdot \infty$).[5]

If the functions f, g, taking values in the intervals θ_1 and $\theta_2 \subset \bar{\mathbb{R}}$, are lower semicontinuous at x_0, and if φ is an increasing and lower semicontinuous function (in the sense of definition (3.3.3)) in $\theta_1 \times \theta_2$ then the function $\varphi(f, g)$ is lower semicontinuous at x_0.

We will show this by using (3.3.5). Let

$$a < \varphi(f(x_0), g(x_0)).$$

Take a sequence $\alpha_n \in \theta_1$, $\alpha_n \to f(x_0)$, such that $\alpha_n < f(x_0)$, provided $f(x_0)$ is not the minimum of θ_1, in which case take $\alpha_n = f(x_0)$. In either case we have $f(x) \geqslant \alpha_n$ in some neighbourhood of x_0, depending on n. Let β_n be a similar sequence for $g(x_0)$. Then

$$\liminf_{n\to\infty} \varphi(\alpha_n, \beta_n) \geqslant \varphi(f(x_0), g(x_0)) > a$$

so that $\varphi(\alpha_N, \beta_N) > a$ for some N. Hence

$$\varphi(f(x), g(x)) \geqslant \varphi(\alpha_N, \beta_N) > a$$

in some neighbourhood of x_0.

From this it follows, for example, that the maximum and the minimum of two lower semicontinuous functions (at x_0) are lower semicontinuous functions (at x_0).

Let $E_a = \{x : f(x) > a\}$. Then, condition (3.3.5) can be formulated as follows:

$$x_0 \in E_a \quad \text{implies that} \quad x_0 \in \operatorname{Int} E_a.$$

Since this condition is satisfied for all $x_0 \in E_a$ it follows that each set E_a is open. We have thus,

THEOREM 3.3.1. *A necessary and sufficient condition for a function f to be lower semicontinuous on E is that for all a the set* $\{x : f(x) > a\}$ *is open.*

COROLLARY 3.3.1. *A necessary and sufficient condition for a function f to be lower semicontinuous on E is that for all a the set* $\{x : f(x) \leqslant a\}$ *is closed.*

COROLLARY 3.3.2. *A necessary and sufficient condition for a set to be open is that its characteristic function*[6] *be lower semicontinuous.*

THEOREM 3.3.2 (Weierstrass). *A lower semicontinuous function defined on a compact space attains its minimum.*

Proof. Suppose that f is lower semicontinuous on a compact space E. Let $m = \inf_E f$. There is a sequence $\{x_\nu\}$ such that $f(x_\nu) \to m$. From this sequence select a convergent subsequence: $x_{\alpha_\nu} \to x_0$. By (3.3.3) we have

$$f(x_0) \leqslant \liminf_{\nu \to \infty} f(x_{\alpha_\nu}) = m.$$

This, together with the inequality $f(x_0) \geqslant \inf_E f = m$, implies that $f(x_0) = m$. ■

THEOREM 3.3.3. *The upper envelope of a family of lower semicontinuous functions at* x_0 *is a lower semicontinuous function at* x_0.

Proof. Let $\mathscr{R}$ be a family of functions defined on E and lower semicontinuous at x_0. Let $\varphi(x) = \sup_{f \in \mathscr{R}} f(x)$. If $a < \varphi(x_0)$ then $a < \bar{f}(x_0)$ for some $\bar{f} \in \mathscr{R}$, so there exists $\delta > 0$ such that $\bar{f}(x) > a$ for $\rho(x, x_0) < \delta$. Hence, *a fortiori*, $\varphi(x) > a$ for $\rho(x, x_0) < \delta$. ■

COROLLARY 3.3.3. *If* $f_n \to f$ *uniformly on E and if* f_n *are lower semicontinuous at* x_0, *then f is also lower semicontinuous at* x_0.

Indeed, since $\varepsilon_n = \sup_E |f_n(x) - f(x)| \to 0$, we have

$$f_n(x) - \varepsilon_n \leqslant f(x) \quad \text{and} \quad f_n(x) - \varepsilon_n \to f(x)$$

so

$$f(x) = \sup_n (f_n(x) - \varepsilon_n).$$

Hence, it follows that $f(x)$ is lower semicontinuous at x_0.

COROLLARY 3.3.4. *The limit of an increasing sequence of functions lower semicontinuous (at x_0) is a lower semicontinuous function (at x_0).*

In particular, the limit of an increasing sequence of continuous functions on E is a lower semicontinuous function on E. Conversely, we have the following:

THEOREM 3.3.4 (Baire). *Every function f which is lower semicontinuous on E is the limit of an increasing sequence of continuous functions on E (finite, provided $f(x) > -\infty$ on E).*

Proof. Let f be a lower semicontinuous function on E. It suffices to prove that the function

$$g(x) = \frac{2}{\pi} \operatorname{arc\,tan} f(x),$$

which is lower semicontinuous and satisfies $-1 \leqslant g(x) \leqslant 1$, is the limit of an increasing sequence of functions g_n which are continuous and satisfy $-1 \leqslant g_n(x) \leqslant 1$ (or $g_n(x) > -1$ in the case when $g(x) > -1$ on E, for the inequality $g_n(x) < 1$ can always be obtained by replacing the sequence g_n by the sequence $\min[g_n, 1-(1/n)]$).

For fixed n the functions

$$\varphi_{a,n}(x) = g(a) + n\rho(x, a)$$

form an equicontinuous family since they satisfy a Lipschitz condition with common constant n. Then, by Th. 3.2.1, the functions

$$g_n(x) = \inf_{a \in E} \varphi_{a,n}(x)$$

are continuous and also constitute an increasing sequence.

Fix $x \in E$. Since $-1 \leqslant g(a)$ and $\varphi_{x,n}(x) = g(x)$ we have

$$-1 \leqslant g_n(x) \leqslant g(x) \leqslant 1. \tag{3.3.6}$$

Now let $\alpha < g(x)$. By (3.3.5) there exists $\delta > 0$ such that $\alpha < g(a)$ for $\rho(a, x) < \delta$. If, therefore, $\rho(a, x) < \delta$ then

$$\varphi_{a,n}(x) \geqslant g(a) > \alpha.$$

But if $\rho(a, x) \geqslant \delta$ then

$$\varphi_{a,n}(x) \geqslant -1 + n\delta.$$

From this it follows that

$$g_n(x) \geqslant \min(\alpha, -1 + n\delta).$$

(If $g(x) > -1$ then we can choose $-1 < \alpha < g(x)$, so that $g_n(x) > -1$ for all n.) Hence, in the limit

$$\lim_{n \to \infty} g_n(x) \geqslant \alpha.$$

Thus $\lim_{n \to \infty} g_n(x) \geqslant g(x)$ so that, by (3.3.6), $g_n(x) \to g(x)$. ■

LEMMA. *Let $a \leqslant b$ $(a, b \in \bar{\mathbb{R}})$, and let $a_\nu \to a$ be decreasing and $b_\nu \to b$ be increasing. Then, the sequences $\{\alpha_n\}$, $\{\beta_n\}$, defined recursively by the formulae*

$$\alpha_1 = a_1, \quad \beta_1 = \min(a_1, b_1), \quad \alpha_n = \max(\beta_{n-1}, a_n), \quad \beta_n = \min(\alpha_n, b_n) \qquad (3.3.7)$$

converge to a common limit c such that $a \leqslant c \leqslant b$. Also, the sequence $\{\alpha_n\}$ is decreasing and the sequence $\{\beta_n\}$ is increasing.

Proof. By (3.3.7) $\alpha_n \geqslant a_n \geqslant a_{n+1}$ and $\beta_n \leqslant \alpha_n$. Therefore

$$\alpha_{n+1} = \max(\beta_n, a_{n+1}) \leqslant \alpha_n$$

showing that $\{\alpha_n\}$ is decreasing. Similarly, since $\beta_n \leqslant b_n \leqslant b_{n+1}$ and $\alpha_{n+1} \geqslant \beta_n$ we have

$$\beta_{n+1} = \min(\alpha_{n+1}, b_{n+1}) \geqslant \beta_n$$

so that $\{\beta_n\}$ is increasing. Let $\alpha = \lim_{n\to\infty} \alpha_n$, $\beta = \lim_{n\to\infty} \beta_n$. Proceeding to the limit in equalities (3.3.7) we have

$$\alpha = \max(\beta, a), \quad \beta = \min(\alpha, b)$$

from which $\alpha \geqslant a$ and $\beta \leqslant b$. If $\alpha \neq \beta$ then we must have $\alpha = a$, $\beta = b$ so that $a = \max(b, a) = b$ and $\alpha = \beta$. Thus $\alpha = \beta$ and the proof is complete. ■

THEOREM 3.3.5. *Let f be an upper semicontinuous function and let g be a lower semicontinuous function on E such that $f(x) \leqslant g(x)$ on E. Then there exists a function φ continuous on E and such that $f(x) \leqslant \varphi(x) \leqslant g(x)$.*

Proof. By Th. 3.3.4 there exists an increasing sequence of continuous functions g_n converging to g on E and a decreasing sequence of continuous functions f_n converging to f on E. Defining

$$\varphi_1(x) = f_1(x), \quad \psi_1(x) = \min(f_1(x), g_1(x)), \ldots, \varphi_n(x) = \max(\psi_{n-1}(x), f_n(x)),$$
$$\psi_n(x) = \min(\varphi_n(x), g_n(x)), \ldots$$

we obtain sequences $\{\varphi_n\}$, $\{\psi_n\}$ of continuous functions on E. By the lemma these sequences converge to a common limit $\varphi(x)$ such that

$$f(x) \leqslant \varphi(x) \leqslant g(x).$$

Moreover, the first sequence is decreasing while the second is increasing. Thus the function φ is both upper and lower semicontinuous and therefore continuous. ■

3.4 MAXIMUM AND MINIMUM AT A POINT

Let f be a real function on a space E. The numbers

$$M(x) = \limsup_{\rho(z,x)\to 0} f(z), \quad m(x) = \liminf_{\rho(z,x)\to 0} f(z) \qquad (3.4.1)$$

are called the *maximum* and *minimum* respectively *of the function f at the point*

x. We always have the inequalities (see (0.2.3))

$$m(x) \leqslant f(x) \leqslant M(x).$$

Hence, by (3.3.1), lower semicontinuity and upper semicontinuity of f at x are equivalent to the conditions $f(x) = m(x)$ and $f(x) = M(x)$ respectively. Also, continuity is equivalent to the condition $m(x) = M(x)$. If $|f(x)| < \infty$ then the quantity

$$\omega(x) = M(x) - m(x)$$

is called the *oscillation of the function* f at the point x. Thus, the continuity of the function f at the point x_0 is equivalent to the condition $\omega(x_0) = 0$. Next, we have (see (0.2.1)):

$$\begin{aligned} M(x) &= \inf_{\varepsilon > 0} \left\{ \sup_{K(x,\varepsilon)} f(x) \right\} = \lim_{\varepsilon \to 0} \left\{ \sup_{K(x,\varepsilon)} f(x) \right\}, \\ m(x) &= \sup_{\varepsilon > 0} \left\{ \inf_{K(x,\varepsilon)} f(x) \right\} = \lim_{\varepsilon \to 0} \left\{ \inf_{K(x,\varepsilon)} f(x) \right\}. \end{aligned} \tag{3.4.2}$$

where $K(x, \varepsilon)$ denotes the ball with centre x and radius ε. From this it follows that if the function f is finite in a neighbourhood of x, then (see (0.1.14)):

$$\omega(x) = \lim_{\varepsilon \to 0} \left\{ \sup_{u, v \in K(x,\varepsilon)} |f(v) - f(u)| \right\}$$

THEOREM 3.4.1. *If* $\{A_n\}$ *is a sequence of sets such that* $x \in \operatorname{Int} A_n$ *and* $\delta(A_n) \to 0$, *then*

$$M(x) = \lim_{n \to \infty} \left\{ \sup_{A_n} f(x) \right\}, \quad m(x) = \lim_{n \to \infty} \left\{ \inf_{A_n} f(x) \right\}$$

Proof. Let $\delta_n = \delta(A_n)$. We have

$$K(x, \varepsilon_n) \subset A_n \subset K(x, 2\delta_n)$$

for some ε_n $(\varepsilon_n \leqslant 2\delta_n)$, therefore

$$\sup_{K(x,\varepsilon_n)} f(x) \leqslant \sup_{A_n} f(x) \leqslant \sup_{K(x,2\delta_n)} f(x).$$

Hence, by (3.4.2) we have $\sup_{A_n} f(x) \to M(x)$. The proof for $m(x)$ is similar. ∎

THEOREM 3.4.2. *The function* $x \to m(x)$ *is lower semicontinuous and the function* $x \to M(x)$ *is upper semicontinuous. If* $f(x)$ *is finite then the oscillation* $x \to \omega(x)$ *is upper semicontinuous.*

Proof. Let $x_n \to x$ and define $\delta_n = \rho(x_n, x) + 1/n$. Since $K(x_n, \delta_n) \subset K(x, 2\delta_n)$ we have from (3.4.2)

$$m(x_n) \geqslant \inf_{K(x_n,\delta_n)} f(x) \geqslant \inf_{K(x,2\delta_n)} f(x).$$

Hence in the limit: $\liminf_{n\to\infty} m(x_n) \geqslant m(x)$. Similarly for $M(x)$. Hence (if $|f(x)| < \infty$) since $-m$ is upper semicontinuous then $\omega = M - m$ is also upper semicontinuous. ■

THEOREM 3.4.3. *The set of all points of continuity of an arbitrary function $f(x)$ defined on a space E is of type G_δ.*

Proof. Since f and arc tan f are continuous or discontinuous together we may therefore we may therefore confine ourselves to the case of finite functions. In this case the set of points of continuity can be represented in the form

$$\bigcap_{n=1}^{\infty} \left\{ x : \omega(x) < \frac{1}{n} \right\}$$

Since the function ω is upper semicontinuous we see from Th. 3.3.1 that this set is of type G_δ. ■

3.5 FUNCTIONS OF THE FIRST CLASS OF BAIRE

We say that a real function f defined on a metric space E is a *function of the first class* (*of Baire*) if it is the limit of a convergent sequence of functions continuous on E.

A function of the first class is always the limit of a sequence of finite continuous functions on E. Indeed, if f_n are continuous on E and $f_n(x) \to f(x)$ on E, then $\bar{f}_n = \max(-n, \min(n, f_n(x))$ are continuous and finite on E and also $\bar{f}_n(x) \to \max(-\infty, \min(\infty, f(x))) = f(x)$.

It follows from this definition that a restriction of a function of the first class is a function of the first class. Also, a linear combination (never of the form $\infty - \infty$) and a product (never of the form $0 \cdot \infty$) of functions of the first class are also functions of the first class.

If f, g are functions of the first class with values in the intervals θ_1, θ_2 and φ is a continuous function on $\theta_1 \times \theta_2$ then $\varphi(f, g)$ is a function of the first class. For, take sequences $a'_n, b'_n \in \theta_1$, $a''_n, b''_n \in \theta_2$ tending respectively to the end-points of these intervals and let f_n, g_n be continuous functions tending to f, g. Then the functions

$$\bar{f}_n(x) = \max(a'_n, \min(b'_n, f_n(x))), \quad \bar{g}_n(x) = \max(a''_n, \min(b''_n, f_n(x)))$$

are continuous, take values in θ_1, θ_2 and converge to $f(x)$, $g(x)$, so that

$$\varphi(\bar{f}_n(x), \bar{g}_n(x)) \to \varphi(f(x), g(x)).$$

By Th. 3.3.4, semicontinuous functions (upper or lower) are functions of the first class. In the case of functions of a single real variable, monotone functions and, more generally, all functions which have at most a countable number of points of discontinuity,[7] are functions of the first class. Indeed, let f be a finite real function (otherwise consider arc tan f) in $[a, b]$ and let $\{\xi_n\}$ be a sequence of all its points of discontinuity. For each n take a subdivision

$a = x_0^{(n)} < \cdots < x_{k_n}^{(n)} = b$ satisfying

$$\max_i |x_i^{(n)} - x_{i-1}^{(n)}| \leqslant \frac{1}{n}$$

and such that $\xi_1, \ldots, \xi_n$ are points of this subdivision. Let $\varphi_n(x)$ be the polygonal function whose graph is made up of line segments with vertices:

$$(x_i^{(n)}, f(x_i^{(n)})) \qquad (i = 0, \ldots, k_n).$$

If $x = \xi_p$ then $\varphi_n(x) = f(x)$ for $n \geqslant p$, so that

$$\varphi_n(x) \to f(x).$$

If x is a point of continuity of $f(x)$, then since $x_{i_n-1}^{(n)} \leqslant x \leqslant x_{i_n}^{(n)}$ for some i_n therefore $\varphi_n(x) \in [f(x_{i_n-1}^{(n)}), f(x_{i_n}^{(n)})]$ $x_{i_n-1}^{(n)} \to x$, $x_{i_n}^{(n)} \to x$. Hence $f(x_{i_n-1}^{(n)}) \to f(x)$, $f(x_{i_n}^{(n)}) \to f(x)$ and so

$$\varphi_n(x) \to f(x).$$

Thus $\varphi_n(x) \to f(x)$ in $[a, b]$ so that f is of the first class of Baire.

Another equivalent definition of functions of the first class is the following: we say that the real function f defined on a metric space E is a *function of the first class (of Baire)* if for every a the sets

$$\{x : f(x) < a\} \quad \text{and} \quad \{x : f(x) > a\}$$

are of type F_σ. In fact we have:

THEOREM 3.5.1. *The following conditions are mutually equivalent:*

(A) *The function f is of the first class.*

(B) *For every a the sets $\{x : f(x) < a\}$ and $\{x : f(x) > a\}$ are of type F_σ (or, equivalently, $\{x : f(x) \geqslant a\}$ and $\{x : f(x) \leqslant a\}$ are of type G_δ).*

(C) *The function f is both the limit of a decreasing sequence of lower semicontinuous functions and the limit of an increasing sequence of upper semicontinuous functions.*

Proof. *(A)⇒(B).* Let f_n be continuous and let $f_n(x) \to f(x)$ on E. Let $a < \infty$ (if $a = \infty$ then $\{x : f(x) > a\} = \emptyset$) and let $\alpha_k \to a$, $\alpha_k > a$. Now, the condition $f(x) > a$ is equivalent to the condition: there exist k and p such that $f_n(x) \geqslant \alpha_k$ for $n \geqslant p$. Hence

$$\{x : f(x) > a\} = \bigcup_{k=1}^{\infty} \bigcup_{p=1}^{\infty} \bigcap_{n=p}^{\infty} \{x : f_n(x) \geqslant \alpha_k\}.$$

Since $\{x : f_n(x) \geqslant \alpha_k\}$ are closed, therefore $\{x : f(x) > a\}$ is of type F_σ. The case $\{x : f(x) < a\}$ is similar.

(B)⇒(C). Suppose condition (B) is satisfied. Let $\{a_n\}$ be a sequence of all the rational numbers. The sets

$$E_k = \{x : f(x) > a_k\}$$

are of type F_σ, and so each of them is of the form

$$E_k = \bigcup_{l=1}^{\infty} F_{kl}$$

where F_{kl} are closed sets. The functions

$$\chi_{kl}(x) = \begin{cases} a_k & \text{for } x \in F_{kl} \\ -\infty & \text{for } x \in \backslash F_{kl} \end{cases}$$

are upper semicontinuous (for the set $\{x : \chi_{kl}(x) \geqslant \alpha\}$ is always closed) and moreover $\chi_{kl}(x) \leqslant f(x)$ on E. Hence the functions

$$\varphi_n(x) = \max_{k+l \leqslant n} \chi_{kl}(x)$$

are upper semicontinuous and form an increasing sequence with $\varphi_n(x) \leqslant f(x)$ on E.

Take $x \in E$ and let $a < f(x)$. Then $a < a_k < f(x)$ for some k, hence $x \in E_k$ and $x \in F_{kl}$ for some l, that is

$$\chi_{kl}(x) = a_k.$$

If therefore $n \geqslant k + l$ then $\varphi_n(x) \geqslant a_k$, hence $\varphi_n(x) > a$. It follows that

$$\varphi_n(x) \to f(x).$$

In a similar way we can construct a decreasing sequence of lower semicontinuous functions which converge to f on E.

(C)$\Rightarrow$(A). Suppose that $\varphi_n(x) \to f(x)$ and $\psi_n(x) \to f(x)$ on E, where φ_n is an increasing sequence of upper semicontinuous functions and ψ_n is a decreasing sequence of lower semicontinuous functions. Then $\varphi_n(x) \leqslant \psi_n(x)$ and so by Th. 3.3.5, there exists a function f_n continuous on E and such that

$$\varphi_n(x) \leqslant f_n(x) \leqslant \psi_n(x)$$

on E. It follows that

$$f_n(x) \to f(x)$$

on E, so that f is a function of the first class. ■

THEOREM 3.5.2. *The limit of a uniformly convergent sequence of functions of the first class is a function of the first class.*

Proof. Let f_n be of the first class and let $f_n(x) \to f(x)$ uniformly on E. Then

$$\varepsilon_n = \sup_E |f_n(x) - f(x)| \to 0,$$

so that

$$f_n(x) - \varepsilon_n \to f(x) \quad \text{and} \quad f_n(x) - \varepsilon_n \leqslant f.$$

The condition $a < f(x)$ is equivalent to the condition: there exists n such that

$a<f_n(x)-\varepsilon_n$. Hence

$$\{x:a<f(x)\}=\bigcup_{n=1}^{\infty}\{x:a<f_n(x)-\varepsilon_n\}.$$

But, since $\{x:a<f_n(x)-\varepsilon_n\}$ is of type F_σ, therefore so is the set $\{x:a<f(x)\}$.
We can show similarly that the set $\{x:f(x)>a\}$ is of type F_σ. Hence f is of the first class. ■

THEOREM 3.5.3. *The set of all points of discontinuity of an arbitrary function of the first class is a set of the first category.*

Proof. Let f be a function of the first class and let D be the set of its points of discontinuity. We denote by $m(x)$, $M(x)$ the minimum and maximum of this function at the point x. Let $\{\xi_n\}$ be a sequence of all the rational numbers. Then

$$\begin{aligned}D&=\{x:m(x)<f(x)\}\cup\{x:f(x)<M(x)\}\\&=\bigcup_{\xi_i<\xi_j}\{x:m(x)<\xi_i<\xi_j<f(x)\}\cup\bigcup_{\xi_k<\xi_l}\{x:f(x)<\xi_k<\xi_l<M(x)\}\end{aligned}$$

For $\xi_i<\xi_j$ the set

$$\{x:m(x)<\xi_i<\xi_j<f(x)\}=\{x:m(x)<\xi_i\}\cap\{x:\xi_j<f(x)\}$$

is a set of the first category. Indeed, since f, m are of the first class, therefore the set is of type F_σ (for it is the intersection of two sets of type F_σ). Thus, it suffices to show that it does not contain any interior points (for then $F_\sigma=\bigcup_{n=1}^{\infty}F_n$, where F_n are closed sets without any interior points, that is, nowhere dense). If the set had an interior point x_0, then we would have $f(x)>\xi_j$ in a neighbourhood of x_0. Hence

$$m(x_0)=\liminf_{\rho(x,x_0)\to 0} f(x)\geqslant\xi_j$$

which contradicts the inequality $m(x_0)<\xi_i$.
Similarly, for $\xi_k<\xi_l$ the set

$$\{x:f(x)<\xi_k<\xi_l<M(x)\}$$

is of the first category. ■

It follows from this theorem that *when the space E is complete, the set of points of continuity of a function of the first class is a set of the second category and so is a set which is dense in E.*

Such a function, having a set of points of continuity dense in E, is called *pointwise discontinuous.* Thus, in a space E which is complete, every function of the first class is pointwise discontinuous. Moreover, its restriction to any closed set is pointwise discontinuous, for it is also a function of the first class defined on a set which forms a complete space, since it is a closed set of a complete space. It turns out that the converse is true. Namely, we have:

THEOREM 3.5.4 (Baire). *Suppose that E is a complete[8] space. For a function f to be a function of the first class it is necessary and sufficient that its restriction to an arbitrary non-empty closed subset should have points of continuity (or, equivalently, should be pointwise discontinuous).*

Proof. (Under the assumption that E is separable.) Since the necessity of the condition has already been demonstrated we consider sufficiency and suppose that the restriction of f to any closed subset has points of continuity.

Let $\alpha < a$ and define

$$A = \{x : f(x) < a\} \quad \text{and} \quad B = \{x : f(x) \leqslant \alpha\}. \tag{3.5.1}$$

It suffices to prove the existence of a set H of type F_σ such that $B \subset H \subset A$. For then, taking $\alpha_n \to a$, $\alpha_n < a$ (if $a = -\infty$ then $A = \varnothing$), and letting

$$B_n = \{x : f(x) \leqslant \alpha_n\}$$

and H_n a set of type F_σ such that $B_n \subset H_n \subset A$, we will have $A = \bigcup_{n=1}^{\infty} B_n \subset \bigcup_{n=1}^{\infty} H_n \subset A$, so that $A = \bigcup_{n=1}^{\infty} H_n$ is of type F_σ. Taking the function $-f$ in place of f we derive the conclusion that the sets $\{x : f(x) > a\}$ are also of type F_σ. Hence, it follows that f is of the first class.

Let $\mathscr{S}$ be the family of (open) balls K satisfying the condition

$$K \cap B \subset H \subset A \tag{3.5.2}$$

for some set H (depending on K) of type F_σ. Let G be the union of all the balls in the family $\mathscr{S}$. By Lindelöf's theorem $G = \bigcup_{n=1}^{\infty} K_n$, where $K_n \in \mathscr{S}$, that is $K_n \cap B \subset H_n \subset A$ for some H_n of type F_σ. Hence

$$G \cap B \subset \bigcup_{n=1}^{\infty} H_n \subset A \tag{3.5.3}$$

so it suffices to prove that

$$G = E.$$

Suppose that $G \neq E$. Since $\backslash G$ is a non-empty, closed set therefore the function f restricted to $\backslash G$ possesses some point of continuity:

$$x_0 \in \backslash G. \tag{3.5.4}$$

From the inequality $\alpha < a$ we must have $f(x_0) > \alpha$ or $f(x_0) < a$.

Suppose that $f(x_0) > \alpha$. Then there exists $\delta > 0$ such that

$$f(x) > \alpha \qquad \text{if } x \in \backslash G \quad \text{and} \quad \rho(x, x_0) < \delta.$$

Putting $K_0 = K(x_0, \delta)$ we therefore have, by (3.5.1), that $(K_0 \backslash G) \cap B = \varnothing$ so that $K_0 \cap B \subset G \cap B$. That is, by (3.5.3), the ball K_0 satisfies condition (3.5.2). Therefore $K_0 \subset G$ which contradicts (3.5.4).

Suppose that $f(x_0) < a$. Then, there exists $\delta > 0$ such that

$$f(x) < a \qquad \text{if} \quad x \in \backslash G \quad \text{and} \quad \rho(x, x_0) < \delta.$$

Putting $K_0 = K(x_0, \delta)$ we have, by (3.5.1) that $K_0 \backslash G \subset A$. Then, because of the inclusion $K_0 \cap B \subset ((K_0 \backslash G) \cap B) \cup (G \cap B)$, we obtain, by (3.5.3)

$$K_0 \cap B \subset (K_0 \backslash G) \cup \bigcup_{n=1}^{\infty} H_n \subset A.$$

Because the middle term of this inclusion is of type F_σ, therefore K_0 satisfies condition (3.5.2) from which $K_0 \subset G$ contradicting (3.5.4). Therefore we must have $G = E$ as required. ■

The proof for the non-separable case is given below in an optional section.

COROLLARY 3.5.1. *If E is a complete space and if the set of points of discontinuity of the function f is at most countable, then f is a function of the first class.*

For let F be a closed set. If F has isolated points then f is continuous at these points. If the set F does not have isolated points then it is a perfect set which (by the completeness of E) is uncountable and so must have points of continuity of the function f. Hence, by Th. 3.5.4, the function f is of the first class.

**Non-separable case.* In order to prove the theorem without assuming that the space E is separable we need the following

LEMMA. *Let $\varepsilon > 0$ and let $\{F_\nu\}_{\nu \in I}$ be a family of closed sets satisfying the condition $\rho(F_\nu, F_\kappa) \geqslant \varepsilon > 0$ for $\nu \neq \kappa$. Then*

$$F = \bigcup_{\nu \in I} F_\nu$$

is a closed set.

Proof of Lemma. Let $x_n \to x_0$, $x_n \in F$. There exists N such that

$$\rho(x_n, x_m) < \varepsilon \qquad \text{for} \quad n, m \geqslant N.$$

It follows that the points $x_N, x_{N+1}, \ldots$ all belong to the same set F_{ν_0}. Hence $x_0 \in F_{\nu_0}$ so that $x_0 \in F$ and F is closed. ■

Now we return to the proof of Th. 3.5.4 without assuming that the space E is separable. With the notation used in the proof of that theorem, we must demonstrate the existence of a set H of type F_σ such that $B \subset H \subset A$.

With each closed set F we associate a closed set $\varphi(F)$ as follows: if $F = \varnothing$ then $\varphi(F) = \varnothing$; if $F \neq \varnothing$ then the function f restricted to F has a point of continuity $x_0 \in F$. We must have $f(x_0) < a$ or $f(x_0) > \alpha$. Choosing a sufficiently small ball K centred on x_0 we will have $f(x) < a$ in $K \cap F$ or $f(x) > \alpha$ in $K \cap F$, that is (using (3.5.1))

$$K \cap F \subset A \quad \text{or} \quad K \cap F \subset \backslash B.$$

We define $\varphi(F) = F\backslash K$. Then

$$F\backslash\varphi(F) = K \cap F.$$

It follows that the function φ so defined has the following properties:

$$\varphi(F) \subset F \quad \text{and} \quad \varphi(F) \neq F \text{ if } F \neq \varnothing, \tag{3.5.5}$$

$$F\backslash\varphi(F) \subset A \quad \text{or} \quad F\backslash\varphi(F) \subset \backslash B. \tag{3.5.6}$$

Let ζ be a limit ordinal number whose cardinality is greater than the cardinality of the family of all closed sets in E. We define by induction a transfinite sequence $\{F_\xi\}_{\xi<\zeta}$:

$$F_0 = E, \quad F_{\xi+1} = \varphi(F_\xi) \quad \text{and} \quad F_\lambda = \bigcap_{\xi<\lambda} F_\xi,$$

if λ is a limit number ($\xi, \lambda < \zeta$). The terms in this sequence are closed sets (by induction) and by (3.5.6)

$$F_\xi\backslash F_{\xi+1} \subset A \quad \text{or} \quad F_\xi\backslash F_{\xi+1} \subset \backslash B \quad (\xi < \zeta). \tag{3.5.7}$$

By (3.5.5) the sequence $\{F_\xi\}_{\xi<\zeta}$ is decreasing (if $\xi < \eta < \zeta$ then $F_\eta \subset F_\xi$; this is easily proved by transfinite induction on η). If the transfinite sequence were strictly decreasing then the set of terms in this transfinite sequence would have equal cardinality with the set of numbers $\xi < \zeta$ and this would contradict the assumption that the cardinality of ζ is greater than the cardinality of the family of all closed sets in E. Therefore we must have $F_\xi = F_\eta$ for some pair $\xi < \eta < \zeta$. Hence $F_\xi = F_{\xi+1}$ and so, by (3.5.5), $F_\xi = \varnothing$.

Let T be the set of all those $\xi < \zeta$ for which $F_\xi\backslash F_{\xi+1} \subset A$ and define:

$$H = \bigcup_{\xi\in T} (F_\xi\backslash F_{\xi+1}). \tag{3.5.8}$$

Then $H \subset A$. We show that $B \subset H$.

Indeed, if $x \in B$ then $x \notin F_\xi$ for some ξ. Let ξ_0 be the least such ξ. Then $x \in \bigcap_{\xi<\xi_0} F_\xi$, that is (from the definition of the transfinite sequence F_ξ) ξ_0 cannot be a limit number. Hence

$$\xi_0 = \xi' + 1 \quad \text{and} \quad x \in F_{\xi'}\backslash F_{\xi'+1}.$$

Because $x \in B$ therefore, by (3.5.7), we must have $F_{\xi'}\backslash F_{\xi'+1} \subset A$, so that

$$\xi' \in T.$$

Hence, by (3.5.8), $x \in H$.

It therefore suffices to show that H is of type F_σ. To this end consider the open sets

$$G_{\xi_n} = \left\{x : \rho(x, F_{\xi+1}) < \frac{1}{n}\right\}.$$

Then (since $F_{\xi+1}$ is closed) $F_{\xi+1} = \bigcap_{n=1}^\infty G_{\xi_n}$. Hence

$$F_\xi\backslash F_{\xi+1} = \bigcup_{n=1}^\infty (F_\xi\backslash G_{\xi_n})$$

and so by (3.5.8)

$$H = \bigcup_{n=1}^{\infty} \Phi_n \quad \text{where } \Phi_n = \bigcup_{\xi \in T} (F_\xi \backslash G_{\xi_n}). \tag{3.5.9}$$

On the other hand, it follows from the definition of G_{ξ_n} that $\rho(F_{\xi+1}, \backslash G_{\xi_n}) \geqslant 1/n$. If therefore $\xi < \eta$ then $F_\eta \backslash G_{\eta_n} \subset F_\eta \subset F_{\xi+1}$ and $F_\xi \backslash G_{\xi_n} \subset \backslash G_{\xi_n}$, hence *a fortiori*,

$$\rho(F_\eta \backslash G_{\eta_n}, F_\xi \backslash G_{\xi_n}) \geqslant \frac{1}{n}.$$

By the lemma it follows that the Φ_n are closed and so by (3.5.9) H is of type F_σ. ■

NOTES

1. In the case $f(x_0) = \infty$ (resp. $f(x_0) = -\infty$) given $A < \infty$ (resp. $A > -\infty$) there exists $\delta > 0$ such that $f(x) > A$ (resp. $f(x) < A$), when $\rho(x, x_0) < \delta$. In extending the arguments to the case of topological spaces the reader should use definitions and conditions of the Cauchy type.
2. The reader should also find a proof based on Cauchy's definition.
3. The reader should propose an equivalent definition of the Cauchy type and prove Ths: 3.1.8 and 3.1.9 on this basis.
4. In fact each of the conditions (3.3.1) implies equality, for taking a constant sequence $x_\nu = x_0$, $f(x_\nu) \to f(x_0)$.
5. It follows from the next assertion that this assumption is superfluous, for the product xy, as a function defined on the set $[0, \infty] \times [0, \infty]$ is lower semicontinuous at the points $(0, \infty)$, $(\infty, 0)$. The assumption is necessary for the case of upper semicontinuity.
6. See definition after Th. 4.4.9.
7. See Corollary 3.5.1.
8. The assumption of completeness is required for the proof of necessity. In the proof of sufficiency we do not make use of this assumption.

CHAPTER 4

ALGEBRAS OF SETS AND MEASURABLE FUNCTIONS

4.1 ALGEBRAS OF SETS

Let X be an arbitrary set (space). A non-empty class $\mathscr{S}$ of subsets of X is called a *countably additive algebra* if it is closed under the operations of taking countable unions and differences, that is, if the following conditions are satisfied:

(i) if $A_i \in \mathscr{S}$ $(i=1,2,\ldots)$ then $\bigcup_{i=1}^{\infty} A_i \in \mathscr{S}$;
(ii) if $A, B \in \mathscr{S}$ then $A \backslash B \in \mathscr{S}$.

Taking $A=B$ in (ii) we see that the empty set is in $\mathscr{S}$. An intersection can be expressed in terms of unions and differences

$$\bigcap_{i=1}^{\infty} A_i = A_1 \backslash \bigcup_{i=1}^{\infty} (A_1 \backslash A_i);$$

A finite union or intersection can be regarded as countably infinite:

$$A_1 \cup \cdots \cup A_n = A_1 \cup \cdots \cup A_n \cup A_n \cup \cdots.$$

Hence we have

THEOREM 4.1.1. *The union and intersection of an at most countable number of sets from an algebra (countably additive) $\mathscr{S}$ belong to $\mathscr{S}$. The upper and lower bounds of a sequence of sets from $\mathscr{S}$ belong to $\mathscr{S}$.*

If $X \in \mathscr{S}$ then, by (ii), the complement of a set in $\mathscr{S}$ belongs to $\mathscr{S}$. Conversely

THEOREM 4.1.2. *A non-empty class $\mathscr{S}$ of subsets of a space X which satisfies the conditions:*

(i) *if $A_i \in \mathscr{S}$ $(i=1,2,\ldots)$ then $\bigcup_{i=1}^{\infty} A_i \in \mathscr{S}$;*
(ii′) *if $A \in \mathscr{S}$ then $\backslash A \in \mathscr{S}$;*
(iii) $\varnothing \in \mathscr{S}$,

is a countably additive algebra such that $X \in \mathscr{S}$.

Indeed, if $A, B\in\mathscr{S}$ then by (i) and (ii′)

$$A\backslash B=\backslash(\backslash A\cup B\cup B\cup\cdots)\in\mathscr{S}$$

and by (ii′) and (iii) $X\in\mathscr{S}$.

THEOREM 4.1.3. *A non-empty class of subsets of a space X which satisfies the conditions:*

(i′) *if* $A_i\in\mathscr{S}$, $A_i\cap A_j=\varnothing$ *for* $i\neq j$ $(i, j=1,2,\ldots)$ *then* $\bigcup_{i=1}^{\infty}A_i\in\mathscr{S}$;
(ii) *if* $A, B\in\mathscr{S}$ *then* $A\backslash B\in\mathscr{S}$;

is a countably additive algebra.

Proof. We must show that condition (i) is satisfied. From the identity

$$\bigcup_{n=1}^{\infty}A_n=A_1\cup\bigcup_{n=2}^{\infty}\left(A_n\backslash\bigcup_{i=1}^{n-1}A_i\right)$$

which replaces an arbitrary union by a disjoint union, it suffices to show that a finite union of sets from $\mathscr{S}$ belongs to $\mathscr{S}$. But the union of two sets in $\mathscr{S}$ belongs to $\mathscr{S}$ for

$$A\cup B=A\cup(B\backslash A),$$

where A and $B\backslash A$ are disjoint. It follows by induction that a finite union of sets from $\mathscr{S}$ belongs to $\mathscr{S}$. ■

A non-empty class $\mathscr{R}$ of subsets of a space X is called a *finitely additive algebra* if the following conditions are satisfied:

(1) if $A, B\in\mathscr{R}$ then $A\cup B\in\mathscr{R}$;
(2) if $A, B\in\mathscr{R}$ then $A\backslash B\in\mathscr{R}$.

Condition (1) can be replaced by the condition

(1′) if $A, B\in\mathscr{R}$ and $A\cap B=\varnothing$ then $A\cup B\in\mathscr{R}$

(for $A\cup B=A\cup(B\backslash A)$, where A and $B\backslash A$ are disjoint).

A finitely additive algebra is closed under the operations of taking unions and intersections of a finite number of sets in the algebra.

THEOREM 4.1.4. *A finitely additive algebra* $\mathscr{R}$ *which satisfies the condition that the limit of an increasing sequence of sets from* $\mathscr{R}$ *belongs to* $\mathscr{R}$ *is a countably additive algebra.*

This follows from the identity

$$\bigcup_{n=1}^{\infty}A_n=\lim_{n\to\infty}(A_1\cup\cdots\cup A_n),$$

in which the sequence $\{A_1\cup\cdots\cup A_n\}$ is increasing.

THEOREM 4.1.5. *The intersection of any family of countably additive algebras is a countably additive algebra.*

Proof. Let $\{\mathcal{S}_l\}$ be a family of countably additive algebras and let $\mathcal{S} = \bigcap_l \mathcal{S}_l$. If $A_i \in \mathcal{S}$ $(i = 1, 2, \ldots)$ then $A_i \in \mathcal{S}_l$ $(i = 1, 2, \ldots)$ for all l so that $\bigcup_{i=1}^{\infty} A_i \in \mathcal{S}_l$ for each l, hence $\bigcup_{i=1}^{\infty} A_i \in \mathcal{S}$. Condition (ii) can be checked similarly. ■

Let $E \subset X$. Since the class of all subsets of the set E is a countably additive algebra, we have from Th. 4.1.5:

THEOREM 4.1.6. *Let $\mathcal{S}$ be a countably additive algebra. Then, the class $\mathcal{S}_E$ of all those sets in $\mathcal{S}$ which are contained in E is a countably additive algebra.*

The algebra $\mathcal{S}_E$ is called the *restriction of the algebra S to the set E*.
Let $f: X \to Y$. From the properties of pre-images we have the following

THEOREM 4.1.7. *Let $\mathcal{S}$ be a countably additive algebra in X. Then the class of all sets $F \subset Y$ such that $f^{-1}(F) \in \mathcal{S}$, is a countably additive algebra in Y.*

Let any class $\mathcal{K}$ of subsets of a set X be given. The family of all countably additive algebras which contain the class $\mathcal{K}$ is non-empty because it contains, for example, the algebra of all subsets of X. By Th. 4.1.5, the intersection of all the algebras of this family is also a countably additive algebra containing the class $\mathcal{K}$. We call it the *countably additive algebra generated by the class* $\mathcal{K}$, and we denote it by $\mathcal{S}(\mathcal{K})$. The following properties follow from this definition. If $\mathcal{S}$ is a countably additive algebra then

$$\mathcal{K} \subset \mathcal{S} \quad \text{implies} \quad \mathcal{S}(\mathcal{K}) \subset \mathcal{S}; \tag{4.1.1}$$

$$\mathcal{K} \subset \mathcal{K}' \quad \text{implies} \quad \mathcal{S}(\mathcal{K}) \subset \mathcal{S}(\mathcal{K}') \tag{4.1.2}$$

(for $\mathcal{K} \subset \mathcal{K}' \subset \mathcal{S}(\mathcal{K}')$ so by (4.1.1) $\mathcal{S}(\mathcal{K}) \subset \mathcal{S}(\mathcal{K}')$). From these two properties it follows that

$$\mathcal{K} \subset \mathcal{K}' \subset \mathcal{S}(\mathcal{K}) \quad \text{implies} \quad \mathcal{S}(\mathcal{K}') = \mathcal{S}(\mathcal{K}). \tag{4.1.3}$$

We say that the class $\mathcal{M}$ is *monotone* if the limit of a monotone sequence of sets in $\mathcal{M}$ belongs to $\mathcal{M}$. We have the following

THEOREM 4.1.8. *If $\mathcal{R}$ is a finitely additive algebra then $\mathcal{S}(\mathcal{R})$ is the smallest monotone class containing $\mathcal{R}$.*

Proof. Let $\mathcal{N}$ be the smallest monotone class containing $\mathcal{R}$ (its existence follows in the same way as in the case of algebras). We have $\mathcal{N} \subset \mathcal{S}(\mathcal{R})$ for $\mathcal{S}(\mathcal{R})$ is a monotone class. It suffices to show that $\mathcal{N}$ is a finitely additive algebra for then, by Th. 4.1.4, it is countably additive and so $\mathcal{S}(\mathcal{R}) \subset \mathcal{N}$. But for an arbitrary class $\mathcal{L}$, the class

$$\mathcal{J}(\mathcal{L}) = \{E : E \cup F, E \setminus F, F \setminus E \in \mathcal{N} \quad \text{if } F \in \mathcal{L}\}$$

is monotone, for given any monotone sequence E_n tending to E, the sequences $E_n \cup F$, $E_n \backslash F$ and $F \backslash E_n$ are monotone and tend to $E \cup F$, $E \backslash F$ and $F \backslash E$. We note that the conditions $\mathscr{K} \subset \mathscr{J}(\mathscr{L})$ and $\mathscr{L} \subset \mathscr{J}(\mathscr{K})$ are equivalent and mean that $E \cup F, E \backslash F, F \backslash E \in \mathscr{N}$ if $E \in \mathscr{K}$ and $F \in \mathscr{L}$. Hence, using the assumed properties of $\mathscr{R}$ we obtain successively the inclusions:

$$\mathscr{R} \subset \mathscr{J}(\mathscr{R}), \quad \mathscr{N} \subset \mathscr{J}(\mathscr{R}), \quad \mathscr{R} \subset \mathscr{J}(\mathscr{N}), \quad \mathscr{N} \subset \mathscr{J}(\mathscr{N})$$

the last of which means that $\mathscr{N}$ is a finitely additive algebra. ■

4.2 CARTESIAN PRODUCTS

Let $\mathscr{K}_1, \mathscr{K}_2$ be arbitrary classes of sets. We use the notation $\mathscr{K}_1 \bar{\times} \mathscr{K}_2$ for the class of all sets $E_1 \times E_2$ where $E_1 \in \mathscr{K}_1$, $E_2 \in \mathscr{K}_2$.

Let two countably additive algebras, $\mathscr{S}_1$ in space X_1, and $\mathscr{S}_2$ in the space X_2, be given. We call the countably additive algebra $\mathscr{S}_1 \times \mathscr{S}_2 = \mathscr{S}(\mathscr{S}_1 \bar{\times} \mathscr{S}_2)$ in the space $X_1 \times X_2$, the *cartesian product of the algebras* $\mathscr{S}_1$ and $\mathscr{S}_2$.

THEOREM 4.2.1. *The algebra $\mathscr{S}_1 \times \mathscr{S}_2$ is the smallest within the classes $\mathscr{S}'$ which contain $\mathscr{S}_1 \bar{\times} \mathscr{S}_2$, are monotone and satisfy the condition:*

$$\text{if } A, B \in \mathscr{S}', \text{ and } A \cap B = \varnothing \text{ then } A \cup B \in \mathscr{S}'. \tag{4.2.1}$$

Proof. Let $\mathscr{K} = \mathscr{S}_1 \bar{\times} \mathscr{S}_2$. Clearly $\mathscr{S}_1 \times \mathscr{S}_2 = \mathscr{S}(\mathscr{K})$ is a monotone class containing $\mathscr{K}$ and satisfying condition (4.2.1). Let $\mathscr{S}'$ be any such class. It is required to show that $\mathscr{S}_1 \times \mathscr{S}_2 \subset \mathscr{S}'$. By (4.2.1) the class $\mathscr{S}'$ contains the class $\mathscr{R}$ of all finite unions of disjoint sets in $\mathscr{K}$. It suffices to show that $\mathscr{R}$ is a finitely additive algebra for then, by Th. 4.1.8, we have

$$\mathscr{S}' \supset \mathscr{S}(\mathscr{R}) \supset \mathscr{S}(\mathscr{K}) = \mathscr{S}_1 \times \mathscr{S}_2.$$

Clearly $\mathscr{R}$ satisfies condition (1') in the definition of a finitely additive algebra. Thus it suffices to show that $\mathscr{R}$ satisfies condition (2). From the identity

$$(E_1 \times E_2) \backslash (F_1 \times F_2) = ((E_1 \backslash F_1) \times E_2) \cup ((E_1 \cap F_1) \times (E_2 \backslash F_2))$$

it follows that the difference of two sets in $\mathscr{K}$ is the disjoint sum of two sets in $\mathscr{K}$. Hence, if $A \in \mathscr{R}$ and $B \in \mathscr{K}$ then $A \backslash B \in \mathscr{R}$, since $(\bigcup_1^n A_i) \backslash B = \bigcup_1^n (A_i \backslash B)$. Now, if $A, B \in \mathscr{R}$ then $B = B_1 \cup \cdots \cup B_m$, where $B_i \in \mathscr{K}$, therefore

$$A \backslash B = ((A \backslash B_1) \backslash \cdots) \backslash B_m \in \mathscr{R}. \quad ■$$

THEOREM 4.2.2. *If $\mathscr{K}_1$ is a class of sets from X_1 and $\mathscr{K}_2$ is a class of sets from X_2, then*

$$\mathscr{S}(\mathscr{K}_1) \times \mathscr{S}(\mathscr{K}_2) = \mathscr{S}(\mathscr{K}_1 \bar{\times} \mathscr{K}_2).$$

Proof. Since $\mathscr{K}_1 \bar{\times} \mathscr{K}_2 \subset \mathscr{S}(\mathscr{K}_1) \bar{\times} \mathscr{S}(\mathscr{K}_2)$ it suffices, by (4.1.3), to show that

$$\mathscr{S}(\mathscr{K}_1) \bar{\times} \mathscr{S}(\mathscr{K}_2) \subset \mathscr{S}(\mathscr{K}_1 \bar{\times} \mathscr{K}_2).$$

Let $\mathcal{S} = \mathcal{S}(\mathcal{K}_1 \bar{\times} \mathcal{K}_2)$. Since $\mathcal{K}_1 \bar{\times} \mathcal{K}_2 \subset \mathcal{S}$ it suffices to show that if $\mathcal{L}_1 \bar{\times} \mathcal{L}_2$ is contained in $\mathcal{S}$ then also $\mathcal{S}(\mathcal{L}_1) \bar{\times} \mathcal{L}_2$ and $\mathcal{L}_1 \bar{\times} \mathcal{S}(\mathcal{L}_2)$ are contained in $\mathcal{S}$. So, assume that $\mathcal{L}_1 \bar{\times} \mathcal{L}_2 \subset \mathcal{S}$. For every $F \in \mathcal{L}_2$ the class $\{E: E \times F \in \mathcal{S}\}$ is a countably additive[1] algebra containing $\mathcal{L}_1$, and hence (by (4.1.1)) containing $\mathcal{S}(\mathcal{L}_1)$ also. This gives the inclusion $\mathcal{S}(\mathcal{L}_1) \bar{\times} \mathcal{L}_2 \subset \mathcal{S}$. The second inclusion is obtained similarly. ■

Let $A \subset X_1 \times X_2$. The set

$$(A)_{x_1} = \{x_2 : (x_1, x_2) \in A\}$$

is called the *section of the set* A determined by $x_1 \in X_1$. We have

$$(A)_{x_1} = \varphi_{x_1}^{-1}(A) \qquad \text{where} \quad \varphi_{x_1}: x_2 \to (x_1, x_2),$$

from which the following properties of sections derive:

$$(A/B)_{x_1} = (A)_{x_1} \backslash (B)_{x_1}, \quad \left(\bigcup_l A_l\right)_{x_1} = \bigcup_l (A_l)_{x_1}, \quad \left(\bigcap_l A_l\right)_{x_1} = \bigcap_l (A_l)_{x_1} \tag{4.2.2}$$

Similarly, for each $x_2 \in X_2$ we define the section

$$(A)^{x_2} = \{x_1 : (x_1, x_2) \in A\},$$

which has similar properties.

THEOREM 4.2.3. *Let $\mathcal{S}_1$ be a countably additive algebra in X_1 and $\mathcal{S}_2$ a countably additive algebra in X_2. If $A \in \mathcal{S}_1 \times \mathcal{S}_2$, then*

$$(A)_{x_1} \in \mathcal{S}_2 \text{ for all } x_1 \in X_1 \quad \text{and} \quad (A)^{x_2} \in \mathcal{S}_1 \text{ for all } x_2 \in X_2. \tag{4.2.3}$$

Proof. By (4.2.2) the class $\mathcal{S}^*$ of all sets A satisfying (4.2.3) is a countably additive algebra. Since $\mathcal{S}_1 \bar{\times} \mathcal{S}_2 \subset \mathcal{S}^*$ (for always $(E_1 \times E_2)_{x_1} = E_2$ or $\varnothing$) therefore, by (4.1.1), $\mathcal{S}_1 \times \mathcal{S}_2 \subset \mathcal{S}^*$, from which it follows that every set in $\mathcal{S}_1 \times \mathcal{S}_2$ satisfies condition (4.2.3). ■

4.3 BOREL SETS

Let X be a metric space. The countably additive algebra generated by the class of all open sets in X is called the *algebra of Borel sets* and is denoted by $\mathcal{B}(X)$. The sets belonging to this algebra are called *Borel sets* of the space X. The algebra of Borel sets for the space $\mathbb{R}^n$ will be denoted by $\mathcal{B}_n$.

The algebra $\mathcal{B}(X)$ contains all closed sets (since they are the complements of open sets) and so also, by (4.1.1), the algebra $\mathcal{B}'$ generated by the class of all closed sets. Conversely, $\mathcal{B}'$ contains all open sets and hence it contains $\mathcal{B}(X)$ also. Thus $\mathcal{B}(X) = \mathcal{B}'$; that is, *$\mathcal{B}(X)$ is the algebra generated by the class of all closed sets in X.*

Similarly, $\mathcal{B}_n$ contains all intervals (n-dimensional, closed and open) and so it contains also the algebra $\mathcal{B}'_n$ generated by the class of all closed intervals.

Conversely, $\mathscr{B}'_n$ contains all open sets, for every set D open in $\mathbb{R}^n$ is the union of a countable number of closed intervals (namely, the set of all intervals contained in D whose vertices have rational coordinates). Thus $\mathscr{B}'_n$ contains $\mathscr{B}_n$ and we have

THEOREM 4.3.1. *The algebra $\mathscr{B}_n$ is generated by the class of all n-dimensional closed intervals.*[2]

Let $\mathscr{K}$ denote the class of open intervals of the form $(a, \infty]$ with $|a| < \infty$ or $\bar{\mathbb{R}}$ itself and let $\mathscr{K}'$ be the class of open sets in $\bar{\mathbb{R}}$, then $\mathscr{K} \subset \mathscr{K}' \subset \mathscr{S}(\mathscr{K})$. For, note that every open set in $\bar{\mathbb{R}}$ is the union of a countable number of open intervals which either belong to $\mathscr{K}$ or are of the form $(-\infty, \infty] = \bigcup_1^\infty (-n, \infty]$, or $[-\infty, d) = \bar{\mathbb{R}} \setminus \bigcap_1^\infty (d_n, \infty]$, where $d > d_n \to d$, or $(c, d) = [-\infty, d) \cap (c, \infty]$. Hence we have

THEOREM 4.3.2. *The algebra $\mathscr{B}(\bar{\mathbb{R}})$ is generated by the class of all intervals of the form $(a, \infty]$, where $|a| < \infty$, or $\bar{\mathbb{R}}$ itself.*

THEOREM 4.3.3. *Let $E \in \mathscr{B}(X)$, Then the algebra of Borel sets in E (as a space) is the class of all Borel sets in X which are contained in E:*

$$\mathscr{B}(E) = \mathscr{B}(X)_E.$$

Proof. Since the open sets of E are sets of the form $E \cap D$, where D is open (in X), therefore the algebra $\mathscr{B}(X)_E$ contains all open sets of E and hence $\mathscr{B}(E) \subset \mathscr{B}(X)_E$. The reverse inclusion arises from the fact that every set of $\mathscr{B}(X)$ contained in E is the inverse image of itself by the continuous mapping $x\ (\in E) \to x \in X$ and so belongs to $\mathscr{B}(E)$ by the following theorem. ■

THEOREM 4.3.4. *The inverse image of a Borel set by a continuous mapping is a Borel set.*

Proof. Let f be a continuous mapping of a metric space X to a metric space Y. The class $\mathscr{B}'$ of all sets $F \subset Y$ such that

$$f^{-1}(F) \in \mathscr{B}(X)$$

is, by Th. 4.1.7, a countably additive algebra containing the open sets in Y. Hence $\mathscr{B}(Y) \subset \mathscr{B}'$ which means that if $F \in \mathscr{B}(Y)$ then $f^{-1}(F) \in \mathscr{B}(X)$. ■

Let X_1 and X_2 be metric spaces. By Th. 4.2.2, the countably additive algebra $\mathscr{B}(X_1) \times \mathscr{B}(X_2)$ is generated by the class $\mathscr{K}$ of all cartesian products of open sets:

$$\mathscr{B}(X_1) \times \mathscr{B}(X_2) = \mathscr{S}(\mathscr{K}).$$

Let $\mathscr{G}$ be the class of all open sets in $X_1 \times X_2$. Then $\mathscr{K} \subset \mathscr{G}$, so that

$$\mathscr{S}(\mathscr{K}) \subset \mathscr{S}(\mathscr{G}) = \mathscr{B}(X_1 \times X_2).$$

Hence we have the inclusion

$$\mathscr{B}(X_1) \times \mathscr{B}(X_2) \subset \mathscr{B}(X_1 \times X_2) \tag{4.3.1}$$

In particular, we have

THEOREM 4.3.5. *The cartesian product of Borel sets is a Borel set.*

Suppose that the spaces X_1, X_2 are separable. Then, every open set D in $X_1 \times X_2$ is a countable union of sets in $\mathscr{K}$ (indeed, D is the union of all sets in $\mathscr{K}$ which are contained in D and so, by Lindelöf's theorem, D is the union of a countable number of such sets). Hence $\mathscr{K} \subset \mathscr{G} \subset \mathscr{S}(\mathscr{K})$ and so, by (4.1.3),

$$\mathscr{S}(\mathscr{K}) = \mathscr{S}(\mathscr{G}) = \mathscr{B}(X_1 \times X_2).$$

We have therefore

THEOREM 4.3.6. *If the spaces X_1, X_2 are separable, then $\mathscr{B}(X_1) \times \mathscr{B}(X_2) = \mathscr{B}(X_1 \times X_2)$. In particular*

$$\mathscr{B}_{n+m} = \mathscr{B}_n \times \mathscr{B}_m. \tag{4.3.2}$$

THEOREM 4.3.7. *Every section of a Borel set in $X_1 \times X_2$ is a Borel set in X_1 or X_2 respectively.*

Indeed, the section $(A)_{x_1}$ is the inverse image of A under the continuous mapping

$$x_2(\in X_2) \to (x_1, x_2) \in X_1 \times X_2.$$

Theorem 4.3.7 is therefore a corollary of Th. 4.3.4.

Classification of Borel Sets

Let Φ be a class of sets. We denote by Φ_σ the class of all sets which are countable unions of sets in Φ. We denote by Φ_δ the class of all sets which are countable intersections of sets in Φ. Clearly

$$\Phi \subset \Phi_\sigma \quad \text{and} \quad \Phi \subset \Phi_\delta.$$

The class Φ_σ is closed under the operation of taking countable unions. The class Φ_δ is closed under countable intersections. Thus

$$\Phi_{\sigma\sigma} = \Phi_\sigma \quad \text{and} \quad \Phi_{\delta\delta} = \Phi_\delta.$$

If the class Φ is *finitely additive* and *finitely multiplicative* that is, if Φ is closed under the operations of taking finite unions and finite intersections, then the classes Φ_σ and Φ_δ have the same property.

Let $\mathscr{B} = \mathscr{B}(X)$ be the algebra of Borel sets for the space X. Let F denote the class of all closed sets and let G denote the class of all open sets. Then $F_\delta = F$

and $G_\sigma = G$. But the inclusion $F \subset F_\sigma$ is usually a strict inclusion and

$$F \subset F_\sigma \subset F_{\sigma\delta} \subset F_{\sigma\delta\sigma} \subset \cdots$$

can be shown, in general, to be a sequence which is strictly increasing.[3] These are obviously classes of Borel sets. The union of all the classes in this sequence is not necessarily closed under countable unions (or under countable intersections) so that these operations can be applied further to obtain classes which are, in general, strictly wider. The situation is similar for the sequence

$$G \subset G_\delta \subset G_{\delta\sigma} \subset \cdots.$$

In this way we can define, by transfinite induction, sequences $\{F_\xi\}_{\xi<\Omega}$ and $\{G_\xi\}_{\xi<\Omega}$:

$$F_0 = F, F_\alpha = \begin{cases} \left(\bigcup_{\xi<\alpha} F_\xi\right)_\sigma & \text{if } \alpha \text{ is odd,} \\ \left(\bigcup_{\xi<\alpha} F_\xi\right)_\delta & \text{if } \alpha \text{ is even,} \end{cases} \tag{4.3.3}$$

$$G_0 = G, G_\alpha = \begin{cases} \left(\bigcup_{\xi<\alpha} G_\xi\right)_\delta & \text{if } \alpha \text{ is odd,} \\ \left(\bigcup_{\xi<\alpha} G_\xi\right)_\sigma & \text{if } \alpha \text{ is even.} \end{cases} \tag{4.3.4}$$

In particular

$$F_1 = F_\sigma, F_2 = F_{\sigma\delta}, \ldots, G_1 = G_\delta, G_2 = G_{\delta\sigma}, \ldots$$

It follows from the definition that these sequences are increasing (in general strictly[3]).

The complements of sets in F_α are sets in G_α. Indeed, the complements of sets in F_0 (closed sets) are sets in G_0 (open sets). Choose an arbitrary $\alpha < \Omega$ and suppose that the assertion is true for $\xi < \alpha$. If α is odd and if $E \in F_\alpha$ then from (4.3.3)

$$E = \bigcup_{i=1}^{\infty} E_i \qquad \text{where } E_i \in F_{\xi_i}, \quad \xi_i < \alpha.$$

Hence $\backslash E = \bigcap_{i=1}^{\infty} (\backslash E_i)$ and $\backslash E_i \in G_{\xi_i}$ so that by (4.3.4), $\backslash E \in G_\alpha$. The proof is similar when α is even. Hence, the assertion is true for all $\alpha < \Omega$. In the same way the complements of sets in G_α are sets in F_α.

Finally, we have the inclusions

$$F_\alpha \subset G_{\alpha+1} \quad \text{and} \quad G_\alpha \subset F_{\alpha+1} \qquad (\alpha < \Omega).^4 \tag{4.3.5}$$

Indeed, $F_0 \subset G_1$, for every closed set is a member of G_δ.[4] Take $\alpha < \Omega$ and assume that $F_\xi \subset G_{\xi+1}$ for $\xi < \alpha$. If α is odd and if $E \in F_\alpha$ then

$$E = \bigcup_{i=1}^{\infty} E_i \qquad \text{where } E_i \in F_{\xi_i}, \quad \xi_i < \alpha.$$

Hence

$$E_i \in G_{\xi_i+1} \subset \bigcup_{\xi<\alpha+1} G_\xi \qquad \text{so that } E \in G_{\alpha+1}.$$

We have thus shown that $F_\alpha \subset G_{\alpha+1}$. The proof is similar in the case in which α is even.

From definitions (4.3.3), (4.3.4) we obtain by induction the inclusions

$$F_\alpha \subset \mathscr{B}, \quad G_\alpha \subset \mathscr{B}.$$

In fact, the classes F_α, G_α contain all Borel sets. We have namely

THEOREM 4.3.8. $\mathscr{B} = \bigcup_{\xi<\Omega} F_\xi = \bigcup_{\xi<\Omega} G_\xi$.

Proof. It follows from (4.3.5) that both unions are equal. It suffices to show that they form a countably additive algebra for then they must contain $\mathscr{B}$. Since the complement of a set in F_ξ is a set in G_ξ it follows that these unions are closed under the operations of taking complements.

Let

$$E_n \in \bigcup_{\xi<\Omega} F_\xi \qquad (n = 1, 2, \ldots)$$

then $E_n \in F_{\xi_n}$, $\xi_n < \Omega$. As there exists an odd $\alpha < \Omega$ such that $\xi_n < \alpha$ for all n, then by (4.3.3)

$$\bigcup_{n=1}^{\infty} E_n \in F_\alpha.$$

Thus the class $\bigcup_{\xi<\Omega} F_\xi$ is closed also under countable unions and so is a countably additive algebra. ■

From other properties of F_α and G_α it is easily proved (by transfinite induction) that inverse images, cartesian products and sections of sets in the class F_α are sets in the class F_α and that the class F_α in the subspace $E \in F_\alpha$ of the space X is identical to the class of sets of the form $A \cap E$ where $A \in F_\alpha$ (in the space X). Similar properties hold for the class G_α.

4.4 MEASURABLE FUNCTIONS

Let $\mathscr{S}$ be a countably additive algebra on a set (space) X. The real function f defined on the set $E \in \mathscr{S}$ is called *measurable relatively to the algebra* $\mathscr{S}$ or ($\mathscr{S}$)-*measurable* if

$$f^{-1}((a, \infty]) = \{x : f(x) > a\} \in \mathscr{S} \qquad \text{for all finite } a. \tag{4.4.1}$$

Because of the identities

$$\{x : f(x) \geqslant a\} = \bigcap_{n=1}^{\infty} \left\{x : f(x) > a - \frac{1}{n}\right\}, \{x : f(x) < a\} = E \backslash \{x : f(x) \geqslant a\},$$

$$\{x : f(x) \leqslant a\} = \bigcap_{n=1}^{\infty} \left\{x : f(x) < a + \frac{1}{n}\right\}, \{x : f(x) > a\} = E \backslash \{x : f(x) \leqslant a\},$$

the condition (4.4.1) implies the first of the conditions

$$\begin{aligned} &\{x : f(x) \geqslant a\} \in \mathscr{S}, \qquad \text{if } |a| < \infty, \\ &\{x : f(x) < a\} \in \mathscr{S}, \qquad \text{if } |a| < \infty, \\ &\{x : f(x) \leqslant a\} \in \mathscr{S}, \qquad \text{if } |a| < \infty, \end{aligned} \tag{4.4.2}$$

which in turn implies the second, which implies the third, which implies (4.4.1). Each of these conditions is therefore an equivalent definition of the measurability of a function. If the function f is measurable, then

$$\{x : f(x) = a\} = \{x : f(x) \geqslant a\} \cap \{x : f(x) \leqslant a\} \in \mathscr{S},$$

$$\{x : f(x) = \infty\} = \bigcap_{n=1}^{\infty} \{x : f(x) \geqslant n\} \in \mathscr{S},$$

$$\{x : f(x) = -\infty\} = \bigcap_{n=1}^{\infty} \{x : f(x) \leqslant -n\} \in \mathscr{S},$$

$$\{x : a < f(x) < b\} = \{x : a < f(x)\} \bigcap \{x : f(x) < b\} \in \mathscr{S}, \text{ etc.}$$

In general,

THEOREM 4.4.1. *If the function f is measurable then $f^{-1}(A) \in \mathscr{S}$ for all $A \in \mathscr{B}(\bar{\mathbb{R}})$.*

Proof. The class $\mathscr{M}$ of all sets $A \subset \mathbb{R}$ for which $f^{-1}(A) \in \mathscr{S}$ is, by Th. 4.1.7, a countably additive algebra. By (4.4.1), the algebra $\mathscr{M}$ contains all intervals $(a, \infty]$ for $|a| < \infty$ and $\bar{\mathbb{R}}$, for the domain of the function belongs to $\mathscr{S}$ (by definition). Hence, by Th. 4.3.2, $\mathscr{B}(\bar{\mathbb{R}}) \subset \mathscr{M}$. ■

If E is the domain of a function f then $\mathscr{S}$-measurable is the same as $\mathscr{S}_E$-measurable (see Th. 4.1.6).

From the identity

$$\{x : f_A(x) > a\} = A \cap \{x : f(x) > a\}$$

we have

THEOREM 4.4.2. *The restriction of an $\mathscr{S}$-measurable function to a set $A \in \mathscr{S}$ is an $\mathscr{S}$-measurable function.*

We say that a function f is *measurable on a set A* belonging to $\mathscr{S}$ if f_A is measurable.

If $A = \bigcup_n A_n$, then

$$\{x : f_A(x) > a\} = \bigcup_n \{x : f_{A_n}(x) > a\}.$$

Hence

THEOREM 4.4.3. *A function which is measurable on every set in a finite or countable sequence of sets $A_1, A_2, \ldots$ ($A_n \in \mathscr{S}$) is also measurable on the union of these sets.*

THEOREM 4.4.4. *If φ is a function of a real variable which is $\mathscr{B}(\bar{\mathbb{R}})$-measurable, that is, a Baire function on $A \in B(\bar{\mathbb{R}})$,*[5] *and if f is a measurable function, then the composite function $\varphi \circ f$ is also measurable.*

Proof. We have

$$\varphi^{-1}((a, \infty]) \in \mathscr{B}(\bar{\mathbb{R}}),$$

and so by Th. 4.4.1

$$\{x : \varphi(f(x)) > a\} = f^{-1}(\varphi^{-1}((a, \infty])) \in \mathscr{S}$$

for every finite a. ■

In particular φ can be a continuous function on a set $A \in \mathscr{B}(\bar{\mathbb{R}})$, for then $\varphi^{-1}((a, \infty])$ is an open set in A and so a Borel set in X.

It follows from this theorem that if f is measurable then so are the functions f^2, $|f|$, $1/f$, $af + b$ etc.

THEOREM 4.4.5. *If the functions f and g are defined and measurable on a set $E \in \mathscr{S}$, then the sets*

$$\{x : f(x) > g(x)\}, \quad \{x : f(x) \geqslant g(x)\}, \quad \{x : f(x) = g(x)\}$$

belong to $\mathscr{S}$.

This follows from the identities

$$\{x : f(x) > g(x)\} = \bigcup_{n,m} \left\{x : f(x) > \frac{m}{n}\right\} \cap \left\{x : \frac{m}{n} > g(x)\right\},$$
$$\{x : f(x) \geqslant g(x)\} = E \backslash \{x : g(x) > f(x)\},$$
$$\{x : f(x) = g(x)\} = \{x : f(x) \leqslant g(x)\} \cap \{x : f(x) \geqslant g(x)\}.$$

THEOREM 4.4.6. *A linear combination and a product of functions defined and measurable on a set $E \in \mathscr{S}$ are measurable functions.*

Proof. The measurability of a linear combination of measurable functions follows from the measurability of their sum. By Th. 4.4.3, it suffices to demonstrate the measurability of a sum of measurable functions f, g in the case of finite functions, for it is easy to check that the set

$$\{x : |f(x)| = \infty \quad \text{or} \quad |g(x)| = \infty\}$$

is a finite union of measurable sets on which $f + g$ is constant. Thus, the theorem follows from Th. 4.4.5 by the identity

$$\{x : f(x) + g(x) > a\} = \{x : f(x) > a - g(x)\} \in \mathscr{S}.$$

In the case of products we can again confine ourselves to the case of finite functions. The measurability of products is then a consequence of the identity

$$fg = \tfrac{1}{4}(f + g)^2 - \tfrac{1}{4}(f - g)^2. \quad ■$$

THEOREM 4.4.7. *If every function in a finite or countable sequence of functions $f_1, f_2, \ldots$ is defined and measurable on a set E, then the functions*

$$\varphi(x) = \inf_n f_n(x) \quad \textit{and} \quad \psi(x) = \sup_n f_n(x)$$

are also measurable. In particular, the maximum and minimum of measurable functions are measurable functions.

This follows from the identities

$$\{x:\varphi(x) < a\} = \bigcup_n \{x: f_n(x) < a\} \quad \text{and} \quad \{x:\psi(x) > a\} = \bigcup_n \{x: f_n(x) > a\}.$$

Since also

$$\liminf_{n\to\infty} f_n = \sup_n \left(\inf_{\nu \geqslant n} f_\nu \right) \quad \text{and} \quad \limsup_{n\to\infty} f_n = \inf_n \left(\sup_{\nu \geqslant n} f_\nu \right)$$

we also have

THEOREM 4.4.8. *The* lim inf, lim sup *and* lim *(if it exists) of a sequence of measurable functions are measurable functions.*

For an arbitrary function f we introduce the notation

$$f_+(x) = \max(f(x), 0) = \begin{cases} f(x) & \text{if } f(x) \geqslant 0 \\ 0 & \text{if } f(x) < 0, \end{cases}$$
$$f_-(x) = \max(-f(x), 0) = \begin{cases} -f(x) & \text{if } f(x) \leqslant 0, \\ 0 & \text{if } f(x) > 0. \end{cases} \tag{4.4.3}$$

Then

$$f(x) = f_+(x) - f_-(x), \quad |f(x)| = f_+(x) + f_-(x). \tag{4.4.4}$$

The function f_+ is the *positive part* of f and f_- is the *negative part* of f. From (4.4.3) and (4.4.4) we have

THEOREM 4.4.9. *A function is measurable if and only if both its positive and its negative parts are measurable.*

The function $\chi_E(x)$ which equals 1 on E and 0 on $\setminus E$ is called the *characteristic function of the set E*. A necessary and sufficient condition for χ_E to be measurable is that $E \in \mathscr{S}, \setminus E \in \mathscr{S}$.

A *simple function* is a finite function which assumes only a finite number of distinct values $\alpha_1 < \cdots < \alpha_k$. Let E_i be the set of all x for which the function takes the value α_i, then the function may be written

$$\sum_{i=1}^{k} \alpha_i \chi_{E_i}.$$

A necessary and sufficient condition for it to be measurable is $E_i \in \mathscr{S}$ $(i = 1, \ldots, k)$. We have the following

THEOREM 4.4.10. *Every measurable function f is the limit of a sequence of simple, measurable functions. We can also require that the sequence be increasing with terms $\geqslant 0$ if $f(x) \geqslant 0$, and that the sequence be uniformly convergent if f is bounded.*

Proof. It suffices to take

$$f_n(x) = \begin{cases} n & \text{if } f(x) \geqslant n, \\ \dfrac{k}{2^n} & \text{if } -n \leqslant \dfrac{k}{2^n} \leqslant f(x) < \dfrac{k+1}{2^n} \leqslant n, \\ -n & \text{if } f(x) < -n. \end{cases}$$

Then $f_n(x) \to f(x)$. Indeed, in the case $|f(x)| < \infty$ we have $|f_n(x) - f(x)| \leqslant (1/2^n)$ when $n \geqslant |f(x)|$. When $f(x) = \pm\infty$ then $f_n(x) = \pm n$.

If $|f(x)| \leqslant M$, M a constant, for all x, then for $n > M$ we have

$$|f_n(x) - f(x)| \leqslant \frac{1}{2^n}$$

so that the sequence is uniformly convergent.

If $f(x) \geqslant 0$ for all x, then

$$f_n(x) = \sup\{w \in Z_n : w \leqslant f(x)\},$$

where

$$Z_n = \left\{0, \frac{1}{2^n}, \ldots . \frac{n \cdot 2^n}{2^n} = n\right\}$$

form an increasing sequence of sets. Hence the sequence f_n is increasing. ■

Directly from the definition of measurability we have

THEOREM 4.4.11. *If $\mathscr{S}, \mathscr{S}'$ are countably additive algebras in X and $\mathscr{S} \subset \mathscr{S}'$ then every $\mathscr{S}$-measurable function is $\mathscr{S}'$-measurable.*

THEOREM 4.4.12. *Let $\mathscr{S}_1$ be a countably additive algebra in X_1 and let $\mathscr{S}_2$ be a countably additive algebra in X_2. If the function f is $(\mathscr{S}_1 \times \mathscr{S}_2)$-measurable then the function $f_{x_1}(x_2) = f(x_1, x_2)$ is $\mathscr{S}_2$-measurable for each x_1 and $f^{x_2}(x_1) = f(x_1, x_2)$ is $\mathscr{S}_1$-measurable for each x_2.*

Indeed, by the definition of the section of a set and from Th. 4.2.3, we have

$$\{x_2 : f_{x_1}(x_2) > a\} = (\{(x_1, x_2) : f(x_1, x_2) > a\})_{x_1} \in \mathscr{S}_2$$

for every finite a.

THEOREM 4.4.13. *If the function φ is $\mathscr{S}_1$-measurable then the function $(x_1, x_2) \to \varphi(x_1)$, where x_2 lies in a set $E_2 \in \mathscr{S}_2$ is also $(\mathscr{S}_1 \times \mathscr{S}_2)$-measurable.*

This is because

$$\{(x_1, x_2): \varphi(x_1) > a\} = \{x_1: \varphi(x_1) > a\} \times E_2 \in \mathscr{S}_1 \times \mathscr{S}_2$$

for all finite a.

4.5 BAIRE FUNCTIONS

Let X be a metric space. We denote by Φ_0 the set of all continuous functions in X and by Φ_1 the set of all functions of the first class of Baire, that is, functions which are the limits of convergent sequences of continuous functions. Φ_2 denotes the set of all functions of the second class of Baire, that is, functions which are the limits of convergent sequences of functions of the first class, etc. Let $\alpha < \Omega$ and suppose that we have already defined classes Φ_ξ for $\xi < \alpha$. Then Φ_α denotes the class of all functions (Baire functions of class α) which are the limits of convergent sequences of functions in $\bigcup_{\xi<\alpha} \Phi_\xi$.

The *class of Baire functions* is the smallest class of functions which

(1) contains all finite continuous functions,
(2) contains the limit of any convergent sequence of functions in the class.

We denote this class by Φ; (similarly as in the case of the algebra generated by a given class of sets (Th. 4.1.5)) it is equal to the intersection of all classes of functions which satisfy the conditions (1) and (2). We have the following

THEOREM 4.5.1. $\Phi = \bigcup_{\alpha<\Omega} \Phi_\alpha$.

Proof. It follows from conditions (1) and (2) by transfinite induction that $\Phi_\alpha \subset \Phi$ for $\alpha < \Omega$,[6] so

$$\bigcup_{\alpha<\Omega} \Phi_\alpha \subset \Phi.$$

It is clear that $\bigcup_{\alpha<\Omega} \Phi_\alpha$ satisfies condition (1). But, it also satisfies condition (2) for if f_n is a convergent sequence of functions from this class then

$$f_n \in \Phi_{\xi_n}, \qquad \text{with } \xi_n < \Omega,$$

so there exists $\alpha < \Omega$ such that $\xi_n < \alpha$ for all n, and then

$$\lim f_n \in \Phi_\alpha \subset \bigcup_{\alpha<\Omega} \Phi_\alpha.$$

So we must have

$$\Phi \subset \bigcup_{\alpha<\Omega} \Phi_\alpha. \quad \blacksquare$$

THEOREM 4.5.2. *The class of Baire functions is identical with the class of measurable functions on X with respect to the algebra of Borel sets in X.*

Proof. Denote by Φ^* the class of all $\mathscr{B}(X)$-measurable functions. This class satisfies condition (1) for the set $\{x:f(x)>a\}$ is open and hence belongs to $\mathscr{B}(X)$ for every a when f is continuous. Thus f is $\mathscr{B}(X)$-measurable. By Th. 4.4.8, the class Φ^* also satisfies condition (2). We therefore have the inclusion

$$\Phi \subset \Phi^*.$$

Now note that Φ has the following property: if f, g are finite and $f, g \in \Phi$ then $\alpha f + \beta g \in \Phi$ and $fg \in \Phi$ (4.5.1). Indeed, if $g \in \Phi$ is finite then the set Φ_g of all functions f such that $\alpha f + \beta g \in \Phi$, $fg \in \Phi$ satisfies conditions (1) and (2). For this is certainly true if g is continuous, hence $\Phi \subset \Phi_g$ in this case. If therefore $g \in \Phi$ is finite and h is finite and continuous, then $g \in \Phi_h$, thus $h \in \Phi_g$. It follows that $\Phi \subset \Phi_g$ which means that condition (4.5.1) holds.

Now, let us prove the inclusion $\Phi^* \subset \Phi$. By Th. 4.4.10, it suffices to show that every simple, $\mathscr{B}(X)$-measurable function belongs to Φ. But because every such function is a linear combination of characteristic functions of sets in $\mathscr{B}(X)$, it suffices, by (4.5.1), to show that the characteristic function of any set in $\mathscr{B}(X)$ belongs to Φ. If, therefore, we denote by $\mathscr{B}'$ the class of all sets A such that $\chi_A \in \Phi$ then we must show the inclusion

$$\mathscr{B}(X) \subset \mathscr{B}'.$$

But $\mathscr{B}'$ contains all open sets, for the characteristic function of an open set is lower semicontinuous and therefore (Th. 3.3.4) of the first class of Baire and so belongs to Φ. Thus, it suffices to show that $\mathscr{B}'$ is a countably additive algebra.

If $E, F \in \mathscr{B}'$, then, by (4.5.1) $\chi_{E \setminus F} = \chi_E - \chi_E \chi_F \in \Phi$, so that $E \setminus F \in \mathscr{B}'$. If $E, F \in \mathscr{B}'$ and $E \cap F = \varnothing$ then by (4.5.1) $\chi_{E \cup F} = \chi_E + \chi_F \in \Phi$, so that $E \cup F \in \mathscr{B}'$. Hence $\mathscr{B}'$ is a finitely additive algebra. If E_n is an expanding sequence of sets in $\mathscr{B}'$ and if $E = \lim_{n\to\infty} E_n$, then, by property (2), $\chi_E(x) = \lim_{n\to\infty} \chi_E(x) \in \Phi$ so that $\lim E_n \in \mathscr{B}'$. Hence, by Th. 4.1.4, it follows that $\mathscr{B}'$ is a countably additive algebra. ■

THEOREM 4.5.3. *If g is a continuous mapping of a space X to a space Y and if φ is a Baire function on Y, then $\varphi \circ g$ is a Baire function on X.*

For then $\varphi^{-1}((a,\infty]) \in \mathscr{B}(Y)$ and so by Th. 4.3.4, we have

$$\{x:\varphi(g(x))>a\} = g^{-1}(\varphi^{-1}((a,\infty])) \in \mathscr{B}(X)$$

for every finite a.

The classes Φ_α can be characterized by inverse images using the classes F_α, G_α of Borel sets. One can show, namely, using transfinite induction and by a similar method to that used in the proof of Th. 4.5.1 and Th. 3.5.1, the following

THEOREM 4.5.4 (Lebesgue–Hausdorff). *Let Φ_α^* be the class of all function f such that for all a the sets*

$$\{x:f(x)<a\} \quad \textit{and} \quad \{x:f(x)>a\}$$

belong to F_α (α-odd) *or to* G_α (α-even). *Then*

$$\Phi_\alpha = \begin{cases} \Phi_\alpha^* & \text{if } \alpha \text{ is finite} \\ \Phi_{\alpha+1}^* & \text{if } \alpha \text{ is infinite.} \end{cases}$$

In particular, functions of the second class are characterized by the conditions

$$\{x : f(x) < a\}, \{x : f(x) > a\} \in G_{\delta\sigma} \quad \text{for all } a.$$

NOTES

1. Note that $(\bigcup_1^\infty E_n) \times F = \bigcup_1^\infty (E_n \times F)$ and $(E_1 \backslash E_2) \times F = (E_1 \times F) \backslash (E_2 \times F)$.
2. Clearly, the closed intervals could be replaced by open intervals (or, for example, by intervals open on the right).
3. This holds, for example, for $X = \mathbb{R}^n$.
4. This is not true in a general topological space.
5. Compare with Ths 4.3.3 and 4.5.2.
6. $\Phi_0 \subset \Phi$ since every continuous function f is the limit of the sequence of continuous and finite functions

$$f_n(x) = \max(-n, \min(n, f(x))).$$

CHAPTER 5

MEASURE AND MEASURABLE FUNCTIONS

5.1 MEASURE

Let $\mathscr{S}$ be a countably additive algebra in the space X. A real-valued function μ defined on $\mathscr{S}$ is called a *measure* if

(1) $\mu(E) \geqslant 0$ for $E \in \mathscr{S}$,
(2) $\mu(\varnothing) = 0$,
(3) $\mu(\bigcup_{i=1}^{\infty} E_i) = \sum_{i=1}^{\infty} \mu(E_i)$ if the E_i are disjoint sets

in $\mathscr{S}$ (countable additivity).
The sets for which μ is defined, that is, the sets belonging to $\mathscr{S}$, are called *μ-measurable.*

The restriction of the measure μ to a countably additive algebra $\mathscr{S}_0 \subset \mathscr{S}$ is also a measure (on $\mathscr{S}_0$). In particular, if $E \subset X$ we can restrict the measure μ to the algebra $\mathscr{S}_E$.

From the definition of measure we have the following properties:

THEOREM 5.1.1. *$\mu(E_1 \cup \cdots \cup E_n) = \mu(E_1) + \cdots + \mu(E_n)$ where $E_1, \ldots, E_n$ are disjoint sets in $\mathscr{S}$ (finite additivity).*

Indeed, by (3) and (2) we have

$$\mu(E_1 \cup \cdots \cup E_n) = \mu(E_1 \cup \cdots \cup E_n \cup \varnothing \cup \cdots) = \mu(E_1) + \cdots + \mu(E_n).$$

If $E \subset F$ and $E, F \in \mathscr{S}$ then

$$\mu(F) = \mu(E \cup (F \backslash E)) = \mu(E) + \mu(F \backslash E),$$

since E and $F \backslash E$ are disjoint. By (1) we therefore have

THEOREM 5.1.2. *If $E \subset F$ $(E, F \in \mathscr{S})$ then $\mu(E) \leqslant \mu(F)$.*

THEOREM 5.1.3. *If $E \subset F$ $(E, F \in \mathscr{S})$ and if $\mu(F) < \infty$ then $\mu(F \backslash E) = \mu(F) - \mu(E)$.*

THEOREM 5.1.4. *Let $E \in \mathscr{S}$ and let $E_1, E_2, \ldots,$ be a finite or countable sequence*

of sets in $\mathscr{S}$. If $E \subset \bigcup_i E_i$ then

$$\mu(E) \leqslant \sum_i \mu(E_i).$$

For we have $\bigcup_i E_i = \bigcup_i F_i$, where $F_1 = E_1, \ldots, F_i = E_i \backslash (E_1 \cup \cdots \cup E_{i-1}), \ldots$ are disjoint and $F_i \subset E_i$, hence

$$\mu(E) \leqslant \mu\left(\bigcup_i E_i\right) = \mu\left(\bigcup_i F_i\right) = \sum_i \mu(F_i) \leqslant \sum_i \mu(E_i).$$

THEOREM 5.1.5. *If $E_1 \subset E_2 \subset \cdots$ is an increasing sequence of sets in $\mathscr{S}$, then*

$$\mu\left(\lim_{n\to\infty} E_n\right) = \lim_{n\to\infty} \mu(E_n). \tag{5.1.1}$$

Proof. By property (3) we have

$$\mu\left(\lim_{n\to\infty} E_n\right) = \mu\left(E_1 \cup \bigcup_{i=2}^{\infty} (E_i \backslash E_{i-1})\right) = \mu(E_1) + \sum_{i=2}^{\infty} \mu(E_i \backslash E_{i-1}).$$

On the other hand, by Th. 5.1.1,

$$\mu(E_n) = \mu\left(E_1 \cup \bigcup_{i=2}^{n} (E_i \backslash E_{i-1})\right) = \mu(E_1) + \sum_{i=2}^{n} \mu(E_i \backslash E_{i-1})$$

and so (5.1.1) follows. ■

THEOREM 5.1.6. *If $E_1 \supset E_2 \supset \cdots$ is a decreasing sequence of sets in $\mathscr{S}$ and if $\mu(E_m) < \infty$ for some m, then*

$$\mu\left(\lim_{n\to\infty} E_n\right) = \lim_{n\to\infty} \mu(E_n). \tag{5.1.2}$$

Proof. Since the sequence $E_m \backslash E_\nu$ is increasing (with ν) then, by Ths 5.1.5 and 5.1.3, we have

$$\begin{aligned}\mu\left(\lim_{\nu\to\infty} (E_m \backslash E_\nu)\right) &= \lim_{\nu\to\infty} \mu(E_m \backslash E_\nu) = \lim_{\nu\to\infty} [\mu(E_m) - \mu(E_\nu)] \\ &= \mu(E_m) - \lim_{\nu\to\infty} \mu(E_\nu).\end{aligned}$$

But

$$\mu\left(\lim_{\nu\to\infty} (E_m \backslash E_\nu)\right) = \mu\left(E_m \backslash \lim_{\nu\to\infty} E_\nu\right) = \mu(E_m) - \mu\left(\lim_{\nu\to\infty} E_\nu\right)$$

from which we obtain (5.1.2). ■

Sets of Measure Zero

From Th. 5.1.2 and from property (1) and Th. 5.1.1 it follows that every measurable subset of a set of measure zero has zero measure and that the union of at most a countable number of sets of measure zero has measure zero.

The measure μ is called *complete* if every subset of any set of measure zero is measurable (and so has measure zero).

Let $w(x)$ be a condition on $x \in X$ and let $E \subset X$. We say that the condition $w(x)$ holds *almost everywhere* (a.e.) in E if the exceptional set Z_0 of all points of E at which this condition is not satisfied is contained in a set of measure zero (or, has measure zero, if μ is complete), in other words, if $w(x)$ holds in a set $F \subset E$ such that $\mu(E \setminus F) = 0$ or in a set $E \setminus Z$ where $\mu(Z) = 0$.

If $w_1(x), w_2(x), \ldots$ is a finite or countable sequence of conditions each of which holds a.e. in E then all these conditions hold simultaneously a.e. in E (for the exceptional set for the simultaneous occurrence of all the conditions is the union of the exceptional sets for each individual condition).

If condition $w(x)$ holds a.e. in E then it holds a.e. in $F \subset E$, for the exceptional set for $w(x)$ in F is contained in the exceptional set for $w(x)$ in E. Also, any consequence $u(x)$ of condition $w(x)$ holds a.e. in E.

We say that the sets $A, B \in \mathscr{S}$ *differ by a set of measure zero* or are *μ-equivalent* if

$$\mu(A \setminus B) = \mu(B \setminus A) = 0. \tag{5.1.3}$$

We then write

$$A \approx B.$$

This is an equivalence relation. Indeed, reflexivity and symmetry are obvious. Also, if $A \approx B$ and $B \approx C$ then $A \approx C$ for, since $A \setminus C \subset (A \setminus B) \cup (B \setminus C)$ we therefore have $\mu(A \setminus C) = 0$ and similarly $\mu(C \setminus A) = 0$.

The implication

$$(A \setminus Z_1) \cup Z_2 \approx A \quad \text{if} \quad \mu(Z_1) = 0 = \mu(Z_2)^1$$

follows since $((A \setminus Z_1) \cup Z_2) \setminus A \subset Z_2$ and $A \setminus ((A \setminus Z_1) \cup Z_2) \subset Z_1$.

THEOREM 5.1.7. *If $A \approx B$ then*

$$\mu(A) = \mu(B).$$

In particular, if $A, Z \in \mathscr{S}$ and $\mu(Z) = 0$ then

$$\mu(A \setminus Z) = \mu(A) = \mu(A \cup Z).$$

For, by (5.1.3), we have

$$\mu(A) \leqslant \mu((A \setminus B) \cup B) \leqslant \mu(A \setminus B) + \mu(B) = \mu(B)$$

and similarly $\mu(B) \leqslant \mu(A)$.

THEOREM 5.1.8. *μ-equivalence is preserved under difference, union and intersection of at most a countable number of sets: if $E_i \approx F_i$ then*

$$E_1 \setminus E_2 \approx F_1 \setminus F_2, \quad \bigcup_i E_i \approx \bigcup_i F_i \quad \text{and} \quad \bigcap_i E_i \approx \bigcap_i F_i.$$

Proof. Let $E_i \approx F_i$. Since

$$(E_1 \backslash E_2) \backslash (F_1 \backslash F_2) \subset (E_1 \backslash F_1) \cup (F_2 \backslash E_2)$$

we have $\mu((E_1 \backslash E_2) \backslash (F_1 \backslash F_2)) = 0$. Similarly $\mu((F_1 \backslash F_2) \backslash (E_1 \backslash E_2)) = 0$.

Since

$$\bigcup_i E_i \backslash \bigcup_i F_i \subset \bigcup_i (E_i \backslash F_i),$$

we have $\mu(\bigcup_i E_i \backslash \bigcup_i F_i) = 0$. Similarly, $\mu(\bigcup_i F_i \backslash \bigcup_i E_i) = 0$.

The result for intersection follows from the result for unions and differences by using the identity

$$\bigcap_i E_i = E_1 \backslash \bigcup_i (E_1 \backslash E_i). \quad \blacksquare$$

THEOREM 5.1.9. *If $E_1, E_2, \ldots$ is a finite or countable sequence of sets in $\mathscr{S}$ such that $\mu(E_i \cap E_j) = 0$ for $i \neq j$, then*

$$\mu\left(\bigcup_i E_i\right) = \sum_i \mu(E_i).$$

Proof. We have $\bigcup_i E_i = \bigcup_i F_i$, where $F_1 = E_1, \ldots, F_i = E_i \backslash (E_1 \cup \cdots \cup E_{i-1}) = E_i \backslash (E_i \cap E_1) \backslash \cdots \backslash (E_i \cap E_{i-1}), \ldots$ are disjoint and $E_i \approx F_i$, therefore

$$\mu\left(\bigcup_i E_i\right) = \mu\left(\bigcup_i F_i\right) = \sum_i \mu(F_i) = \sum_i \mu(E_i). \quad \blacksquare$$

We now show that any measure μ can be extended to a complete measure, that is, a measure such that all subsets of sets of measure zero are measurable.

To do this, let $\mathscr{N}$ denote the class of all subsets of sets of measure zero and let $\bar{\mathscr{S}}$ be the algebra generated by $\mathscr{S} \cup \mathscr{N}$. We have the following

THEOREM 5.1.10. *There exists exactly one measure $\bar{\mu}$ on $\bar{\mathscr{S}}$ such that $\bar{\mu} = \mu$ on $\mathscr{S}$. This measure is complete.*

The measure $\bar{\mu}$ is called the *completion of the measure* μ and the algebra $\bar{\mathscr{S}}$ is the *completion of the algebra* $\mathscr{S}$ with respect to the measure μ.

Proof. First note that any subset of a set in $\mathscr{N}$ belongs to $\mathscr{N}$ and that the union of at most a countable number of sets in $\mathscr{N}$ belongs to $\mathscr{N}$. From this, as for μ-equivalence, it follows that the relation $E \dot{\approx} F$ defined between sets $E \subset X$ and $F \subset X$ by the condition

$$E \backslash F \in \mathscr{N} \quad \text{and} \quad F \backslash E \in \mathscr{N} \tag{5.1.4}$$

is an equivalence relation which is preserved under difference, intersection and union of at most a countable number of sets. This relation, restricted to $\mathscr{S}$, coincides with μ-equivalence.

We denote by $\mathscr{S}'$ the class of all sets E satisfying the condition $E \dot{\approx} F$ for some $F \in \mathscr{S}$. We show that

$$\bar{\mathscr{S}} = \mathscr{S}'.$$

For, let $E \in \mathscr{S}'$, then $E \dot{\approx} F$ where $F \in \mathscr{S}$, and since $E = (F \backslash (F \backslash E)) \cup (E \backslash F)$ we have, by (5.1.4), that $E \in \bar{\mathscr{S}}$. Hence

$$\mathscr{S}' \subset \bar{\mathscr{S}}.$$

To obtain the inclusion

$$\bar{\mathscr{S}} \subset \mathscr{S}'$$

it suffices, since $\mathscr{S} \cup \mathscr{N} \subset \mathscr{S}'$, to show that $\mathscr{S}'$ is a countably additive algebra. But, if $A, B \in \mathscr{S}'$ then $A \dot{\approx} E$, $B \dot{\approx} F$ where $E, F \in \mathscr{S}$, therefore $A \backslash B \dot{\approx} E \backslash F \in \mathscr{S}$ and so $A \backslash B \in \mathscr{S}'$. Also, if $A_i \in \mathscr{S}'$ then $A_i \dot{\approx} E_i \in \mathscr{S}$ and so $\bigcup_{i=1}^{\infty} A_i \dot{\approx} \bigcup_{i=1}^{\infty} E_i \in \mathscr{S}$. Hence $\bigcup_{i=1}^{\infty} A_i \in \mathscr{S}'$.

If $A \in \bar{\mathscr{S}}$ then $A \approx E$ for some $E \in \mathscr{S}$. Define

$$\bar{\mu}(A) = \mu(E).$$

This does not depend upon the choice of E, for if $A \dot{\approx} E' \in \mathscr{S}$ then $E \approx E'$ and so $\mu(E) = \mu(E')$. Clearly $\bar{\mu}(E) = \mu(E)$ for $E \in \mathscr{S}$. The function $\bar{\mu}$ satisfies the measure conditions (1) and (2) and is countably additive. For, if $A_1, A_2, \ldots$ is a sequence of disjoint sets in $\bar{\mathscr{S}}$ then $A_i \dot{\approx} E_i \in \mathscr{S}$ and so for $i \neq j$ we have $E_i \cap E_j \dot{\approx} A_i \cap A_j = \varnothing$. Hence

$$\mu(E_i \cap E_j) = 0.$$

But

$$\bigcup_{i=1}^{\infty} A_i \dot{\approx} \bigcup_{i=1}^{\infty} E_i$$

and so, by Th. 5.1.9, we have

$$\bar{\mu}\left(\bigcup_{i=1}^{\infty} A_i\right) = \mu\left(\bigcup_{i=1}^{\infty} E_i\right) = \sum_{i=1}^{\infty} \mu(E_i) = \sum_{i=1}^{\infty} \bar{\mu}(A_i).$$

Hence $\bar{\mu}$ is a measure on $\bar{\mathscr{S}}$ which satisfies the given conditions.

If $\bar{\bar{\mu}}$ is another measure with the same properties then $\bar{\bar{\mu}} = \bar{\mu}$. Indeed, we must have $\bar{\bar{\mu}} = 0$ on $\mathscr{N}$. Let $\mathscr{A} \in \bar{\mathscr{S}}$. Then $A \dot{\approx} E \in \mathscr{S}$ and so by (5.1.4) we have

$$\bar{\bar{\mu}}(A \backslash E) = \bar{\bar{\mu}}(E \backslash A) = 0.$$

Then, by Th. 5.1.7,

$$\bar{\bar{\mu}}(A) = \bar{\bar{\mu}}((E \backslash (E \backslash A)) \cup (A \backslash E)) = \bar{\bar{\mu}}(E) = \mu(E) = \bar{\mu}(A).$$

The completeness of $\bar{\mu}$ follows from the fact that $\bar{\mu}(A) = 0$ implies that $A \in \mathscr{N}$. This is because $A \subset E \cup (A \backslash E)$ where $A \dot{\approx} E \in \mathscr{S}$. ■

5.2 OUTER MEASURE

A real function μ^* defined on the class of all subsets of a space X is called an *outer measure* on X, if

(1) $\mu^*(E) \geqslant 0$ for $E \subset X$,
(2) $\mu^*(\varnothing) = 0$,
(3) $\mu^*(E) \leqslant \mu^*(F)$ if $E \subset F$,
(4) $\mu^*(\bigcup_{i=1}^{\infty} E_i) \leqslant \sum_{i=1}^{\infty} \mu^*(E_i)$.

Just as for the case of a measure, conditions (2) and (4) imply the inequality

$$\mu^*\left(\bigcup_{i=1}^{n} E_i\right) \leqslant \sum_{i=1}^{n} \mu^*(E_i).$$

We say that the set $E \subset X$ is *μ^*-measurable* if it satisfies *Caratheodory's condition*:

$$\mu^*(A) = \mu^*(A \cap E) + \mu^*(A \setminus E) \qquad \text{for all } A \subset X. \tag{C}$$

THEOREM 5.2.1. *The class Λ of all μ^*-measurable sets, that is, sets satisfying Caratheodory's condition, is a countably additive algebra and μ^* restricted to Λ is a measure.*

Proof. Let $E, F \in \Lambda$ and $A \subset X$, then

$$\mu^*(A) = \mu^*(A \cap E) + \mu^*(A \setminus E) = \mu^*(A \cap E) + \mu^*((A \setminus E) \cap F) + \mu^*((A \setminus E) \setminus F).$$

But since $A \cap E = A \cap (E \cup F) \cap E$ and $(A \setminus E) \cap F = (A \cap (E \cup F)) \setminus E$, therefore

$$\begin{aligned}\mu^*(A) &= \mu^*(A \cap (E \cup F) \cap E) + \mu^*(A \cap (E \cup F) \setminus E) + \mu^*(A \setminus (E \cup F)) \\ &= \mu^*(A \cap (E \cup F)) + \mu^*(A \setminus (E \cup F)).\end{aligned}$$

Hence $E \cup F \in \Lambda$. So we have

$$\text{if} \quad E, F \in \Lambda \quad \text{then} \quad E \cup F \in \Lambda. \tag{5.2.1}$$

If $E \in \Lambda$ then by the symmetry of condition (C) we have $\setminus E \in \Lambda$. Thus since $E \setminus F = \setminus((\setminus E) \cup F)$, therefore

$$\textit{if} \quad E, F \in \Lambda \quad \text{then} \quad E \setminus F \in \Lambda. \tag{5.2.2}$$

Let E, F be disjoint sets in Λ. Then for any A

$$\begin{aligned}\mu^*(A \cap (E \cup F)) &= \mu^*(A \cap (E \cup F) \cap E) + \mu^*(A \cap (E \cup F) \setminus E) \\ &= \mu^*(A \cap E) + \mu^*(A \cap F).\end{aligned}$$

If now $E_1, E_2, \ldots$ is a sequence of disjoint sets in Λ then by induction and by (5.2.1) we obtain

$$\mu^*\left(A \cap \bigcup_{i=1}^{n} E_i\right) = \sum_{i=1}^{n} \mu^*(A \cap E_i).$$

Thus, by (3)

$$\mu^*\left(A \cap \bigcup_{i=1}^{\infty} E_i\right) \geqslant \sum_{i=1}^{n} \mu^*(A \cap E_i)$$

and in the limit

$$\mu^*\left(A\cap\bigcup_{i=1}^{\infty}E_i\right)\geqslant\sum_{i=1}^{\infty}\mu^*(A\cap E_i).$$

But property (4) gives the reverse inequality. So we obtain

$$\mu^*\left(A\cap\bigcup_{i=1}^{\infty}E_i\right)=\sum_{i=1}^{\infty}\mu^*(A\cap E_i) \tag{5.2.3}$$

when $A\subset X$ and E_i are disjoint sets in Λ.

Under the same assumptions about A and E_i we have, by (5.2.1)

$$\bigcup_{i=1}^{n}E_i\in\Lambda,$$

and so

$$\begin{aligned}\mu^*(A)&=\mu^*\left(A\cap\bigcup_{i=1}^{n}E_i\right)+\mu^*\left(A\Big\backslash\bigcup_{i=1}^{n}E_i\right)\\&\geqslant\sum_{i=1}^{n}\mu^*(A\cap E_i)+\mu^*\left(A\Big\backslash\bigcup_{i=1}^{\infty}E_i\right),\end{aligned}$$

whence, in the limit,

$$\mu^*(A)\geqslant\sum_{i=1}^{\infty}\mu^*(A\cap E_i)+\mu^*\left(A\Big\backslash\bigcup_{i=1}^{\infty}E_i\right)=\mu^*\left(A\cap\bigcup_{i=1}^{\infty}E_i\right)+\mu^*\left(A\Big\backslash\bigcup_{i=1}^{\infty}E_i\right)$$

Since property (4) gives the reverse inequality, we have

$$\mu^*(A)=\mu^*\left(A\cap\bigcup_{i=1}^{\infty}E_i\right)+\mu^*\left(A\Big\backslash\bigcup_{i=1}^{\infty}E_i\right).$$

Hence $\bigcup_{i=1}^{\infty}E_i\in\Lambda$.

Thus, we have shown that

$$\bigcup_{i=1}^{\infty}E_i\in\Lambda \text{ if the } E_i \text{ are disjoint sets in } \Lambda. \tag{5.2.4}$$

From properties (5.2.2), (5.2.4) and Th. 4.1.3, it follows that Λ is a countably additive algebra. Also, from properties (1), (2) and equation (5.2.3) with $A=X$ it follows that $(\mu^*)_\Lambda$ is a measure. ∎

THEOREM 5.2.2. *Sets E for which $\mu^*(E)=0$ are μ^*-measurable. The outer measure μ^* restricted to Λ is a complete measure.*

Proof. Let $\mu^*(E)=0$. By (3) we have $\mu^*(A\cap E)=0$ for all A and so $\mu^*(A)\leqslant\mu^*(A\cap E)+\mu^*(A\backslash E)\leqslant\mu^*(A)$. Hence $\mu^*(A)=\mu^*(A\cap E)+\mu^*(A\backslash E)$. Thus $E\in\Lambda$. It follows that μ^* is a complete measure on Λ. ∎

Now suppose that X is a metric space. The outer measure μ^* is called a *metric outer measure* if the following condition is satisfied:

(5) $\mu^*(E \cup F) = \mu^*(E) + \mu^*(F)$ when $\rho(E, F) > 0$.

From this condition we obtain by induction that

$$\mu^*\left(\bigcup_{i=1}^{n} E_i\right) = \sum_{i=1}^{n} \mu^*(E_i) \qquad \text{if } \rho(E_i, E_j) > 0 \text{ for } i \neq j. \tag{5.2.5}$$

THEOREM 5.2.3. *If μ^* is a metric outer measure then all Borel sets in X are μ^*-measurable. Thus μ^*, restricted to $\mathscr{B}(X)$, is a measure on $\mathscr{B}(X)$.*

Proof. Since Λ is a countably additive algebra it follows that to prove the inclusion $\mathscr{B}(X) \subset \Lambda$ it suffices to show that every open set is μ^*-measurable.

So, let G be an open set. Let

$$G_n = \left\{x : \rho(x, \backslash G) > \frac{1}{n}\right\};$$

then

$$\rho(G_n, \backslash G) \geqslant \frac{1}{n} > 0. \tag{5.2.6}$$

We put

$$D_n = \left\{x : \frac{1}{n+1} < \rho(x, \backslash G) \leqslant \frac{1}{n}\right\};$$

then, as is easily shown, we have

$$\rho(D_i, D_j) \geqslant \frac{1}{i+1} - \frac{1}{j} > 0 \qquad \text{if } i + 2 \leqslant j, \tag{5.2.7}$$

and, noting that $G = \{x : \rho(x, \backslash G) > 0\}$, we obtain

$$G \backslash G_n = \bigcup_{i=n}^{\infty} D_i. \tag{5.2.8}$$

Take an arbitrary set $A \subset X$. We have to show that

$$\mu^*(A) = \mu^*(A \cap G) + \mu^*(A \backslash G).$$

Since $\mu^*(A) \leqslant \mu^*(A \cap G) + \mu^*(A \backslash G)$ it suffices to prove the reverse inequality. This is obvious if $\mu^*(A) = \infty$, hence assume that $\mu^*(A) < \infty$.

By (5.2.5) and (5.2.7), we have

$$\begin{aligned} &\mu^*(A \cap D_1) + \mu^*(A \cap D_3) + \cdots + \mu^*(A \cap D_{2n-1}) \\ &\quad = \mu^*(A \cap (D_1 \cup D_3 \cup \cdots \cup D_{2n-1})) \leqslant \mu^*(A) \end{aligned}$$

and

$$\begin{aligned} &\mu^*(A \cap D_2) + \mu^*(A \cap D_4) + \cdots + \mu^*(A \cap D_{2n}) \\ &\quad = \mu^*(A \cap (D_2 \cup D_4 \cup \cdots \cup D_{2n})) \leqslant \mu^*(A). \end{aligned}$$

Hence

$$\sum_{i=1}^{\infty} \mu^*(A \cap D_i) \leqslant 2\mu^*(A) < \infty.$$

But since, by (5.2.8)

$$\mu^*(A \cap (G \backslash G_n)) \leqslant \sum_{i=n}^{\infty} \mu^*(A \cap D_i)$$

therefore

$$\mu^*(A \cap (G \backslash G_n)) \to 0 \quad \text{as } n \to \infty. \tag{5.2.9}$$

By property (5) and (5.2.6) we have

$$\mu^*(A \cap G_n) + \mu^*(A \backslash G) = \mu^*((A \cap G_n) \cup (A \backslash G)) \leqslant \mu^*(A).$$

Hence

$$\begin{aligned}\mu^*(A \cap G) + \mu^*(A \backslash G) &\leqslant \mu^*(A \cap G_n) + \mu^*(A \cap (G \backslash G_n)) + \mu^*(A \backslash G) \\ &\leqslant \mu^*(A) + \mu^*(A \cap (G \backslash G_n)).\end{aligned}$$

Thus in the limit, by (5.2.9),

$$\mu^*(A \cap G) + \mu^*(A \backslash G) \leqslant \mu^*(A). \quad \blacksquare$$

An example of a metric outer measure in an arbitrary metric space X is given by the *Hausdorff α-dimensional measure.* For every set E and number $\alpha > 0$ we put

$$\lambda_\alpha^\varepsilon(E) = \inf_{E = \bigcup_{i=1}^{\infty} E_i, \delta(E_i) < \varepsilon} \sum_{i=1}^{\infty} [\delta(E_i)]^\alpha.$$

The number $\mu_\alpha^*(E) = \sup_{\varepsilon > 0} \lambda_\alpha^\varepsilon(E) = \lim_{\varepsilon \to 0} \lambda_\alpha^\varepsilon(E)$ is called the α-dimensional Hausdorff measure of the set E. It is not difficult to show that μ_α^* is a metric outer measure and hence a measure on an algebra containing the Borel sets. It can be shown that when $X = \mathbb{R}^n$, μ_n^* differs from Lebesgue outer measure on $\mathbb{R}^n$ by a constant factor: $\mu_n^*(E) = \kappa_n m_n^*(E)$, where κ_n equals the ratio of the volume of the n-dimensional cube to the volume of the n-dimensional ball inscribed in the cube.

THEOREM 5.2.4. *If λ is a function defined on the class of open sets and satisfying conditions* (1)–(5), *then*

$$\mu^*(E) = \inf_{E \subset G} \lambda(G) \tag{5.2.10}$$

is a metric outer measure.

Proof. It follows from (5.2.10) that μ^* satisfies conditions (1)–(3). For a sequence of sets E_ν and for $\varepsilon > 0$ take G_ν open such that

$$E_\nu \subset G_\nu \quad \text{and} \quad \lambda(G_\nu) \leqslant \mu^*(E_\nu) + \frac{\varepsilon}{2^\nu}.$$

Then

$$\mu^*\left(\bigcup_{\nu=1}^{\infty} E_\nu\right) \leqslant \lambda\left(\sum_{\nu=1}^{\infty} G_\nu\right) \leqslant \sum_{\nu=1}^{\infty} \lambda(G_\nu) \leqslant \sum_{\nu=1}^{\infty} \mu^*(E_\nu) + \varepsilon.$$

Hence

$$\mu^*\left(\bigcup_{\nu=1}^{\infty} E_\nu\right) \leqslant \sum_{\nu=1}^{\infty} \mu^*(E_\nu)$$

so that μ^* satisfies condition (4).

If $\eta=\rho(E,F)>0$ then

$$G=\{x:\rho(x,E)<\tfrac{1}{3}\eta\},\quad H=\{x:\rho(x,F)<\tfrac{1}{3}\eta\}$$

are open sets containing E and F respectively. Also $\rho(G,H)\geqslant\frac{1}{3}\eta>0$. Let D be any open set containing $E\cup F$. We have

$$\begin{aligned}\mu^*(E)+\mu^*(F)&\leqslant\lambda(G\cap D)+\lambda(H\cap D)\\&=\lambda((G\cap D)\cup(H\cap D))\leqslant\lambda(D).\end{aligned}$$

Hence, taking the infimum over D we obtain

$$\mu^*(E)+\mu^*(F)\leqslant\mu^*(E\cup F).$$

Now, using property (4), it follows that μ^* satisfies property (5) which completes the proof. ■

Remark. Replacing condition (5) *by the conditions*

(5a) $\lambda(G\cup H)=\lambda(G)+\lambda(H)$ *if* G *and* H *are disjoint,*
(5b) $\lambda(G)=\sup_{\bar{D}\subset G}\lambda(D)$,

(it follows from (1)–(4) that μ^* is an outer measure), *we obtain the measurability of the Borel sets without using* Th. 5.2.3.[2]).

Indeed, it suffices to prove the inequality

$$\mu^*(A\cap G)+\mu^*(A\backslash G)\leqslant\mu^*(A)$$

for every open G and arbitrary A. To do this take, successively, arbitrary open sets H, C, D such that

$$H\supset A,\quad \bar{C}\subset H\cap G\quad\text{and}\quad \bar{D}\subset H\backslash\bar{C}.$$

Taking a supremum (by (5b)) and using the inequality $\lambda(C)+\lambda(D)\leqslant\lambda(H)$ we obtain

$$\lambda(C)+\mu^*(A\backslash G)\leqslant\lambda(C)+\lambda(H\backslash\bar{C})\leqslant\lambda(H)$$

and

$$\mu^*(A\cap G)+\mu^*(A\backslash G)\leqslant\lambda(H\cap G)+\mu^*(A\backslash G)\leqslant\lambda(H)$$

from which the required inequality follows.

THEOREM 5.2.5. *If μ is a measure on $\mathscr{B}(X)$, then*

$$\mu(E)=\inf_{E\subset G_{\text{open}}}\mu(G)$$

in the countably additive algebra $\mathscr{B}_\mu$ generated by the open sets with finite μ-measure.

If X is the union of a sequence of open sets with finite μ-measure, then

$$\mathscr{B}_\mu=\mathscr{B}(X).$$

Thus any two such measures which are equal on the class of open sets must be identical.[3]

Proof. By Ths 5.2.4 and 5.2.3,

$$\mu^*(E) = \inf_{E \subset G_{\text{open}}} \mu(G)$$

is a measure on $\mathscr{B}(X)$, where clearly $\mu^*(G) = \mu(G)$, when G is open and (taking an infimum) $\mu(E) \leqslant \mu^*(E)$ for $E \in \mathscr{B}(X)$. Also, if E is a Borel subset of an open set G of finite measure, then we also have the reverse inequality, for $\mu(G \backslash E) \leqslant \mu^*(G \backslash E)$, hence

$$\mu(E) = \mu^*(E).$$

Thus, this equality holds also on the class of countable unions of such sets (by taking the limit in $\mu(\bigcup_1^n E_i) = \mu^*(\bigcup_1^n E_i)$) which is a countably additive algebra containing $\mathscr{B}_\mu$.[4] The second part of the theorem follows from the fact that in this case every open set is the union of a sequence of open sets of finite measure. ■

Outer Measure on a Countably Additive Ideal

Let X be an arbitrary set (space). A non-empty class $\mathscr{I}$ of subsets of the space X is called a *countably additive ideal* if the following conditions are satisfied:

(a) if $A \in \mathscr{I}$ and $B \subset A$ then $B \in \mathscr{I}$,
(b) if $A_i \in \mathscr{I}$ $(i = 1, 2, \ldots)$ then $\bigcup_{i=1}^{\infty} A_i \in \mathscr{I}$.

An *outer measure* on a countably additive ideal $\mathscr{I}$ is a function μ^* defined on $\mathscr{I}$ which satisfies conditions (1)–(4). Then Caratheodory's condition is that equation (C) holds for every $A \in \mathscr{I}$ and Th. 5.2.1 then takes the form (proof unchanged):

The class Λ of all sets satisfying Caratheodory's condition is a countably additive algebra and the restriction of μ^* to the class $\Lambda \cap \mathscr{I}$ (which is also a countably additive algebra) is a measure.

We make use of this theorem in the following section.

Extension of a Finitely Additive Function to a Measure

Let $\mathscr{R}$ be a finitely additive algebra on X and let F be a finite, non-negative and finitely-additive function on $\mathscr{R}$, that is:

(1′) $0 \leqslant F(A) < \infty$ for $A \in \mathscr{R}$,
(2′) $F(A \cup B) = F(A) + F(B)$ if A, B are disjoint sets in $\mathscr{R}$. Clearly in this case $F(\varnothing) = 0$ and

$$F\left(\bigcup_{i=1}^{k} A_i\right) = \sum_{i=1}^{k} F(A_i) \qquad \text{for disjoint } A_i \in \mathscr{R}.$$

In addition

$$F(A) \leqslant F(B) \quad \text{for} \quad A \subset B \quad (\text{in } \mathscr{R}).$$

The class $\mathscr{I}$ of sets which can be covered by a countable number of sets in $\mathscr{R}$ is a countably additive ideal. Also, the function

$$F^*(E) = \inf_{E \subset \cup A_i} \sum_{i=1}^{\infty} F(A_i) \qquad \text{for } E \in \mathscr{I}$$

is an outer measure on $\mathscr{I}$.

Indeed, conditions (1)–(3) are clearly satisfies and condition (4) can be verified by taking (for arbitrary $\varepsilon > 0$) a covering $\{A_{i\nu}\}$ of the set E_i such that $\sum_\nu F(A_{i\nu}) \leqslant F^*(E_i) + 2^{-i}\varepsilon$, from which $F^*(E) \leqslant \sum_i F^*(E_i) + \varepsilon$.

If F is countably additive, that is $F(\bigcup_{i=1}^{\infty} E_i) = \sum_{i=1}^{\infty} F(E_i)$ for disjoint $E_i \in \mathscr{R}$ whose union is in $\mathscr{R}$, or, equivalently

(2°) $F(B_n) \to 0$ if $B_n \in \mathscr{R}$ and $B_1 \supset B_2 \supset \cdots \to \varnothing$,

then

$$F^* = F \quad \text{on } \mathscr{R}.$$

For, if $E \in \mathscr{R}$ then the inequality $F^*(E) \leqslant F(E)$ is obvious, and the reverse inequality is obtained by taking an arbitrary covering $\{E_i\}$ from $\mathscr{R}$ of the set E, which is then the disjoint union of the sets $F_i = E \cap (E_i \backslash \bigcup_{j=1}^{i-1} E_j) \in \mathscr{R}$, from which $F(E) = \sum_1^\infty F(F_i) \leqslant \sum_1^\infty F(E_i)$. Thus we have the following

THEOREM 5.2.6. *Condition (2°) is necessary and sufficient in order that a finite, non-negative and finitely additive function F can be extended to a measure. Then, the restriction of F^* to $\Lambda \cap \mathscr{I}$ (where Λ is the algebra of sets satisfying Caratheodory's condition) is a measure which extends F.*

Proof. The condition is obviously necessary. To prove sufficiency it is enough to obtain the inclusion $\mathscr{R} \subset \Lambda \cap \mathscr{I}$, that is (since $\mathscr{R} \subset \mathscr{I}$) to show that every set $E \in \mathscr{R}$ satisfies condition (C). So let $A \in \mathscr{I}$. By (4) it suffices to show that

$$F^*(A \cap E) + F^*(A \backslash E) \leqslant F^*(A).$$

But, for any covering $\{E_i\}$ from $\mathscr{R}$ of the set A, the sequences $\{E_i \cap E\}$ and $\{E_i \backslash E\}$ are respectively coverings of the sets $A \cap E$ and $A \backslash E$. Therefore

$$F^*(A \cap E) + F^*(A \backslash E) \leqslant \sum_1^\infty F(E_i \cap E) + \sum_1^\infty F(E_i \backslash E) = \sum_1^\infty F(E_i),$$

from which the desired inequality follows. ■

5.3 CONTENT OF AN INTERVAL

An *n-dimensional interval* (in $\mathbb{R}^n$) is the Cartesian product of n finite real intervals. In this section we will be concerned with closed intervals

$$P = \Delta^{(1)} \times \cdots \times \Delta^{(n)} = [a^{(1)}, b^{(1)}] \times \cdots \times [a^{(n)}, b^{(n)}]. \tag{5.3.1}$$

We say that the intervals P and $P_1 = \Delta_1^{(1)} \times \cdots \times \Delta_1^{(n)}$ *do not overlap* if they have no interior points in common. This occurs if and only if $\Delta^{(k)}$ and $\Delta_1^{(k)}$ do not overlap for at least one k (for $P \cap P_1 = (\Delta^{(1)} \cap \Delta_1^{(1)}) \times \cdots \times (\Delta^{(n)} \cap \Delta_1^{(n)})$).

We say that two real intervals are *adjoining* if they have a common endpoint and do not overlap: for example $[a,b]$ and $[b,c]$ $(a \leqslant b \leqslant c)$ are adjoining intervals. The union of two adjoining intervals is an interval. We say that the intervals P, P_1 are *adjoining* if $\Delta^{(k)}$ and $\Delta_1^{(k)}$ are adjoining for some k and $\Delta^{(i)} = \Delta_1^{(i)}$ for $i \neq k$. Thus

$$[a^{(1)}, b^{(1)}] \times \cdots \times [a^{(k)}, b^{(k)}] \times \cdots \times [a^{(n)}, b^{(n)}]$$

and

$$[a^{(1)}, b^{(1)}] \times \cdots \times [b^{(k)}, c^{(k)}] \times \cdots \times [a^{(n)}, b^{(n)}]$$

are adjoining intervals. The union of adjoining intervals is an interval:

$$\begin{aligned}(\Delta^{(1)} \times \cdots \times \Delta^{(k)} \times \cdots \times \Delta^{(n)}) &\cup (\Delta^{(1)} \times \cdots \times \Delta_1^{(k)} \times \cdots \times \Delta^{(n)}) \\ = \Delta^{(1)} \times \cdots \times (\Delta^{(k)} \cup \Delta_1^{(k)}) &\times \cdots \times \Delta^{(n)}.\end{aligned}$$

The *content* of the interval P defined by (5.3.1) is given by the number

$$|P| = |\Delta^{(1)}| \cdots |\Delta^{(n)}| = (b^{(1)} - a^{(1)}) \cdots (b^{(n)} - a^{(n)}).^5 \tag{5.3.2}$$

Content is an additive function of intervals, that is, if P and P_1 are adjoining intervals then

$$|P \cup P_1| = |P| + |P_1|. \tag{5.3.3}$$

Indeed

$$|P \cup P_1| = |\Delta^{(1)}| \cdots |\Delta^{(k)} \cup \Delta_1^{(k)}| \cdots |\Delta^{(n)}| = |P| + |P_1|,$$

for $|\Delta^{(k)} \cup \Delta_1^{(k)}| = c^{(k)} - a^{(k)} = (b^{(k)} - a^{(k)}) + (c^{(k)} - b^{(k)}) = |\Delta^{(k)}| + |\Delta_1^{(k)}|$.

In general, a finite function $F(P)$ defined on the class of intervals is called an *additive interval function* if for any pair of adjoining intervals P and Q

$$F(P \cup Q) = F(P) + F(Q). \tag{5.3.4}$$

A *normal subdivision* of a real interval $\Delta = [a,b]$ is a system of intervals

$$\theta_i = [x_{i-1}, x_i], \qquad i = 1, \ldots, k \text{ where } a = x_0 < \cdots < x_k = b.$$

These intervals do not overlap and $\Delta = \theta_1 \cup \cdots \cup \theta_k$. If $0 \leqslant p < q \leqslant k$ and $\Delta_0 = [x_p, x_q]$ then the intervals $\theta_i \subset \Delta_0$ constitute a (normal) subdivision of Δ_0.

A *normal subdivision* of the interval P given by (5.3.1) is a system of intervals

$$P_{\kappa_1, \ldots, \kappa_n} = \theta_{\kappa_1}^{(1)} \times \cdots \times \theta_{\kappa_n}^{(n)} \qquad (\kappa_1 = 1, \ldots, p_1; \ldots; \kappa_n = 1, \ldots, p_n), \tag{5.3.5}$$

where $\theta_1^{(k)}, \ldots, \theta_{p_k}^{(k)}$ is a normal subdivision of $\Delta^{(k)}$, that is, $\theta_i^{(k)} = [x_{i-1}^{(k)}, x_i^{(k)}]$ $(i = 1, \ldots, p_k)$, where $a^{(k)} = x_0^{(k)} < \cdots < x_{p_k}^{(k)} = b^{(k)}$. The intervals $P_{\kappa_1, \ldots, \kappa_n}$ do not overlap and $P = \cup P_{\kappa_1, \ldots, \kappa_n}$.

If $0 \leqslant q_k \leqslant q_k' \leqslant p_k$, $\Delta_0^{(k)} = [x_{q_k}^{(k)}, x_{q_k'}^{(k)}]$ and $P_0 = \Delta_0^{(1)} \times \cdots \times \Delta_0^{(n)}$, then the intervals $P_{\kappa_1, \ldots, \kappa_n} \subset P_0$ constitute a normal subdivision of P_0.

LEMMA 5.3.1. *If F is an additive interval function then for the normal subdivision (5.3.5) of the interval P we have*

$$F(P) = \sum F(P_{\kappa_1, \ldots, \kappa_n}). \tag{5.3.6}$$

Proof. We will use induction on the number $N = p_1 \cdots p_n$ of intervals of the subdivision. For $N = 1$ the equation (5.3.6) holds. Suppose that $N > 1$ and that (5.3.6) is true for every normal subdivision into less than N intervals. We must have $p_k > 1$ for some k. Let

$$\Delta_0^{(k)} = [x_0^{(k)}, x_1^{(k)}], \quad \Delta_1^{(k)} = [x_1^{(k)}, x_{p_k}^{(k)}]$$

and

$$P_0 = \Delta^{(1)} \times \cdots \times \Delta_0^{(k)} \times \cdots \times \Delta^{(n)}, \quad P_1 = \Delta^{(1)} \times \cdots \times \Delta_1^{(k)} \times \cdots \times \Delta^{(n)}.$$

Then, the intervals $P_{\kappa_1,\ldots,\kappa_n} \subset P_0$, numbering less than N, make up a normal subdivision of P_0, so that

$$F(P_0) = \sum_{\kappa_k = 1} F(P_{\kappa_1,\ldots,\kappa_n}).$$

Similarly

$$F(P_1) = \sum_{\kappa_k > 1} F(P_{\kappa_1,\ldots,\kappa_n}).$$

But, since P_0 and P_1 are adjoining, therefore

$$F(P) = F(P_0) + F(P_1) = \sum F(P_{\kappa_1,\ldots,\kappa_n}). \quad \blacksquare$$

Let there be given a system of intervals

$$P_1, \ldots, P_s. \tag{5.3.7}$$

A system of non-overlapping intervals $R_1, \ldots, R_m$ is called a *normal system* for the system (5.3.7) if, for every j, the intervals $R_\nu \subset P_j$ make up a normal subdivision of P_j. Such a system always exists; for let

$$P_k = [a_j^{(1)}, b_j^{(1)}] \times \cdots \times [a_j^{(n)}, b_j^{(n)}] \quad (j = 1, \ldots, s)$$

and let

$$P = [a^{(1)}, b^{(1)}] \times \cdots \times [a^{(n)}, b^{(n)}]$$

be an interval containing all the intervals (5.3.7). Take a normal subdivision (5.3.5) of the interval P such that for every k the sequence $x_1^{(k)}, \ldots, x_{p_k}^{(k)}$ contains each of the numbers $a_1^{(k)}, \ldots, a_s^{(k)}, b_1^{(k)}, \ldots, b_s^{(k)}$. Then, for any j the intervals

$$P_{\kappa_1,\ldots,\kappa_n} \subset P_j$$

will constitute a normal subdivision of P_j. Thus, the system of intervals (5.3.5) will be a normal system for the system (5.3.7).

LEMMA 5.3.2. *Let $R_1, \ldots, R_m$ be a normal system for $P_1, \ldots, P_s$. If $R_p \subset P_{\alpha_1} \cup \cdots \cup P_{\alpha_q}$ then $R_p \subset P_{\alpha_r}$ for some r.*

Proof. We have $P_{\alpha_i} = \bigcup_{R_\nu \subset P_{\alpha_i}} R_\nu$, then

$$R_p \subset \bigcup_{i=1}^{q} \bigcup_{R_\nu \subset P_{\alpha_i}} R_\nu.$$

Let $x \in \operatorname{Int} R_p$. We have $x \in R_\nu \subset P_{\alpha_r}$ for some ν and r. Hence R_p and R_ν must have common interior points, so that $R_p = R_\nu$ and $R_p \subset P_{\alpha_r}$. $\blacksquare$

Let F be an additive interval function. Let $Q, P_1, \ldots, P_s$ be intervals. Choosing a normal system for this system of intervals, we will have

$$F(Q) = \sum_{R_v \subset Q} F(R_v) \quad \text{and} \quad F(P_i) = \sum_{R_v \subset P_i} F(R_v) \tag{5.3.8}$$

for $i = 1, \ldots, s$.

THEOREM 5.3.1. *If $Q = \bigcup_{i=1}^{s} P_i$ where the P_i do not overlap, then*

$$F(Q) = \sum_{i=1}^{s} F(P_i).$$

Proof. We have

$$\sum_{R_v \subset \bigcup_{i=1}^{s} P_i} F(R_v) = \sum_{i=1}^{s} \sum_{R_v \subset P_i} F(R_v)$$

since, by Lemma 5.3.2, every term in the first sum occurs exactly once in the second sum (for the P_i do not overlap) and conversely. By (5.3.8), this is the desired equality. ■

THEOREM 5.3.2 *If $F \geqslant 0$, the intervals $P_1, \ldots, P_s$ do not overlap and are contained in Q, then*

$$\sum_{i=1}^{s} F(P_i) \leqslant F(Q).$$

In particular, $F(P) \leqslant F(Q)$ if $P \subset Q$.

For in this case, as before, we have

$$\sum_{i=1}^{s} \sum_{R_v \subset P_i} F(R_v) = \sum_{R_v \subset \bigcup_{i=1}^{s} P_i} F(R_v) \leqslant \sum_{R_v \subset Q} F(R_v).$$

By (5.3.8), this gives the desired inequality. ■

THEOREM 5.3.3. *If $F \geqslant 0$ and $Q \subset \bigcup_{i=1}^{s} P_i$, then*

$$F(Q) \leqslant \sum_{i=1}^{s} F(P_i).$$

For we have

$$\sum_{R_v \subset Q} F(R_v) \leqslant \sum_{R_v \subset \bigcup_{i=1}^{s} P_i} F(R_v) \leqslant \sum_{i=1}^{s} \sum_{R_v \subset P_i} F(R_v)$$

since, by Lemma 5.3.2, every term in the second sum occurs at least once in the third. The inequality in the theorem then follows by (5.3.8). ■

Since content is an additive, non-negative, interval function, we can apply Ths 5.3.1–5.3.3 to this case.

We also consider the question of whether an additive, non-negative interval function can be extended to a measure. It is easy to verify that the class $\mathscr{R}$ of finite unions of intervals open on the right (but closed on the left) is a finitely additive algebra.

Let F be an additive, non-negative interval function. From Th. 5.3.1 it follows that the following formula uniquely defines a function $\tilde{F}$ on $\mathscr{R}$:

$$\tilde{F}(E) = \sum_i F(P_i),$$

where E is a finite union of disjoint intervals P_i which are open on the right. This function is finite, non-negative and additive, so that, by Th. 5.2.6, it can be extended to a measure if and only if condition (2°) is satisfied. Let us show that this condition is equivalent to the following:

$$F([a_1, b_1 - \delta] \times \cdots \times [a_n, b_n - \delta]) \to F([a_1, b_1] \times \cdots \times [a_n, b_n]) \quad \text{as } \delta \to 0+. \tag{2*}$$

Clearly (2°) implies (2*). The reverse implication with be proved indirectly. Suppose that for some sequence $B_k \in \mathscr{R}$ such that $B_1 \supset B_2 \supset \cdots \to \varnothing$ we have

$$\tilde{F}(B_k) \geqslant \varepsilon > 0.$$

Using condition (2*) we can find $A_k \in \mathscr{R}$ such that

$$\bar{A}_k \subset B_k \quad \text{and} \quad \tilde{F}(B_k \backslash A_k) \leqslant \varepsilon/2^k,$$

but then, since $B_k \backslash (A_1 \cap \cdots \cap A_k) \subset (B_1 \backslash A_1) \cup \cdots \cup (B_k \backslash A_k)$, we have

$$\tilde{F}(B_k \backslash (A_1 \cap \cdots \cap A_k)) < \varepsilon,$$

and so $A_1 \cap \cdots \cap A_k \neq \varnothing$. Then, by Cantor's theorem, $\varnothing \neq \bigcap_1^\infty \bar{A}_k \subset \bigcap_1^\infty B_k$, contrary to hypothesis.

Hence, we have the following

THEOREM 5.3.4. *If F is an additive and non-negative interval function, then, for the function of intervals open on the right $P \to F(\bar{P})$ to be extendable to a measure, it is necessary and sufficient that condition (2*) is satisfied.*[6]

5.4 LEBESGUE MEASURE

Let a subset E of the space $\mathbb{R}^n$ be given. The infimum of sums of contents of closed intervals constituting a finite or countable covering of a set E

$$m_n^*(E) = \inf_{E \subset \bigcup P_i} \sum |P_i| \quad (P_i\text{—closed intervals}) \tag{5.4.1}$$

is called the *outer (n-dimensional) Lebesgue measure* of the set E. In this definition we can replace closed intervals by open intervals. For, let $Q_1, Q_2, \ldots,$ be a finite or countable sequence of open intervals covering a set E, then $E \subset \cup \bar{Q}_i$ and so

$$m_n^*(E) \leqslant \sum |\bar{Q}_i| = \sum |Q_i|.$$

Also, if $\varepsilon > 0$, then taking a sequence of closed intervals P_ν covering E such that

$$\sum |P_\nu| < m_n^*(E) + \tfrac{1}{2}\varepsilon,$$

and a sequence of open intervals Q_ν such that

$$P_\nu \subset Q_\nu \quad \text{and} \quad |Q_\nu| < |P_\nu| + \frac{\varepsilon}{2^{\nu+1}},$$

we will have

$$E \subset \cup Q_\nu \quad \text{and} \quad \sum |Q_\nu| \leqslant m_n^*(E) + \varepsilon.$$

Hence

$$m_n^*(E) = \inf_{E \subset \cup Q_\nu} \sum |Q_\nu| \quad (Q_\nu\text{-open intervals}) \tag{5.4.2}$$

Equally, we can confine ourselves to closed (or open) intervals whose diameters are less than a given number $\eta > 0$ and which have points in common with the set E:

$$m_n^*(E) = \inf_{\substack{E \subset \cup P_\nu \\ E \cap P_\nu \neq \varnothing, \delta(P_\nu) < \eta}} \sum |P_\nu| \tag{5.4.3}$$

For, $m_n^*(E) \leqslant \sum |P_\nu|$ where the sequence $P_1, P_2, \ldots$ satisfies the conditions under the inf sign. But, if $\varepsilon > 0$, then taking a sequence of closed intervals P_ν covering E and such that $\sum |P_\nu| \leqslant m_n^*(E) + \varepsilon$, and taking normal subdivisions $P_\nu = \bigcup_\sigma P_{\nu\sigma}$ such that $\delta(P_{\nu\sigma}) < \eta$ we will have

$$E \subset \bigcup_{P_{\nu\sigma} \cap E \neq \varnothing} P_{\nu\sigma} \quad \text{and} \quad \sum_{P_{\nu\sigma} \cap E \neq \varnothing} |P_{\nu\sigma}| \leqslant \sum |P_{\nu\sigma}| = \sum P_\nu \leqslant m_n^*(E) + \varepsilon.$$

Finally, we can confine ourselves to coverings with n-cubes (open or closed).[7]

THEOREM 5.4.1. *m_n^* is a metric outer measure on $\mathbb{R}^n$.*

Proof. Conditions (1) and (2) of §5.2 are clearly satisfied. Condition (3) follows from (5.4.1), for if $E \subset F$ then the class of coverings of E contains the class of coverings of F, so that $m_n^*(E) \leqslant m_n^*(F)$.

Now let $\{E_i\}$ be a sequence of sets. Let $\varepsilon > 0$ and choose a covering of E_i by a sequence of closed intervals $P_{i1}, P_{i2}, \ldots$ such that

$$\sum_\nu |P_{i\nu}| \leqslant m_n^*(E_i) + \frac{\varepsilon}{2^i}.$$

We then have

$$\bigcup_{i=1}^{\infty} E_i \subset \bigcup_{i,\nu} P_{i\nu} \quad \text{and} \quad m_n^*\left(\bigcup_{i=1}^{\infty} E_i\right) \leqslant \sum_{i=1}^{\infty} \sum_\nu |P_{i\nu}| \leqslant \sum_{i=1}^{\infty} m_n^*(E_i) + \varepsilon.$$

Hence

$$m_n^*\left(\bigcup_{i=1}^{\infty} E_i\right) \leqslant \sum_{i=1}^{\infty} m_n^*(E_i),$$

that is, condition (4) is satisfied. We have shown that m_n^* is an outer measure.

It remains to show that this is a metric outer measure, i.e. that condition (5) is satisfied. Let E, F be sets such that

$$\eta = \rho(E, F) > 0.$$

Take any covering of $E \cup F$ by a sequence of closed intervals $P_1, P_2, \ldots$ such that $(E \cup F) \cap P_\nu \neq \varnothing$ and $\delta(P_\nu) < \eta$. Then

$$E \subset \bigcup_{P_\nu \cap E \neq \varnothing} P_\nu \quad \text{and} \quad F \subset \bigcup_{P_\nu \cap F \neq \varnothing} P_\nu$$

and we cannot have both $P_\nu \cap E \neq \varnothing$ and $P_\nu \cap F \neq \varnothing$ simultaneously. Hence

$$m_n^*(E) + m_n^*(F) \leqslant \sum_{P_\nu \cap E \neq \varnothing} |P_\nu| + \sum_{P_\nu \cap F \neq \varnothing} |P_\nu| = \sum_\nu |P_\nu|$$

It follows, by (5.4.3), that

$$m_n^*(E) + m_n^*(F) \leqslant m_n^*(E \cup F)$$

and, since the reverse inequality is always satisfied by an outer measure, therefore

$$m_n^*(E \cup F) = m_n^*(E) + m_n^*(F). \quad \blacksquare$$

Sets E which are m_n^*-measurable, i.e. which satisfy Caratheodory's condition

$$m_n^*(A) = m_n^*(A \cap E) + m_n^*(A \backslash E) \quad \text{for any } A \subset \mathbb{R}^n \tag{C}$$

are called (*n-dimensionally*) *measurable in the sense of Lebesgue*. We will denote the class of such sets by $\mathscr{L}_n$. By Th. 5.2.1, *$\mathscr{L}_n$ is a countably additive algebra and the restriction m_n of the outer measure m_n^* to $\mathscr{L}_n$ is a measure on $\mathscr{L}_n$. This measure is called* (n-dimensional) *Lebesgue measure. By Th. 5.2.2, this measure is complete: if $m_n^*(E) = 0$, then E is measurable ($E \in \mathscr{L}_n$). By Th. 5.2.3, all Borel sets are measurable.*[8]

The Lebesgue measure of any interval (open or closed) P is equal to its content:

$$m_n(P) = |P|. \tag{5.4.4}$$

For, since $P \subset \bar{P}$, we have

$$m_n(P) = m_n^*(P) \leqslant |\bar{P}| = |P|.$$

Let $P \subset \bigcup_\nu Q_\nu$, where Q_ν are open intervals. Also, let $\varepsilon > 0$ and let P_0 be a closed interval contained in P and such that

$$|P| - \varepsilon < |P_0|.$$

Since $P_0 \subset \bigcup_\nu Q_\nu$ we have, by the Borel–Lebesgue theorem,

$$P_0 \subset Q_1 \cup \cdots \cup Q_N$$

for some N and so, by Th. 5.3.3,

$$|P| - \varepsilon < |P_0| \leqslant |Q_1| + \cdots + |Q_N| \leqslant \sum_\nu |Q_\nu|.$$

Hence, by (5.4.2), $|P| - \varepsilon \leqslant m_n^*(P)$ and so

$$|P| \leqslant m_n^*(P) = m_n(P).$$

For every natural number p the space $\mathbb{R}^n$ is a disjoint union of a system $\mathscr{S}_p$ of cubes open on the right

$$\left[\frac{k_1}{2^p}, \frac{k_1+1}{2^p}\right) \times \cdots \times \left[\frac{k_n}{2^p}, \frac{k_n+1}{2^p}\right),$$

where $k_1, \ldots, k_n$ take all integer values.

Let an open set $G \subset \mathbb{R}^n$ be given. We define a seqence of sets $\{H_p\}$, by induction. Let H_1 be the union of those cubes in the system $\mathscr{S}_1$ which are contained in G. Let H_p be the union of the cubes of $\mathscr{S}_p$ contained in $G \backslash (\bigcup_{i=1}^{p-1} H_i)$. Let us show that

$$\bigcup_{p=1}^{\infty} H_p = G.$$

Clearly

$$\bigcup_{p=1}^{\infty} H_p \subset G.$$

Let $x \in G$. For every p there exists exactly one cube P_p of the system $\mathscr{S}_p$ which contains x. Since $\rho(x, \backslash G) > 0$, therefore

$$P_p \subset G$$

for p sufficiently large; let p denote the smallest such value. Then, either $p = 1$, so $x \in P_1 \subset H_1$, or $p > 1$ and $P_i \not\subset G$ for $i \leqslant p-1$ so that $P_p \cap \bigcup_{i=1}^{p-1} H_i = \varnothing$. But since $P_p \subset P_{p-1}$, therefore $P_p \subset G \backslash \bigcup_{i=1}^{p-1} H_i$ and so $x \in P_p \subset H_p$.

Thus

$$G \subset \bigcup_{p=1}^{\infty} H_p.$$

Since the H_p are disjoint we have proved the following

LEMMA 5.4.1. *Every open set is the union of at most a countable number of disjoint cubes open on the right.*

LEMMA 5.4.2. *Every measurable set is the union of at most a countable number of disjoint sets which are measurable and bounded (and hence of finite measure).*

For, we can take a system of disjoint intervals P_ν such that $\mathbb{R}^n = \bigcup_{\nu=1}^{\infty} P_\nu$. Then, for any measurable set E,

$$E = \bigcup_{\nu=1}^{\infty} E \cap P_\nu.$$

LEMMA 5.4.3. *Let $E \subset \mathbb{R}^n$. For every $\varepsilon > 0$ there exists an open set G such that*

$$E \subset G \quad \text{and} \quad m_n(G) \leqslant m_n^*(E) + \varepsilon.$$

There exists a set H of type G_δ such that

$$E \subset H \quad \text{and} \quad m_n(H) = m_n^*(E).$$

Indeed, it suffices to take $G=\bigcup_\nu Q_\nu$, where Q_ν are open intervals such that $E\subset\bigcup_\nu Q_\nu$ and $\sum|Q_\nu|\leqslant m_n^*(E)+\varepsilon$, for then $m_n(G)\leqslant\sum|Q_\nu|$.

Next, taking G_k open such that $E\subset G_k$ and $m_n(G_k)\leqslant m_n^*(E)+1/k$ it suffices to take $H=\bigcap_{k=1}^\infty G_k$, for then $E\subset H$ and

$$m_n^*(E)\leqslant m_n(H)\leqslant m_n(G_k)\leqslant m_n^*(E)+1/k$$

for any k. Thus $m_n(H)=m_n^*(E)$. ■

THEOREM 5.4.2. *A necessary and sufficient condition for a set E to be* $\mathscr{L}_n$-*measurable is given by each of the following conditions:*

(A) *for any* $\varepsilon>0$ *there exists an open set G such that*

$$E\subset G \quad \textit{and} \quad m_n^*(G\backslash E)\leqslant\varepsilon,$$

(B) *there exists a set H of type* G_δ *such that*

$$E\subset H \quad \textit{and} \quad m_n^*(H\backslash E)=0.$$

Proof. Recall that the definition of measurability is given by condition (C).

(C)⇒(*A*). If the set E is measurable ($\mathscr{L}_n$) then, by Lemma 5.4.2,

$$E=\bigcup_\nu E_\nu \quad \text{and} \quad m_n(E_\nu)<\infty,$$

where the sets E_ν are measurable. By Lemma 5.4.3, we can choose G_ν open and such that

$$E_\nu\subset G_\nu, \quad m_n(G_\nu)\leqslant m_n(E_\nu)+\varepsilon/2^\nu$$

and put $G=\bigcup_\nu G_\nu$. We then have $E\subset G$ and, noting that $G\backslash E\subset(G_\nu\backslash E_\nu)$ we obtain

$$m_n^*(G\backslash E)=m_n(G\backslash E)\leqslant\sum_\nu m_n(G_\nu\backslash E_\nu)=\sum_\nu(m_n(G_\nu)-m_n(E_\nu))\leqslant\varepsilon.$$

(A)⇒(B). Choose G_k open such that

$$E\subset G_k, \quad m_n^*(G_k\backslash E)\leqslant 1/k$$

and put $H=\bigcap_{k=1}^\infty G_k$. Then $E\subset H$ and $m_n^*(H\backslash E)\leqslant m_n^*(G_k\backslash E)\leqslant 1/k$, so that

$$m_n^*(H\backslash E)=0.$$

(B)⇒(C). The set $Z=H/E$ is $\mathscr{L}_n$-measurable since $m_n^*(Z)=0$. Hence $E=H\backslash Z$ is $\mathscr{L}_n$-measurable. ■

Remark 1. It follows from the above argument the $\mathscr{L}_n$ is the only countably additive algebra containing the intervals on which m_n^* is a complete measure. Thus, in this theorem (and in Lemma 5.4.2) we could replace $\mathscr{L}_n$ by any algebra with the mentioned properties.

Remark 2. It follows from Th. 5.4.2 that the algebra $\mathscr{L}_n$ is the completion of the algebra $\mathscr{B}_n$ with respect to the measure m_n. For, on the one hand $\mathscr{L}_n$ contains $\mathscr{B}_n$ and all subsets of Borel sets of measure zero, so that $\bar{\mathscr{B}}_n \subset \mathscr{L}_n$. But on the other hand, by (B) every set in $\mathscr{L}_n$ is of the form $H \backslash Z$ where $H \in \mathscr{B}_n$ and $m_n^*(Z) = 0$. Thus, by Lemma 5.4.3, Z is a subset of a Borel set of measure zero so that $H \backslash Z \in \bar{\mathscr{B}}_n$. Therefore $\mathscr{L}_n = \bar{\mathscr{B}}_n$.

Since a set E is measurable if and only if $\backslash E$ is measurable, we can take complements in Th. 5.4.2 and obtain

THEOREM 5.4.2′. *A necessary and sufficient condition for a set E to be measurable is given by each of the following conditions*:

(A′) *for any* $\varepsilon > 0$ *there exists a closed set* F *such that* $F \subset E$ *and* $m_n^*(E \backslash F) \leqslant \varepsilon$,
(B′) *there exists a set* J *of type* F_σ *such that* $J \subset E$ *and* $m_n^*(E \backslash J) = 0$.

Let E be a bounded set. By Lemma 5.4.3, the outer measure of E is given by

$$m^*(E) = \inf_{E \subset G} m(G) \qquad (G \text{ open}).$$

If we define the *inner measure* of a set E in an analogous way:

$$m_*(E) = \sup_{F \subset E} m(F) \qquad (F \text{ closed}),$$

then, from Ths 5.4.2 and 5.4.2′ (by properties (A) and (A′)) we obtain

$$m_*(E) = m^*(E)$$

as a necessary and sufficient condition for a bounded set E to be measurable.

THEOREM 5.4.3. *Let* $E \subset \mathbb{R}^n$ *be* $\mathscr{L}_n$*-measurable and let* $F \subset \mathbb{R}^p$ *be* $\mathscr{L}_p$*-measurable. Then the set* $E \times F \subset \mathbb{R}^{n+p}$ *is* $\mathscr{L}_{n+p}$*-measurable and*

$$m_{n+p}(E \times F) = m_n(E) m_p(F). \tag{5.4.5}$$

Proof. The formula (5.4.5) is true when E and F are intervals. If E and F are open sets then by Lemma 5.4.1

$$E = \bigcup_\nu P_\nu, \quad F = \bigcup_\sigma Q_\sigma,$$

where $\{P_\nu\}$, $\{Q_\sigma\}$ are sequences of disjoint intervals. Hence $E \times F = \bigcup_\nu \bigcup_\sigma P_\nu \times Q_\sigma$, whence (since the union is disjoint)

$$m_{n+p}(E \times F) = \sum_\nu \sum_\sigma m_{n+p}(P_\nu \times Q_\sigma) = \sum_\nu \sum_\sigma m_n(P_\nu) m_p(Q_\sigma)$$
$$= \left(\sum_\nu m_n(P_\nu) \right) \left(\sum_\sigma m_p(Q_\sigma) \right) = m_n(E) m_p(F)$$

so that (5.4.5) is true in this case.

If E and F are bounded and of type G_δ then they are the limits of decreasing sequences of open bounded sets $\{G_\nu\}$ and $\{D_\nu\}$. Thus $E \times F = \lim_{\nu \to \infty} G_\nu \times D_\nu$

and hence

$$m_{n+p}(E \times F) = \lim_{\nu \to \infty} m_{n+p}(G_\nu \times D_\nu) = \left(\lim_{\nu \to \infty} m_n(G_\nu) \right)\left(\lim_{\nu \to \infty} m_p(D_\nu) \right) = m_n(E) m_p(F).$$

so that (5.4.5) holds.

If E and F are of type G_δ then they are the limits of increasing sequences of bounded sets of type G_δ (e.g. $E = \lim_{\nu \to \infty} E \cap G_\nu$, where G_ν is an increasing sequence of bounded open sets such that $\bigcup_{\nu=1}^{\infty} G_\nu = \mathbb{R}^n$). Thus (5.4.5) holds in this case, just as in the previous paragraph.[9]

If E and F are measurable and $m_n(E) = 0$ (or $m_p(F) = 0$) then, by Lemma 5.4.3, there exists a set H_1 of type G_δ such that $E \subset H_1$ and $m_m(H_1) = 0$. Hence

$$E \times F \subset H_1 \times \mathbb{R}^p,$$

but

$$m_{n+p}(H_1 \times \mathbb{R}^p) = m_n(H_1) \cdot m_p(\mathbb{R}^p) = 0$$

so that $m_{n+p}(E \times F) = 0$ and (5.4.5) holds.

If E and F are measurable then by Th. 5.4.2 there exist sets H_1 and H_2 of type G_δ such that $E \subset H_1$, $F \subset H_2$ and $m_n(Z_1) = m_p(Z_2) = 0$, where $Z_1 = H_1 \backslash E$, $Z_2 = H_2 \backslash F$. Then

$$H_1 \times H_2 = (E \cup Z_1) \times (F \cup Z_2) = (E \times F) \cup (E \times Z_2) \cup (Z_1 \times F) \cup (Z_1 \times Z_2)$$

is a disjoint union whose last three terms have measure zero. Thus $E \times F$ is $\mathscr{L}_{n+p}$-measurable and

$$m_{n+p}(E \times F) = m_{n+p}(H_1 \times H_2) = m_n(E) m_p(F). \quad \blacksquare$$

THEOREM 5.4.4. *$\mathscr{L}_n$ is the largest countably additive algebra which contains all intervals and on which m_n^* is a measure.*[10]

Proof. Let $\mathscr{S}$ be a countably additive algebra containing all intervals and on which m_n^* is a measure. Then $\mathscr{B}_n \subset \mathscr{S}$. Let $E \in \mathscr{S}$ and let $A \subset \mathbb{R}^n$. By Lemma 5.4.3, there exists a set H of type G_δ such that $A \subset H$ and $m_n^*(A) = m_n^*(H)$. Then, since $H \cap E$, $H \backslash E \in \mathscr{S}$, we have

$$m_n^*(A) \leqslant m_n^*(A \cap E) + m_n^*(A \backslash E) \leqslant m_n^*(H \cap E) + m_n^*(H \backslash E) = m_n^*(H) = m_n^*(A).$$

Hence

$$m_n^*(A) = m_n^*(A \cap E) + m_n^*(A \backslash E).$$

Thus E satisfies condition (C) and so $E \in \mathscr{L}_n$. Thus $S \subset \mathscr{L}_n$ as required. $\blacksquare$

THEOREM 5.4.5. *On any additive algebra $\mathscr{S}$ such that $\mathscr{B}_n \subset \mathscr{S} \subset \mathscr{L}_n$, the only measure which equals content on intervals*[11] *is m_n.*

Proof.[12] Let $\bar{m}$ be such a measure on $\mathscr{S}$. Clearly

$$\bar{m}(E) \leqslant m_n(E) \qquad \text{for } E \in \mathscr{S}$$

for, given any covering by intervals, $E \subset \cup P_\nu$ we have $\bar{m}(E) \leqslant \sum_\nu \bar{m}(P_\nu) = \sum_\nu |P_\nu|$. For any bounded $E \in \mathscr{S}$ we also have the reverse inequality, for, taking an interval P containing E we have $\bar{m}(P \backslash E) \leqslant m_n(P \backslash E)$ so that

$$\bar{m}(E) = m_n(E).$$

Hence we obtain equality for any $E \in \mathscr{S}$ by noting that $E = \bigcup_{\nu=1}^{\infty} E \cap Q_\nu$, where Q_ν is an increasing sequence of bounded Borel sets whose union is $\mathbb{R}^n$. ■

We note that in the space $\mathbb{R}^n$ a set is of type F_σ if and only if it is *σ-compact*, that is, the set is the union of a sequence of compact sets.[13] Since the image of a σ-compact set by a continuous map is σ-compact and by Th. 5.4.2′, every measurable set is the union of a set of type F_σ and a set of measure zero (and conversely), we have therefore the following

THEOREM 5.4.6. *Let T be a continuous map of a measurable set $E \subset \mathbb{R}^n$ to $\mathbb{R}^p$. If the images of sets of $\mathscr{L}_n$-measure zero are sets of $\mathscr{L}_p$-measure zero (Luzin's condition) then the image of an $\mathscr{L}_n$-measurable set is an $\mathscr{L}_p$-measurable set.*

LEMMA 5.4.4. *Let T be a map of a measurable set $G \subset \mathbb{R}^n$ to $\mathbb{R}^n$ which satisfies a Lipsichitz condition.*[14] *Then, the image of a set of measure zero has measure zero and the image of a measurable set is measurable.*

Proof. Let $E \subset G$, $m_n(E) = 0$. Take $\varepsilon > 0$ and let $E \subset \cup Q_\nu$, $\sum |Q_\nu| < \varepsilon$, where Q_ν are cubes and $Q_\nu \cap E \neq \varnothing$. Let $x_\nu \in Q_\nu \cap E$ and let Q_ν^* be a cube with centre $T(x_\nu)$ and side-length $2Mr_\nu\sqrt{n}$, where M is the Lipschitz constant and r_ν is the side-length of Q_ν. We have

$$T(Q_\nu) \subset Q_\nu^*.$$

Hence $T(E) \subset \cup Q_\nu^*$ and

$$\sum |Q_\nu^*| = (2M\sqrt{n})^n \sum r_\nu^n \leqslant (2M\sqrt{n})^n \varepsilon.$$

Hence $m_n(T(E)) = 0$. ■

For $a \in \mathbb{R}^n$ and $H \subset \mathbb{R}^n$ we write $a + H = \{x : x - a \in H\}$. This is the image of the set H under the map $x \to a + x$. We then have

$$(c_1, \ldots, c_n) + ([a_1, b_1] \times \cdots \times [a_n, b_n]) = [c_1 + a_1, c_1 + b_1] \times \cdots \times [c_n + a_n, c_n + b_n]$$

Always $m_n^*(a + E) = m_n^*(E)$; if $H \in \mathscr{B}_n$ then $a + H \in \mathscr{B}_n$; if $E \in \mathscr{L}_n$ then $a + E \in \mathscr{L}_n$.

LEMMA 5.4.5. *Every interval function F which is additive, non-negative and satisfies $F(a + P) = F(P)$ (for any $a \in \mathbb{R}^n$ and any interval P) is of the form*

$$F(P) = C|P|, \qquad \text{where } C \geqslant 0.$$

Proof. If $n = 1$ then $F(\Delta) = F([0, 1]).|\Delta|$. For, putting $g(s) = F([0, s])$ with $s > 0$, we have

$$g(s+t)=F([0,s+t])=F([0,s])+F([s,s+t])$$
$$=g(s)+g(t).$$

Thus, g is increasing and $g(s)=s\cdot g(1)$.[15]

If n is arbitrary, the function $F_1(\Delta)=F(\cdots\times\Delta\times\cdots)$ satisfies the hypotheses of the lemma, hence

$$F(\cdots\times\Delta\times\cdots)=F(\cdots\times[0,1]\times\cdots)\cdot|\Delta|.$$

Thus $F(\Delta_1\times\cdots\times\Delta_n)=F([0,1]^n).|\Delta_1|\cdots|\Delta_n|$, that is $F(P)=F([0,1]^n)\cdot|P|$. ■

THEOREM 5.4.7. *If μ is a measure on $\mathscr{L}_n$ (on $\mathscr{B}_n$) such that $\mu\neq 0$, $\mu(E)<\infty$ for bounded E (in $\mathscr{L}_n$ or $\mathscr{B}_n$ respectively) and $\mu(a+E)=\mu(E)$ for $E\in\mathscr{L}_n$ $(E\in\mathscr{B}_n)$ and $a\in\mathbb{R}^n$ (in which case we say that μ is invariant), then $\mu=Cm_n$, where $C>0$.*

Proof. The interval function F defined by $F(\bar{P})=\mu(P)$, where P is an interval open on the right, satisfies the hypotheses of Lemma 5.4.5, therefore

$$\mu(P)=C\cdot m_n(P),\qquad \text{where } C\geqslant 0.$$

But $C>0$, for otherwise $\mu=0$. Thus

$$\mu/C=m_n$$

on the class of intervals open on the right. Hence, by Th. 5.4.5,

$$\mu=Cm_n\qquad \text{on } \mathscr{L}_n \text{ (on } \mathscr{B}_n\text{)}. \quad ■$$

A Characterization of the Modulus of a Determinant

If c is a map from the set of non-singular matrices (with n rows and n columns) to $(0,\infty)$ such that

$$c(AB)=c(A)c(B)\quad \textit{and}\quad c(sI)=s^n \textit{ for } s>0$$

(where I is the identity matrix), then

$$c(A)=|\det A|.$$

Proof. Denote by $A_i(s)$ the diagonal matrix whose ith diagonal element is $-s$, and whose other diagonal elements are equal to s. Since $A_i(s)^2=s^2I$ we have $c(A_i(s))^2=s^{2n}$. Hence

$$c(A_i(s))=|s^n|=|\det A_i(s)|.$$

Let $B_{kl}(s)$ $(k\neq l)$ denote the matrix $[a_{ij}]$ for which $a_{kl}=s$, $a_{ii}=1$ $(i=1,\ldots,n)$ and all other elements are zero. Multiplication on the right by $B_{kl}(s)$ is equivalent to multiplying the kth-column by s and adding it to the lth-column. Left multiplication by $B_{kl}(s)$ is equivalent to multiplying the lth-row by s and adding it to the kth-row. Using these operations we can transform an arbitrary non-singular matrix A into the matrix sI or $A_n(s)$, where $s=\sqrt[n]{|\det A|}$,[16] that is, A is a product of matrices of the above type. Hence, since $\det B_{ij}(s)=1$ it suffices to prove that

$$c(B_{ij}(s))=1.$$

But $B_{ij}(-s)=A_i(1)B_{ij}(s)A_i(1)$ so that $c(B_{ij}(-s))=c(B_{ij}(s))$. But $B_{ij}(s)B_{ij}(-s)=I$, hence $c(B_{ij}(s))^2=I$. Thus $c(B_{ij}(s))=1$.[17] ■

THEOREM 5.4.8. *Let A be the matrix of a non-singular linear*[18] *mapping L from $\mathbb{R}^n$ to $\mathbb{R}^n$, then*

$$m_n(L(E))=|\det A|\cdot m_n(E) \qquad \textit{for all } E\in\mathcal{L}_n.$$

Proof. We can assume that L is homogeneous. By Lemma 5.4.4, $\mu(E)=m_n(L(E))$ is a measure on $\mathcal{L}_n$, $\mu\neq 0$, $\mu(E)<\infty$ for bounded E and, since $L(a+E)=L(a)+L(E)$, $\mu(a+E)=\mu(E)$. Thus, by Th. 5.4.7,

$$m_n(L(E))=c(A)m_n(E) \qquad \text{for } E\in\mathcal{L}_n.$$

In this way $c(A)$ is defined for any non-singular matrix A. Taking homogeneous linear mappings L_1, L_2 with matrices A_1, A_2 we have

$$c(A_1A_2)m_n(E)=m_n(L_1(L_2(E)))=c(A_1)c(A_2)m_n(E) \qquad \text{for } E\in\mathcal{L}_n.$$

Hence

$$c(A_1A_2)=c(A_1)c(A_2).$$

Taking $L(x)=sx$, where $s>0$, we have (from the definition of measure)

$$s^n m_n(E)=m_n(L(E))=c(sI)m_n(E) \qquad \text{for } E\in\mathcal{L}_n$$

so that $c(sI)=s^n$. Hence $c(A)=|\det A|$. ■

From Lemma 5.4.3 and the fact that a hyperplane in $\mathbb{R}^n$ has measure zero (since it is the image of the set $\{x:x_n=0\}$ under a non-singular map) we easily obtain

COROLLARY 5.4.1. *For any linear map L from $\mathbb{R}^n$ to $\mathbb{R}^n$ with matrix A,*

$$m_n^*(L(E))=|\det A|m_n^*(E) \qquad \textit{for } E\subset\mathbb{R}^n.$$

COROLLARY 5.4.2. *From Th.* 5.4.8 (*resp. Cor.* 5.4.1) *it follows that Lebesgue measure* (*resp. outer measure*) *is invariant under orthogonal maps.*

We say that a function g defined in the neighbourhood of a point $a\in\mathbb{R}^n$ has a *differential* at a if there exist constants $A_1,\dots,A_n$ such that

$$\lim_{z\to 0}\left(g(a+z)-g(a)-\sum_{i=1}^{n}A_iz_i\right)\Big/|z|=0$$

(where $z=(z_1,\dots,z_n)$, $|z|=\sqrt{(z_1^2+\cdots+z_n^2)}$). Then, the partial derivatives of g

exist at a and

$$A_i = \frac{\partial g}{\partial x_i}(a) \qquad (i = 1, \ldots, n)$$

Note that a linear combination of functions which are finite in the neighbourhood of a point $a \in \mathbb{R}^n$ and which have a differential at a, also has a differential at a.

We say that the map $u \to g(u) = (g_1(u_1, \ldots, u_n), \ldots, g_k(u_1, \ldots, u_n))$ defined in a neighbourhood of a point $a \in \mathbb{R}^n$ to $\mathbb{R}^k$ has a *differential* at a if each of the functions g_j has a differential at a, that is, if

$$\lim_{z \to 0} (g(a+z) - g(a) - Az)/|z| = 0$$

for some matrix A (with k rows and n columns: $Az = \left(\sum_{j=1}^{n} A_{1j} z_j, \ldots, \sum_{j=1}^{n} A_{kj} z_j \right)$, where $A = [A_{ij}]$). Then

$$A = \left[\frac{\partial g_i}{\partial u_j}(a) \right]$$

that is, A is the Jacobian matrix of the map g at the point a.

LEMMA 5.4.6. *Let g be a homeomorphism of an open set $\Omega \subset \mathbb{R}^n$ onto an open set in $\mathbb{R}^n$ and let g satisfy a Lipschitz condition*[19] *and have a differential at some point $a \in \Omega$. Then*

$$\lim_{r \to 0} \frac{m_n(g(Q_r))}{|Q_r|} = |\Delta(a)|$$

where $Q_r = \{u : |u_i - a_i| \leqslant \frac{1}{2} r,\ i = 1, \ldots, n\}$[20] *and $\Delta(a)$ is the Jacobian of g at a.*

Proof. The map $v \to g_r(v) = (1/r)(g(a + rv) - g(a))$ is the composite of the mappings $v \to a + rv$ and $x \to (1/r)(g(x) - g(a))$ and is therefore a homeomorphism of the open set

$$\Omega_r = \{v : a + rv \in \Omega\} \subset \mathbb{R}^n$$

onto an open set in $\mathbb{R}^n$ which satisfies a Lipschitz condition. Given $\varepsilon > 0$ there exists $\delta > 0$ such that if $a + z \in \Omega$ and $|z| < \delta$, then

$$|g(a+z) - g(a) - Az| \leqslant \varepsilon |z|$$

where A is the Jacobian matrix of g at a. Putting

$$g_0(v) = Av$$

and

$$Q = \{v : |v_i| < \tfrac{1}{2}, i = 1, \ldots, n\}, \quad \bar{Q} = \{v : |v_i| \leqslant \tfrac{1}{2}, i = 1, \ldots, n\}$$

we have that $\bar{Q} \subset \Omega_r$ for sufficiently small $r > 0$ and

$$\lim_{r \to 0} |g_r(v) - g_0(v)| = 0 \quad \text{uniformly in } \bar{Q}.$$

Since $g_r(\bar{Q})$ is the image of $g(\bar{Q}_r)$ under the map $x \to (1/r)(x - g(a))$, therefore

$$m_n(g_r(\bar{Q})) = \frac{1}{r^n} m_n(g(Q_r)) = \frac{1}{|Q_r|} m_n(g(Q_r)).$$

Also (Th. 5.4.8)

$$m_n(g_0(\bar{Q})) = |\det A| m_n(\bar{Q}) = |\Delta(a)|.$$

Thus, it suffices to show that

$$\lim_{r \to 0} m_n(g_r(\bar{Q})) = m_n(g_0(\bar{Q})).$$

By the compactness of $g_0(\bar{Q})$, for any open $G \supset g_0(\bar{Q})$ we have

$$\varepsilon = \rho(g_0(\bar{Q}), \backslash G) > 0$$

and then, for sufficiently small $r > 0$ we have $|g_r(v) - g_0(v)| < \varepsilon$ in $\bar{Q}$, or

$$g_r(\bar{Q}) \subset G.$$

Hence (Lemma 5.4.3)

$$\limsup_{r \to 0} m_n(g_r(\bar{Q})) \leqslant m_n(g_0(\bar{Q})).$$

Thus, in the case where $\Delta(a) = 0$ the proof is complete.

Let $\Delta(a) \neq 0$. Then g_0 is a homeomorphism from $\mathbb{R}^n$ to $\mathbb{R}^n$. Putting $B = \bar{Q} \backslash Q$ (the boundary of Q), we have (Lemma 5.4.4) that $m_n(g_r(B)) = 0$. Hence $m_n(g_r(Q)) = m_n(g_r(\bar{Q}))$. Similarly $m_n(g_0(Q)) = m_n(g_0(\bar{Q}))$.

Thus it suffices to show that

$$m_n(g_0(Q)) \leqslant \liminf_{r \to 0} m_n(g_r(Q)).$$

Let F be a closed subset of the set $g_0(Q)$. Then, since $g_0(B)$ is compact,

$$\delta = \rho(F, g_0(B)) > 0.$$

For sufficiently small $r > 0$ we have $\bar{Q} \subset \Omega_r$ and $|g_r(v) - g_0(v)| < \frac{1}{2}\delta$ in $\bar{Q}$. Hence

$$\rho(F, g_r(B)) \geqslant \tfrac{1}{2}\delta.$$

But for such r

$$F \subset g_r(Q)$$

for otherwise we would have $g_0(v_0) \in F \backslash g_r(Q)$ for some $v_0 \in Q$ and hence, since $g_r(v_0) \in g_r(Q)$, the segment connecting $g_0(v_0)$ and $g_r(v_0)$ (of length $< \frac{1}{2}\delta$) would contain a point $y \in$ Boundary of $g_r(Q) = g_r(B)$.[21] Hence, since $|g_0(v_0) - y| < \frac{1}{2}\delta$ we would obtain a contradiction with the previous inequality. Thus (Th. 5.4.2′)

$$m_n(g_0(Q)) \leqslant \liminf_{r \to 0} m_n(g_r(Q)).$$

Non-measurable Sets

The question of whether there exists a measure on the algebra of all subsets of $\mathbb{R}^n$ which equals content on intervals and is invariant under displacements has a negative answer.[22] In fact, Vitali gave an example of a set $M \subset [0, 1]$ for which there exists a sequence of translates $\{M_\nu\}$ which are disjoint and such that

$$[0, 1] \subset \bigcup_{\nu=1}^{\infty} M_\nu \subset [-1, 2]. \tag{5.4.6}$$

If such a measure m existed (for $n = 1$) then we would have $m(M) = m(M_1) = m(M_2) = \cdots$ and by (5.4.6)

$$1 \leqslant \sum_{\nu=1}^{\infty} m(M_\nu) \leqslant 3$$

which is impossible. For the same reason, the set M is also an example of a set which is not Lebesgue measurable. In the space $\mathbb{R}^n$ it suffices to use the set $M \times P$ in place of M, where P is an $(n-1)$-dimensional interval.

The question of whether there exists a *finitely additive* function on the class of all subsets of $\mathbb{R}^n$ which equals content on intervals and is invariant under orthogonal transformations was settled positively by Banach in the case $n = 1, 2$ (the solution is not unique) and negatively by Hausdorff in the case $n \geqslant 3$.

We now give the construction of a set M satisfying condition (5.4.6). On the set $[0, 1]$ we introduce a relation $\xi \sim \eta$, meaning that $\xi - \eta$ is a rational number. It is easy to see that this is an equivalence relation. Thus $[0, 1]$ splits up into disjoint equivalence classes. We denote the family of these classes by R. Let $M \subset [0, 1]$ be a set which contains exactly one point from each of the classes in R. Thus the difference between any two distinct numbers in M is an irrational number. We list all the rational numbers in $[-1, 1]$ in a sequence $\{r_\nu\}$ of distinct numbers. The sets

$$M_\nu = r_\nu + M$$

are contained in the interval $[-1, 2]$ and are disjoint, for if $\eta \in M_\nu \cap M_\sigma$ for $\nu \neq \sigma$ then $\eta = x + r_\nu = y + r_\sigma$ where $x, y \in M$, and so

$$x - y = r_\sigma - r_\nu \neq 0$$

would be a rational number, which is impossible. It remains to show that

$$[0, 1] \subset \bigcup_{\nu=1}^{\infty} M_\nu.$$

Let $0 \leqslant x \leqslant 1$. Then x belongs to some equivalence class $A \in R$. Let $x_0 \in A \cap M$. Then $x - x_0$ is a rational number in the interval $[-1, 1]$ and so $x - x_0 = r_\nu$ for some ν. Hence $x \in M_\nu$, as required.

5.5 MEASURABLE FUNCTIONS

Let μ be a measure. A function is called *measurable relative to the measure μ* or *μ-measurable* if it is measurable relatively to the algebra of μ-measurable sets. We can then utilize all the theorems of §4.4.

In the class of measurable functions on a measurable set E, equality almost everywhere (a.e.) is an equivalence relation. Every measurable function which is defined almost everywhere on E determines a unique equivalence class.

If the functions f, g are defined and measurable on a measurable set E and are equal almost everywhere on E, then for every a the sets

$$\{x: f(x) \geqslant a\} \quad \text{and} \quad \{x: g(x) \geqslant a\}$$

are μ-equivalent. For in this case the sets

$$\{x: f(x) \geqslant a\} \backslash \{x: g(x) \geqslant a\} \quad \text{and} \quad \{x: g(x) \geqslant a\} \backslash \{x: f(x) \geqslant a\}$$

are contained in the set $\{x: f(x) \neq g(x)\}$ and so have measure zero. We have similarly $\{x: f(x) \leqslant a\} \approx \{x: g(x) \leqslant a\}$, $\{x: f(x) > a\} \approx \{x: g(x) > a\}$ etc.

THEOREM 5.5.1. *Let μ be a complete measure. If the functions f, g are defined on a measurable set E, f is measurable and $f(x) = g(x)$ almost everywhere on E, then g is also measurable. In particular, every function is measurable on a set of measure zero.*

Indeed, if $Z = \{x: f(x) \neq g(x)\}$ then $\mu(Z) = 0$ and so g is measurable on Z (for $\{x: g_Z(x) > a\}$ is μ-measurable since it is a subset of Z). Also g is measurable on $E \backslash Z$ (for $g_{E \backslash Z} = f_{E \backslash Z}$). Hence g is measurable on $E = Z \cup (E \backslash Z)$.

LEMMA. *Let f be a function which is defined, measurable and finite almost everywhere on a measurable set E of finite measure. Then*

$$\lim_{v \to \infty} \mu(\{x: |f(x)| > v\}) = 0.$$

For note that the sets $\{x: |f(x)| > v\}$ have finite measure and form a decreasing sequence whose limit is the set $\{x: |f(x)| = \infty\}$ of measure zero. The lemma then follows from Th. 5.1.6.

$\mathscr{L}_n$-measurable Functions

Functions which are measurable relatively to the measure m_n, that is $\mathscr{L}_n$-measurable, are called *Lebesgue measurable functions.*

THEOREM 5.5.2 (Luzin). *A function f defined on an $\mathscr{L}_n$-measurable set $E \subset \mathbb{R}^n$ is $\mathscr{L}_n$-measurable if and only if given $\varepsilon > 0$ there exists a closed set $F \subset E$ such that f_F is continuous and $m_n^*(E \backslash F) < \varepsilon$.*

Proof. Sufficiency. Let $E_0 = \{x: f(x) \geqslant a\}$. Given $\varepsilon > 0$ choose a closed set F as in the statement of the theorem. The set E_0 contains the closed set

$$F_0 = \{x: f_F(x) \geqslant a\} = E_0 \cap F.$$

Also, since $E_0 \backslash F_0 \subset E \backslash F$ we have

$$m_n^*(E_0 \backslash F_0) < \varepsilon.$$

Thus, by Th. 5.4.2′, the set E_0 is $\mathscr{L}_n$-measurable and so f is $\mathscr{L}_n$-measurable.

Necessity. It suffices to show that the function $g(x) = \arctan f(x)$, which is measurable and bounded, satisfies the condition in the theorem (for then so does $f = \lambda \circ g$, where $\lambda(x) = \tan x$ in $(-\frac{1}{2}\pi, \frac{1}{2}\pi)$ and $\lambda(\pm\frac{1}{2}\pi) = \pm\infty$).

The condition is satisfied by every simple, measurable function φ, for such a function is constant on measurable sets $E_1, \ldots, E_k$ (with union E). Hence, taking closed sets $F_i \subset E_i$ such that $m_n(E_i \backslash F_i) < \varepsilon/k$ and putting $F = F_1 \cup \cdots \cup F_k$ we have

$$m_n(E \backslash F) < \varepsilon$$

and φ_F is continuous (Th. 3.1.1).

By Th. 4.4.10, there exists a sequence of simple, measurable functions φ_ν which converges to g uniformly on E. But φ_ν is continuous when restricted to some closed set $F_\nu \subset E$ such that

$$m_n(E \backslash F_\nu) < \varepsilon/2^\nu \qquad \text{for a given } \varepsilon > 0.$$

It follows that the function g is continuous when restricted to the set $F = \bigcup_{\nu=1}^{\infty} F_\nu \subset E$ and

$$m_n(E \backslash F) = m_n\left(\bigcup_{\nu=1}^{\infty} (E \backslash F_\nu)\right) < \varepsilon$$

which completes the proof. ■

THEOREM 5.5.3 (Fréchet). *Given a function f which is $\mathscr{L}_n$-measurable on a measurable set E, there exists a sequence of continuous and finite (in $\mathbb{R}^n$) functions which converges to f almost everywhere in E.*

Proof. By Th. 5.5.2, there is a closed set $F_\nu \subset E$ such that f_{F_ν} is continuous and

$$m_n(E \backslash F_\nu) < 1/\nu.$$

The sets $\Phi_\nu = F_1 \cup \cdots \cup F_\nu$ are closed and form an increasing sequence. Also f_{Φ_ν} is continuous and for $\Phi = \bigcup_{\nu=1}^{\infty} \Phi_\nu$ we have

$$m_n(E \backslash \Phi) = 0$$

(for $m_n(E \backslash \Phi) \leqslant m_n(E \backslash F_\nu) < 1/\nu$). Let f_ν be a continuous (finite) extension of the continuous finite function $\max[\min(f_{\Phi_\nu}, \nu), -\nu]$ on $\mathbb{R}^n$. If $x \in \Phi$ then

$$f_\nu(x) = \max[\min(f(x), \nu), -\nu]$$

for ν sufficiently large, so that

$$f_\nu(x) \to \max[\min(f(x), \infty), -\infty] = f(x)$$

as required. ■

THEOREM 5.5.4 (Vitali). *Every $\mathscr{L}_n$-measurable function is equal almost everywhere to a function of the second class of Baire.*

Proof. We form an increasing sequence of closed sets Φ_ν as in the proof of

Th. 5.5.3. Let

$$h_\nu(x) = \begin{cases} f_{\Phi_\nu}(x) & \text{for } x \in \Phi_\nu \\ \infty & \text{for } x \in \backslash \Phi_\nu \end{cases}$$

The functions h_ν are lower semicontinuous on $\mathbb{R}^n$ (for $\{x : h_\nu(x) \leqslant a\}$ is always closed) and so are of the first class. Also

$$h_\nu(x) \to h(x) = \begin{cases} f(x) & \text{for } x \in \Phi \\ \infty & \text{for } x \in \backslash \Phi \end{cases}$$

Thus h is a function of the second class which equals f on Φ, that is, almost everywhere on E. ■

5.6 SEQUENCES OF MEASURABLE FUNCTIONS

Let μ be a measure and let f_ν be a sequence of functions defined and measurable on a measurable set E.

If the sequence f_ν is convergent a.e. on E, then its limit (defined a.e. on E) is a measurable function. If f is defined on E and $f_\nu(x) \to f(x)$ a.e. on E, then f is measurable, provided μ is a complete measure. Let f, g, g_ν be functions defined and measurable on E. If $f_\nu \to f$ a.e. on E, $f_\nu = g_\nu$ a.e. on E and $f = g$ a.e. on E, then $g_\nu \to g$ a.e. on E. If $f_\nu \to f$ a.e. on E and $f_\nu \to g$ a.e. on E, then $f = g$ a.e. on E.

LEMMA. *Let f, f_ν be functions defined and measurable on a measurable set E. Suppose that $\mu(E) < \infty$ and that $f(x)$ is finite almost everywhere on E. If $f_\nu(x) \to f(x)$ almost everywhere on E then for any $\varepsilon > 0$, $\lim_{k\to\infty} \mu(\bigcup_{\nu=k}^{\infty} E_\nu(\varepsilon)) = 0$, where $E_\nu(\varepsilon) = \{x : |f_\nu(x) - f(x)| \geqslant \varepsilon\}$.*

Proof. $\bigcup_{\nu=k}^{\infty} E_\nu(\varepsilon)$ is a decreasing sequence of sets of finite measure, hence (Th. 5.1.6)

$$\lim_{k\to\infty} \mu\left(\bigcup_{\nu=k}^{\infty} E_\nu(\varepsilon)\right) = \mu\left(\lim_{k\to\infty} \bigcup_{\nu=k}^{\infty} E_\nu(\varepsilon)\right) = \mu\left(\limsup_{\nu\to\infty} E_\nu(\varepsilon)\right).$$

Thus it suffices to show that

$$\limsup_{\nu\to\infty} E_\nu(\varepsilon) \subset \{x : f_\nu(x) \nrightarrow f(x)\} \cup \{x : |f| = \infty\},$$

for the right-hand term has measure zero. Let $x \in \limsup_{\nu\to\infty} E_\nu(\varepsilon)$, then

$$|f_\nu(x) - f(x)| \geqslant \varepsilon$$

for infinitely many ν. Hence

$$f_\nu(x) \nrightarrow f(x) \quad \text{or} \quad |f(x)| = \infty. \quad ■$$

An immediate corollary of the lemma is

THEOREM 5.6.1 (Lebesgue). *Under the hypotheses of the lemma ($\mu(E)<\infty$, f finite almost everywhere in E), if $f_\nu \to f$ almost everywhere in E, then for any $\varepsilon>0$*

$$\lim_{\nu\to\infty} \mu(\{x:|f_\nu(x)-f(x)|\geqslant\varepsilon\})=0.$$

THEOREM 5.6.2 (Egorov). *Under the hypotheses of the lemma ($\mu(E)<\infty$, f finite almost everywhere on E), if $f_\nu \to f$ almost everywhere on E, then for any $\varepsilon>0$ there is a measurable set F (closed in the case of Lebesgue measure), such that*

$$F\subset E,\quad \mu(E\backslash F)<\varepsilon \quad \textit{and} \quad f_\nu\to f \textit{ uniformly on } F. \tag{5.6.1}$$

Proof. Let $\varepsilon>0$. By the lemma, given any positive integer k there exists N_k such that

$$\mu\left(\bigcup_{\nu=N_k}^{\infty} E_\nu\left(\frac{1}{k}\right)\right)<\varepsilon/2^{k+1}.$$

The set $A=\{x:|f(x)|=\infty\}$ has measure zero. Put

$$F=E\backslash A\backslash \bigcup_{k=1}^{\infty}\bigcup_{\nu=N_k}^{\infty} E_\nu\left(\frac{1}{k}\right)$$

Then

$$\mu(E\backslash F)<\tfrac{1}{2}\varepsilon, \tag{5.6.2}$$

and f is finite on F.

Let us show that $f_\nu\to f$ uniformly on F. So, let $\eta>0$ and take k such that $1/k<\eta$. If $x\in F$ then

$$x\in \bigcap_{\nu=N_k}^{\infty} (E\backslash E_\nu(1/k)),$$

that is $|f_\nu(x)-f(x)|<1/k$ for $\nu\geqslant N_k$. Hence $|f_\nu(x)-f(x)|<\eta$ on F for $\nu\geqslant N_k$, that is, $f_\nu\to f$ uniformly on F.

In the case where μ is Lebesgue measure, Th. 5.4.2 guarantees the existence of a closed set $F_0\subset F$ such that

$$\mu(F\backslash F_0)<\tfrac{1}{2}\varepsilon,$$

whence, by (5.6.2)

$$\mu(E\backslash F_0)<\varepsilon.$$

Also, clearly $f_\nu\to f$ uniformly on F_0. ■

The necessity of the hypothesis $\mu(E)<\infty$ is shown by the example of the sequence $f_\nu(x)=x/\nu$ which converges to zero on $E=(-\infty,\infty)$, ($\mu=m_1$). This sequence does not converge uniformly on any set of infinite measure.

Let f, f_ν be functions defined and measurable on a measurable set E and which are finite almost everywhere on E. We say that $f_\nu\to f$ *almost uniformly* on E if the conditions of Egorov's theorem are satisfied; that is, if given $\varepsilon>0$

there exists a measurable set F satisfying (5.6.1). Then

$$f_\nu \to f \qquad \text{almost everywhere on } E.$$

Indeed, $f_\nu \to f$ on $H = \bigcup_{\nu=1}^{\infty} F_\nu$, where F_ν is a measurable set satisfying (5.6.1) with $\varepsilon = 1/\nu$, hence $\mu(E \backslash H) = 0$.

It follows therefore from Egorov's theorem that, when $\mu(E) < \infty$, convergence almost everywhere is equivalent to almost uniform convergence (of functions finite a.e. to functions finite a.e.).

If g, g_ν are defined and measurable on E, $f_\nu = g_\nu$ a.e. on E, $f = g$ a.e. on E and $f_\nu \to f$ almost uniformly on E then it is easily shown that $g_\nu \to g$ almost uniformly on E.

Let f, f_ν be functions defined, measurable and finite a.e. on a measurable set E. We say that the sequence f_ν is *μ-convergent to f* (on E) if the condition in Lebesgue's theorem is satisfied; that is, if

$$\lim_{\nu \to \infty} \mu(\{x : |f_\nu(x) - f(x)| \geqslant \varepsilon\}) = 0 \qquad \text{for all } \varepsilon > 0. \tag{5.6.3}$$

We then write

$$f_n \xrightarrow{(\mu)} f.$$

If g, g_ν are defined and measurable on E, $g_\nu = f_\nu$ a.e. on E and $f = g$ a.e. on E, then also $g_\nu \xrightarrow{(\mu)} g$. For in this case the sets $\{x : |f_\nu(x) - f(x)| \geqslant \varepsilon\}$ and $\{x : |g_\nu(x) - g(x)| \geqslant \varepsilon\}$ are μ-equivalent.

If $f_\nu \xrightarrow{(\mu)} f$ then also $f_{\alpha_\nu} \xrightarrow{(\mu)} f$.

A sequence which is convergent relative to a measure may be divergent everywhere. For instance, in the case $\mu = m_1$ and $E = [0, 1)$, the sequence of characteristic functions of the intervals

$$[0, \tfrac{1}{2}), [\tfrac{1}{2}, 1), [0, \tfrac{1}{3}), [\tfrac{1}{3}, \tfrac{2}{3}), [\tfrac{2}{3}, 1), \ldots, \left[0, \frac{1}{n}\right), \ldots, \left[\frac{n-1}{n}, 1\right), \ldots$$

is convergent relative to measure to 0 but is divergent at every point of the interval $[0, 1)$.

THEOREM 5.6.3. *If $f_\nu \xrightarrow{(\mu)} f$ and $f_\nu \xrightarrow{(\mu)} g$ then $f = g$ almost everywhere on E.*

For from the inclusion

$$\{x : |f(x) - g(x)| \geqslant \varepsilon\} \subset \{x : |f(x) - f_\nu(x)| \geqslant \tfrac{1}{2}\varepsilon\} \cup \{x : |g(x) - f_\nu(x)| \geqslant \tfrac{1}{2}\varepsilon\}$$

it follows from (5.6.3) that $\mu(\{x : |f(x) - g(x)| \geqslant \varepsilon\}) = 0$ for any ε. Hence, from the equality

$$\{x : f(x) \neq g(x)\} = \bigcup_{k=1}^{\infty} \{x : |f(x) - g(x)| \geqslant 1/k\}$$

we have $\mu(\{x : f(x) \neq g(x)\}) = 0$ so that $f = g$ a.e. ■

By Th. 5.6.1, in the case $\mu(E)<\infty$, convergence almost everywhere (of functions almost everywhere finite to an almost everywhere finite function) implies convergence in measure. But if we do not assume that $\mu(E)<\infty$ we have the following

THEOREM 5.6.4. *Almost uniform convergence implies convergence in measure.*

Proof. Suppose that $f_\nu \to f$ almost uniformly on E. Let $\eta>0$, $\varepsilon>0$. There exists a set F which is measurable and satisfies condition (5.6.1), that is

$$|f_\nu(x)-f(x)|<\eta \quad \text{on } F \text{ for } \nu \geqslant N$$

for some N. It follows that for $\nu \geqslant N$ we have

$$\{x:|f_\nu(x)-f(x)|\geqslant \eta\} \subset E\backslash F \quad \text{so that} \quad \mu(\{x:|f_\nu(x)-f(x)|\geqslant \eta\})<\varepsilon.$$

Hence $\lim_{\nu\to\infty} \mu(\{x:|f_\nu(x)-f(x)|\geqslant \eta\})=0$ for any $\eta>0$. ∎

THEOREM 5.6.5 (Riesz). *Let f_ν be a sequence of functions which converges in measure to f in E. Then, there is a subsequence f_{α_ν} which converges to f almost uniformly and hence also almost everywhere on E.*[23]

Proof. Let $f_\nu \xrightarrow{(\mu)} f$ on E. By (5.6.3), for all ν there exists α_ν such that

$$\mu(\{x:|f_{\alpha_\nu}(x)-f(x)|\geqslant 1/\nu\})\leqslant 1/2^\nu$$

and we can require that the sequence $\{\alpha_\nu\}$ be strictly increasing. Let

$$F_k = E\backslash\{x:|f(x)|=\infty\}\backslash \bigcup_{\nu=k}^{\infty} \{x:|f_{\alpha_\nu}(x)-f(x)|\geqslant 1/\nu\}.$$

Since $\mu(\{x:|f(x)|=\infty\})=0$, therefore

$$\mu(E\backslash F_k)\leqslant \sum_{\nu=k}^{\infty} 1/2^\nu = 1/2^{k-1}.$$

Hence it suffices to show that $f_{\alpha_\nu}\to f$ uniformly on F_k, for every k. But on the set F_k the function f is finite and

$$|f_{\alpha_\nu}(x)-f(x)|<1/\nu \qquad \text{for } \nu \geqslant k.$$

Hence $f_{\alpha_\nu}\to f$ uniformly on F_k. ∎

THEOREM 5.6.6. *A necessary and sufficient condition for $f_\nu \xrightarrow{(\mu)} f$ on E is that from every subsequence f_{α_ν} one can select a subsequence which converges to f almost uniformly on E (or, if $\mu(E)<\infty$, almost everywhere on E).*[23]

Proof. Necessity. This follows from Th. 5.6.5, noting that $f_\nu \xrightarrow{(\mu)} f$ implies that $f_{\alpha_\nu} \xrightarrow{(\mu)} f$.

Sufficiency. Let $\varepsilon > 0$ and let $\varepsilon_\nu = \mu(\{x: |f_\nu(x) - f(x)| \geq \varepsilon\})$. It suffices to show that from every subsequence ε_{α_ν} one can select a subsequence which converges to 0, for then $\varepsilon_\nu \to 0$. But, from the sequence f_{α_ν} we can select a subsequence f_{β_ν} which converges to f almost uniformly on E, and hence (by Th. 5.6.4) which converges to f in measure in E. Hence $\varepsilon_{\beta_\nu} \to 0$. ■

NOTES

1. Conversely, if $B \approx A$, then $B = (A \backslash Z_1) \cup Z_2$ where $Z_1 = A \backslash B$, $Z_2 = B \backslash A$ are sets of measure zero.
2. In this case the argument can be extended to the case of a topological space.
3. Th. 5.2.5, together with its proof, holds in the case of topological spaces under the additional assumption that the measure μ satisfies condition (5b):

   $$\mu(G) = \sup_{\bar{D} \subset G} \mu(D) \text{ for } G, D \text{ open.}$$

   This condition is always satisfied (for any measure on $\mathscr{B}(X)$) when the topological space X is metrizable, for in this case any open set is the union of an increasing sequence of open sets G_n such that $\bar{G}_n \subset G$.
4. In fact, both these algebras are the same.
5. In the case of an open interval (or, an interval open on one side) Q the content $|Q|$ is defined by the same formula.
6. From Th. 7.5.7, it follows that every additive, non-negative interval function, when suitably modified on intervals of discontinuity, must satisfy this condition.
7. This follows from the fact that every interval P, for any $\varepsilon > 0$, can be covered by a finite system of cubes Q_ν such that $\sum |Q_\nu| \leq |P| + \varepsilon$; we can also appeal to Lemmas 5.3.1–5.3.2. An *n-cube* is defined to be an n-dimensional interval which is the cartesian product of real intervals of equal length.
8. There exist $\mathscr{L}_n$-measurable sets which are not Borel sets.
9. One should observe that if $a_n = 0$ and $b_n \to \infty$ then $a_n b_n \to 0. \infty$.
10. See Remark 1 after Th. 5.4.2.
11. It is sufficient for this to hold on intervals of a particular type, provided they are bounded, generate $\mathscr{B}_n$ and are such that (5.4.1) holds.
12. This theorem easily follows also from Lemma 5.4.1 and Th. 5.2.5.
13. This is always the case in a σ-compact space X, for then, given any increasing sequence of closed sets F_n we have $\bigcup_1^\infty F_n = \bigcup_1^\infty (F_n \cap K_n)$, where K_n is an increasing sequence of compact sets whose union is X.
14. It suffices to assume that T satisfies a Lipschitz condition in a neighbourhood of every point of G.
15. For we have the following

   LEMMA. *If the function g is increasing in $(0, \infty)$ and satisfies $g(s + t) = g(s) + g(t)$, then $g(s) = sg(1)$.*

   Indeed, for natural numbers k and n we have $g(ks) = kg(s)$, $ng(k/n) = g(k) = kg(1)$, $g(k/n) = (k/n)g(1)$, so that

   $$g(w) = wg(1) \quad \text{for } w \text{ positive and rational.}$$

For any $s > 0$, take rational w, w' such that $0 < w < s < w'$. We have

$$wg(1) = g(w) \leqslant g(s) \leqslant g(w') = w'g(1)$$

so that in the limit $g(s) = sg(1)$.

16. We transform A as follows: make the second element in the first row different from zero; make the first element in the first row equal to s; make all elements other than the first in the first row and the first column equal to zero, etc. Since $|\det A| = s^n$, the last element in the main diagonal must be $\pm s$.
17. In the last part of the proof we use an idea due to A. B. Turowicz.
18. This means that $L(x) = Ax + b$.
19. It is sufficient if g satisfies Luzin's condition (Th. 5.4.6). See also Lemma 5.4.4.
20. It is not essential for the point a to be the centre of the cube Q; equally

$$\lim_{\substack{\delta(Q)\to 0 \\ a\in Q}} \frac{m_n(g(Q))}{|Q|} = |\Delta(a)|$$

21. For the set $g_r(Q)$ is open and $\overline{g_r(Q)} = g_r(\bar{Q})$ (the inclusion $\supset$ follows from the continuity of g_r, while the reverse inclusion follows from the fact that $g_r(\bar{Q})$ is closed).
22. Using the continuum hypothesis, Banach and Kuratowski proved that on the algebra of all subsets of a set having the power of the continuum, there does not exist a finite measure, not identically zero, which takes zero values on one-point sets.
23. We assume that f, f_ν are defined, measurable and almost everywhere finite on the measurable set E.

CHAPTER 6

INTEGRATION

6.1 INTEGRATION OF NON-NEGATIVE FUNCTIONS

Let μ be a measure. Let f be a measurable non-negative function defined on a measurable set E. The *integral of the function f over the set E relative to the measure μ* is defined as follows:

$$\int_E f(x)\,\mathrm{d}\mu(x) = \int_E f\,\mathrm{d}\mu = \sup_{\substack{E=\cup E_i \\ E_i \text{ measurable, disjoint}}} \sum_{i=1}^{\infty} \left(\inf_{E_i} f\right)\mu(E_i), \tag{6.1.1}$$

(where we put $\inf_{E_i} f = 0$ if E_i is empty). If the measure is Lebesgue measure, $\mu = m_n$, then the integral is called the *Lebesgue integral* of the function f over the set E and we denote it by

$$\int_E f(x)\,\mathrm{d}x = \int_E f(x_1,\ldots,x_n)\mathrm{d}x_1\cdots\mathrm{d}x_n \quad (x=(x_1,\ldots,x_n)).$$

From the properties of bounds it follows immediately that

$$\int_E \alpha f\,\mathrm{d}\mu = \alpha\int_E f\,\mathrm{d}\mu \qquad \text{if } \alpha \geqslant 0. \tag{6.1.2}$$

Also we have

THEOREM 6.1.1. *If f, g are measurable and non-negative on a measurable set E, then*

$$\int_E f\,\mathrm{d}\mu \leqslant \int_E g\,\mathrm{d}\mu, \tag{6.1.3}$$

whenever $f(x) \leqslant g(x)$ on E.

THEOREM 6.1.2 (Mean Value Theorem). *If f is measurable and non-negative on a measurable set E, then*

$$\left(\inf_E f\right)\mu(E) \leqslant \int_E f\,\mathrm{d}\mu \leqslant \left(\sup_E f\right)\mu(E). \tag{6.1.4}$$

For, if $E=\bigcup_{i=1}^{\infty} E_i$ is a disjoint union of measurable sets then (since terms with $E_i=\varnothing$ are zero)

$$\left(\inf_E f\right)\mu(E)=\sum_{i=1}^{\infty}\left(\inf_E f\right)\mu(E_i)\leqslant\sum_{i=1}^{\infty}\left(\inf_{E_i} f\right)\mu(E_i)$$
$$\leqslant\sum_{i=1}^{\infty}\left(\sup_E f\right)\mu(E_i)=\left(\sup_E f\right)\mu(E)$$

and so (6.1.4) follows from (6.1.1). ■

From Th. 6.1.2, we have immediately

$$\int_E c\,\mathrm{d}\mu=c\mu(E)\qquad\text{for } c\geqslant 0 \tag{6.1.5}$$

and

$$\int_E f\,\mathrm{d}\mu=0\qquad\text{if } \mu(E)=0. \tag{6.1.6}$$

THEOREM 6.1.3. *If f is measurable and non-negative on the measurable set E, then*

$$\lambda(A)=\int_A f\,\mathrm{d}\mu$$

is a measure on the algebra of measurable sets contained in E. If $E=\bigcup_i E_i$ is a disjoint union of a finite or countable number of measurable sets, then

$$\int_E f\,\mathrm{d}\mu=\sum_i\int_{E_i} f\,\mathrm{d}\mu\qquad(E_i\cap E_j=\varnothing \text{ for } i\neq j). \tag{6.1.7}$$

Proof. Clearly, we always have $\lambda(A)\geqslant 0$ and $\lambda(\varnothing)=0$. It suffices thus to prove (6.1.7) in the case of a countable union. Let

$$E_i=\bigcup_{\nu=1}^{\infty} A_{i\nu}\qquad(i=1,2,\ldots)$$

be any decomposition into disjoint, measurable sets. Then

$$E=\bigcup_{i=1}^{\infty}\bigcup_{\nu=1}^{\infty} A_{i\nu}$$

is a decomposition into disjoint, measurable sets. Hence

$$\sum_{i=1}^{\infty}\sum_{\nu=1}^{\infty}\left(\inf_{A_{i\nu}} f\right)\mu(A_{i\nu})\leqslant\int_E f\,\mathrm{d}\mu$$

so that

$$\sum_{\nu=1}^{\infty}\left(\inf_{A_{1\nu}} f\right)\mu(A_{1\nu})+\cdots+\sum_{\nu=1}^{\infty}\left(\inf_{A_{k\nu}} f\right)\mu(A_{k\nu})\leqslant\int_E f\mathrm{d}\mu.$$

In this inequality we can take the supremum successively of each of the terms on the left-hand side to obtain

$$\int_{E_1} f\,\mathrm{d}\mu + \cdots + \int_{E_k} f\,\mathrm{d}\mu \leqslant \int_E f\,\mathrm{d}\mu,$$

so that in the limit

$$\sum_{i=1}^{\infty} \int_{E_i} f\,\mathrm{d}\mu \leqslant \int_E f\,\mathrm{d}\mu.$$

It remains to prove the reverse inequality. Let

$$E = \bigcup_{\nu=1}^{\infty} A_\nu$$

be any decomposition into disjoint, measurable sets. Then, since

$$\mu(A_\nu) = \sum_{i=1}^{\infty} \mu(A_\nu \cap E_i)$$

we have

$$\begin{aligned}\sum_{\nu=1}^{\infty}\left(\inf_{A_\nu} f\right)\mu(A_\nu) &= \sum_{i=1}^{\infty}\sum_{\nu=1}^{\infty}\left(\inf_{A_\nu} f\right)\mu(A_\nu \cap E_i)\\ &\leqslant \sum_{i=1}^{\infty}\sum_{\nu=1}^{\infty}\left(\inf_{A_\nu \cap E_i} f\right)\mu(A_\nu \cap E_i)\\ &\leqslant \sum_{i=1}^{\infty}\int_{E_i} f\,\mathrm{d}\mu.\end{aligned}$$

Now, taking the supremum on the left-hand side we obtain

$$\int_E f\,\mathrm{d}\mu \leqslant \sum_{i=1}^{\infty}\int_{E_i} f\,\mathrm{d}\mu. \quad \blacksquare$$

It follows from the above theorem that

$$\int_E f\,\mathrm{d}\mu = \int_E g\,\mathrm{d}\mu \tag{6.1.8}$$

if $f = g$ almost everywhere on E, for then, by (6.1.6),

$$\int_{E_0} f\,\mathrm{d}\mu = 0 = \int_{E_0} g\,\mathrm{d}\mu, \quad \text{where} \quad E_0 = \{x : f(x) \neq g(x)\}.$$

Also,

$$\int_E f\,\mathrm{d}\mu < \infty \quad \text{implies that} \quad f(x) < \infty \text{ a.e. on } E, \tag{6.1.9}$$

for if $\mu(E_\infty) > 0$, where $E_\infty = \{x : f(x) = \infty\}$, then by (6.1.5), we would have $\int_{E_\infty} f\,\mathrm{d}\mu = \infty$.

THEOREM 6.1.4. *Let f be a measurable function on a measurable set E. If $f(x) \geqslant 0$ on E and $\int_E f \, d\mu = 0$ then $f(x) = 0$ almost everywhere on E. If $f(x) > 0$ on E and $\mu(E) > 0$ then $\int_E f \, d\mu > 0$.*

Proof. The second part of the theorem follows from the first part. In turn, the first part follows from the fact that the set

$$\{x : f(x) > 0\} = \bigcup_{n=1}^{\infty} E_n \qquad \text{where } E_n = \left\{x : f(x) \geqslant \frac{1}{n}\right\}$$

is of measure zero. For, by Ths 6.1.2–6.1.3,

$$\frac{1}{n} \mu(E_n) \leqslant \int_{E_n} f \, d\mu = 0$$

and so $\mu(E_n) = 0$. ■

THEOREM 6.1.5 (Lebesgue). *If f_ν is an increasing sequence of measurable and non-negative functions on a measurable set E, then*

$$\lim_{\nu \to \infty} \int_E f_\nu \, d\mu = \int_E f \, d\mu \qquad \textit{where } f(x) = \lim_{\nu \to \infty} f_\nu(x) \textit{ on } E. \qquad (6.1.10)$$

Proof. Since $f_\nu(x) \leqslant f(x)$ on E, therefore (Th. 6.1.1)

$$\int_E f_\nu \, d\mu \leqslant \int_E f \, d\mu \text{ and in the limit } \lim_{\nu \to \infty} \int_E f_\nu \, d\mu \leqslant \int_E f \, d\mu.$$

It therefore suffices to prove the reverse inequality.

If $f(x) < \infty$ on E, let $0 < \theta < 1$. Putting $E_\nu = \{x : f_\nu(x) \geqslant \theta f(x)\}$ we have, by Ths 6.1.1–6.1.3, and using (6.1.2),

$$\int_E f_\nu \, d\mu \geqslant \int_{E_\nu} f_\nu \, d\mu \geqslant \int_{E_\nu} \theta f \, d\mu = \theta \int_{E_\nu} f \, d\mu.$$

But the sequence E_ν is increasing and $\lim_{\nu \to \infty} E_\nu = E$. Thus, in the limit as $\nu \to \infty$ we obtain, by Th. 6.1.3,

$$\lim_{\nu \to \infty} \int_E f_\nu \, d\mu \geqslant \theta \int_E f \, d\mu.$$

Hence, in the limit as $\theta \to 1$ we have

$$\lim_{\nu \to \infty} \int_E f_\nu \, d\mu \geqslant \int_E f \, d\mu.$$

If $f(x) = \infty$ on E, let $A < \infty$. Let $E_\nu = \{x : f_\nu(x) \geqslant A\}$, then we have

$$\int_E f_\nu \, d\mu \geqslant \int_{E_\nu} f_\nu \, d\mu \geqslant A \mu(E_\nu)$$

and as before we obtain

$$\lim_{\nu\to\infty}\int_E f_\nu\,\mathrm{d}\mu \geqslant A\mu(E).$$

Taking the limit as $A\to\infty$ we obtain

$$\lim_{\nu\to\infty}\int_E f_\nu\,\mathrm{d}\mu \geqslant \infty\cdot\mu(E)=\int_E f\,\mathrm{d}\mu.$$

Now in the general case, since $E=E_1\cup E_\infty$, where $E_1=\{x:f(x)<\infty\}$ and $E_\infty=\{x:f(x)=\infty\}$, we have

$$\int_E f_\nu\,\mathrm{d}\mu=\int_{E_1} f_\nu\,\mathrm{d}\mu+\int_{E_\infty} f_\nu\,\mathrm{d}\mu\to\int_{E_1} f\,\mathrm{d}\mu+\int_{E_\infty} f\,\mathrm{d}\mu=\int_E f\,\mathrm{d}\mu$$

which completes the proof. ■

It follows from Th. 6.1.5, that

$$\int_E f_\nu\,\mathrm{d}\mu\to\int_E f\,\mathrm{d}\mu \qquad \text{if } 0\leqslant f_\nu(x)\leqslant f(x) \quad\text{and}\quad f_\nu(x)\to f(x) \tag{6.1.11}$$

on E, for then $g_\nu(x)\leqslant f_\nu(x)\leqslant f(x)$, where $g_\nu(x)=\inf_{\sigma\geqslant\nu} f_\sigma(x)$ is an increasing sequence which converges to $f(x)$ on E.

THEOREM 6.1.6. *If f, g are measurable, non-negative functions on a measurable set E, $\alpha\geqslant 0$ and $\beta\geqslant 0$, then*

$$\int_E(\alpha f+\beta g)\mathrm{d}\mu=\alpha\int_E f\,\mathrm{d}\mu+\beta\int_E g\,\mathrm{d}\mu. \tag{6.1.12}$$

Proof. Since, by Th. 4.4.10, every non-negative measurable function is the limit of an increasing sequence of simple, measurable and non-negative functions, it follows from Th. 6.1.5, that it suffices to prove (6.1.12) for such functions. Also, from (6.1.2), we can confine ourselves to the case $\alpha=\beta=1$.

Thus, let f and g be simple, measurable, non-negative functions on E:

$$f(x)=a_i \text{ on } E_i, \quad g(x)=b_j \text{ on } F_j,$$

where $E=\bigcup_{i=1}^k E_i=\bigcup_{j=1}^l F_j$ (disjoint unions of measurable sets). By (6.1.5), (6.1.7), we have

$$\begin{aligned}\int_E f\,\mathrm{d}\mu+\int_E g\,\mathrm{d}\mu&=\sum_{i=1}^k a_i\mu(E_i)+\sum_{j=1}^l b_j\mu(F_j)\\&=\sum_{i=1}^k a_i\sum_{j=1}^l\mu(E_i\cap F_j)+\sum_{j=1}^l b_j\sum_{i=1}^k\mu(E_i\cap F_j)\\&=\sum_{i=1}^k\sum_{j=1}^l(a_i+b_j)\mu(E_i\cap F_j)=\int_E(f+g)\,\mathrm{d}\mu,\end{aligned}$$

since $f(x)+g(x)=a_i+b_j$ on $E_i \cap F_j$ and $E=\sum_{i=1}^{k}\sum_{j=1}^{l} E_i \cap F_j$ is a disjoint union. ■

From Ths 6.1.5–6.1.6, it follows that if f_ν are measurable and non-negative on a measurable set E, then

$$\sum_{\nu=1}^{\infty} \int_E f_\nu \, d\mu = \int_E \sum_{\nu=1}^{\infty} f_\nu \, d\mu \tag{6.1.13}$$

for in this case the sequence $g_\nu = f_1 + \cdots + f_\nu$ is increasing and $\int_E g_\nu \, d\mu = \int_E f_1 \, d\mu + \cdots + \int_E f_\nu \, d\mu$.

We now show that the properties expressed in Ths 6.1.2–6.1.3, characterize the concept of integral. Thus, suppose that f is a measurable, non-negative function on a measurable set E and let λ be a measure on the algebra of μ-measurable sets contained in E which satisfies the condition

$$\left(\inf_A f\right)\mu(A) \leqslant \lambda(A) \leqslant \left(\sup_A f\right)\mu(A). \tag{6.1.14}$$

Such a measure is given, in particular, by $\lambda(A)=\int_A f \, d\mu$. We will show that we must have

$$\lambda(A) = \int_A f \, d\mu.$$

Let $A \subset E$ be a measurable set of finite measure and such that $f(x) < \infty$ on A. Let $0 = l_0 < l_1 < \cdots < l_\nu \to \infty$ and let $\varepsilon = \sup_\nu (l_\nu - l_{\nu-1}) < \infty$. Let

$$A_i = A \cap \{x : l_{i-1} \leqslant f(x) < l_i\} \quad (i = 1, 2, \ldots).$$

Then, by (6.1.14), we have

$$l_{i-1}\mu(A_i) \leqslant \lambda(A_i) \leqslant l_i \mu(A_i).$$

But $A = \bigcup_{i=1}^{\infty} A_i$ is a disjoint sum, so that

$$\sum_{i=1}^{\infty} l_{i-1}\mu(A_i) \leqslant \lambda(A) \leqslant \sum_{i=1}^{\infty} l_i \mu(A_i) \tag{6.1.15}$$

and in particular

$$\sum_{i=1}^{\infty} l_{i-1}\mu(A_i) \leqslant \int_A f \, d\mu \leqslant \sum_{i=1}^{\infty} l_i \mu(A_i).$$

The left and right sides of this inequality are called a *Lebesgue lower sum* and a *Lebesgue upper sum* respectively for the integral $\int_A f \, d\mu$.

Since

$$\sum_{i=1}^{\infty} l_i \mu(A_i) \leqslant \sum_{i=1}^{\infty} (l_{i-1} + \varepsilon)\mu(A_i) \leqslant \sum_{i=1}^{\infty} l_{i-1}\mu(A_i) + \varepsilon\mu(A)$$

therefore, letting $\varepsilon \to 0$, both sums tend to the integral $\int_A f \, d\mu$ and, simultaneously, from (6.1.15), to $\lambda(A)$.

We have therefore shown that in the case $\mu(A)<\infty$ and $f(x)<\infty$ on A, we must have

$$\lambda(A)=\int_A f\,\mathrm{d}\mu.$$

This equality also holds in the case $\delta\leqslant f(x)<\infty$ on A for some $\delta>0$. For then $\lambda(A)\geqslant\delta\cdot\mu(A)$ and $\int_A f\,\mathrm{d}\mu\geqslant\delta\cdot\mu(A)$. Hence, if $\mu(A)=\infty$, then

$$\lambda(A)=\infty=\int_A f\,\mathrm{d}\mu.$$

In the general case let

$$B_\nu=A\cap\{x:1/\nu\leqslant f(x)<\infty\},$$
$$C=A\cap\{x:f(x)=0\}\quad\text{and}\quad D=A\cap\{x:f(x)=\infty\}.$$

Then $A=C\cup\lim_{\nu\to\infty}B_\nu\cup D$, $\{B_\nu\}$ is an increasing sequence and $\lambda(B_\nu)=\int_{B_\nu}f\,\mathrm{d}\mu$, $\lambda(C)=\int_C f\,\mathrm{d}\mu$, $\lambda(D)=\int_D f\,\mathrm{d}\mu$,[1] from which we deduce $\lambda(A)=\int_A f\,\mathrm{d}\mu$.
Thus we have the following.

THEOREM 6.1.7. *If f is a measurable, non-negative function on a measurable set E, then the unique measure on the algebra of μ-measurable sets contained in E which satisfies condition* (6.1.14) *is given by*

$$\lambda(A)=\int_A f\,\mathrm{d}\mu.$$

The Banach–Vitali Theorem

As an application of Lebesgue's theorem on the integration of increasing sequences (Th. 6.1.5) we prove the following

THEOREM 6.1.8 (Banach–Vitali). *If f is a (finite) function which is continuous in the interval $[a,b]$, then the function $y\to\#f^{-1}(y)$ is measurable and*

$$W_a^b(f)=\int_{\mathbb{R}}\#f^{-1}(y)\,\mathrm{d}y.^2$$

Proof. It follows from the definition of variation that there exists a sequence of subdivisions $a=x_{n0}<\cdots<x_{nk_n}=b$, $n=1,2,\ldots$, such that each subdivision (after the first) is a refinement of the previous one, such that $\delta_n=\max_i|x_{ni}-x_{n,i-1}|\to0$, and $W_n=\sum_{i=1}^{k_n}|f(x_{ni})-f(x_{n,i-1})|\to W_a^b(f)$. Let $\Delta_{ni}=(x_{n,i-1},x_{ni})$, $H_{ni}=f(\bar{\Delta}_{ni})$ and $\varphi_n=\sum_{i=1}^{k_n}\chi_{H_{ni}}$. Then, the sequence φ_n is increasing.[3] Since (by the continuity of f) we have $\operatorname{osc}_{\bar{\Delta}_{ni}}f=m(H_{ni})$, therefore by inequalities (6.1.3), (6.1.4) and Th. 1.3.1, we have $W_n\leqslant\tilde{W}_n=\sum_{i=1}^{k_n}m(H_{ni})\leqslant W_a^b(f)$. Hence

$$\int_{\mathbb{R}}\varphi_n(y)\,\mathrm{d}y=\tilde{W}_n\to W_a^b(f).$$

By Th. 6.1.5,[4] it therefore suffices to show that $\varphi_n(y)\to\#f^{-1}(y)$ almost everywhere in $\mathbb{R}$. So let $y\in\mathbb{R}\backslash Z$, where Z is the set of all points $f(x_{ni})$. We have $\varphi_n(y)\leqslant\#f^{-1}(y)$. But $\varphi_n(y)=\#\Theta$, where $\Theta=\{i:y\in H_{ni}\}$, so for $i\in\Theta$ we have $y=f(\xi_i)$ for some $\xi_i\in\Delta_{ni}$, so that $\#\Theta\leqslant\#f^{-1}(y)$. Now choose any integer $k\leqslant\#f^{-1}(y)$. Then $y=f(\xi_r)$ for some $\xi_1<\cdots<\xi_k$. Now, when n is sufficiently large, $\delta_n<\min(\xi_r-\xi_{r-1})$ and then every ξ_r belongs to a unique Δ_{ni}. Thus $y\in H_{ni}$ for k different i. Hence $\varphi_n(y)\geqslant k$. It follows that $\varphi_n(y)\to\#f^{-1}(y)$. ■

6.2 INTEGRATION OF FUNCTIONS OF ARBITRARY SIGN. SUMMABILITY.

Let f now be any measurable function defined on a measurable set E. Then the functions f_- and f_+ defined by (4.4.3) are also measurable. The *integral of the function f over the set E relative to the measure μ* is then defined by

$$\int_E f\,d\mu=\int_E f_+\,d\mu-\int_E f_-\,d\mu, \tag{6.2.1}$$

if at least one of the integrals on the right-hand side is finite. If both of these integrals are finite, that is, if the integral $\int_E f\,d\mu$ exists and is finite, we say that *the function f is summable over the set E.*

In the case of Lebesgue measure $\mu=m_n$, the integral (6.2.1) is called the *Lebesgue integral* of the function f over the set E and we denote it by

$$\int_E f(x)\,dx=\int_E f(x_1,\ldots,x_n)\,dx_1\cdots dx_n \qquad (x=(x_1,\ldots,x_n)).$$

The definition (6.2.1) of integral is consistent with our previous definition, for if f is a non-negative function we have $f_+(x)=f(x)$ and $f_-(x)=0$. If f is a non-positive function, $f(x)\leqslant 0$, we have $f_+(x)=0$ and $f_-(x)=-f(x)$ on E. Therefore $\int_E f\,d\mu=-\int_E(-f)\,d\mu$ and all the theorems of §6.1 have their correspondents in this case.

From properties (6.1.5), (6.1.6), (6.1.8) we obtain immediately:

$$\int_E c\,d\mu=c\mu(E), \tag{6.2.2}$$

$$\int_E f\,d\mu=0 \qquad \text{if } \mu(E)=0, \tag{6.2.3}$$

$$\int_E f\,d\mu=\int_E g\,d\mu \quad \text{if } f(x)=g(x) \text{ a.e. on } E, \tag{6.2.4}$$

and if one of these integrals exists, then so does the other.

From the last property it follows that the integral $\int_E f\,d\mu$ is uniquely determined if the (measurable) function f is defined almost everywhere on E (measurable), i.e. on a measurable set $E_0\subset E$ such that

$$\mu(E\backslash E_0)=0.$$

For in this case the integrals of all extensions of f to E (measurable extensions if μ is not complete, otherwise all extensions are automatically measurable) exist, or fail to exist, simultaneously and have the same value.

It follows from Ths 6.1.1 and 6.1.6 and the identity $|f(x)| = f_+(x) + f_-(x)$ that we have

THEOREM 6.2.1. *Let f be a measurable function on a measurable set E. Then f is summable if and only if the function $x \to |f(x)|$ on E is summable. If $|f(x)| \leqslant \varphi(x)$ on E, where φ is a summable function on E, then f is summable on E.*

THEOREM 6.2.2. *If f is summable on a measurable set E then it is summable on every measurable subset of E. If f is summable on each of the (measurable) sets $E_1, \ldots, E_k$ then it is summable on their union $\bigcup_{i=1}^k E_i$.*

This follows from the fact that $\int_F |f| \, \mathrm{d}\mu \leqslant \int_E |f| \, \mathrm{d}\mu$ for $F \subset E$ and $\int_{\cup E_i} |f| \, \mathrm{d}\mu \leqslant \sum_i \int_{E_i} |f| \, \mathrm{d}\mu$.

From property (6.1.9) we have

THEOREM 6.2.3. *A summable function on a (measurable) set E is finite almost everywhere on E.*

If $f \leqslant g$ then $f_+ \leqslant g_+$ and $f_- \geqslant g_-$. From Th. 6.1.1, we therefore have

THEOREM 6.2.4. *If $f(x) \leqslant g(x)$ on E, then*

$$\int_E f \, \mathrm{d}\mu \leqslant \int_E g \, \mathrm{d}\mu$$

provided both integrals exist.

This implies the mean-value theorem

$$\left(\inf_E f\right)\mu(E) \leqslant \int_E f \, \mathrm{d}\mu \leqslant \left(\sup_E f\right)\mu(E), \tag{6.2.5}$$

and also, noting the inequality $-|f| \leqslant f \leqslant |f|$,

THEOREM 6.2.5. *If $\int_E f \, \mathrm{d}\mu$ exists, then*

$$\left|\int_E f \, \mathrm{d}\mu\right| \leqslant \int_E |f| \, \mathrm{d}\mu. \tag{6.2.6}$$

THEOREM 6.2.6. *If f is summable on a finite or countable union $E = \bigcup_i E_i$ of disjoint, measurable sets, then*

$$\int_E f \, \mathrm{d}\mu = \sum_i \int_{E_i} f \, \mathrm{d}\mu.^5 \tag{6.2.7}$$

For by Th. 6.1.3,

$$\int_E f\,d\mu = \int_E f_+\,d\mu - \int_E f_-\,d\mu = \sum_i \int_{E_i} f_+\,d\mu - \sum_i \int_{E_i} f_-\,d\mu$$
$$= \sum_i \left(\int_{E_i} f_+\,d\mu - \int_{E_i} f_-\,d\mu \right) = \sum_i \int_{E_i} f\,d\mu.$$

THEOREM 6.2.7. *If f and g are summable on a measurable set E, then $\alpha f + \beta g$ is summable on E and*

$$\int_E (\alpha f + \beta g)\,d\mu = \alpha \int_E f\,d\mu + \beta \int_E g\,d\mu.^6 \tag{6.2.8}$$

Proof. Since

$$\int_E |\alpha f + \beta g|\,d\mu \leqslant \int_E (|\alpha|\,|f| + |\beta|\,|g|)\,d\mu = |\alpha| \int_E |f|\,d\mu + |\beta| \int_E |g|\,d\mu < \infty$$

the function $\alpha f + \beta g$ is summable on E. Next we have

$$\int_E \alpha f\,d\mu = \alpha \int_E f\,d\mu,$$

for, when $\alpha \geqslant 0$, since $(\alpha f)_+ = \alpha f_+$ and $(\alpha f)_- = \alpha f_-$, we have

$$\int_E (\alpha f)_+\,d\mu = \alpha \int_E f_+\,d\mu \quad \text{and} \quad \int_E (\alpha f)_-\,d\mu = \alpha \int_E f_-\,d\mu,$$

and similarly in the case $\alpha \leqslant 0$ (using the relations $(\alpha f)_+ = -\alpha f_-$, $(\alpha f)_- = -\alpha f_+$).

Thus it suffices to confine ourselves to the case $\alpha = \beta = 1$. We have

$$(f+g)_+ - (f+g)_- = f_+ - f_- + g_+ - g_-$$

(where f, g are finite a.e. on E). Therefore

$$(f+g)_+ + f_- + g_- = (f+g)_- + f_+ + g_+$$

almost everywhere on E. Thus

$$\int_E (f+g)_+\,d\mu + \int_E f_-\,d\mu + \int_E g_-\,d\mu = \int_E (f+g)_-\,d\mu + \int_E f_+\,d\mu + \int_E g_+\,d\mu,$$

from which

$$\int_E (f+g)\,d\mu = \int_E f\,d\mu + \int_E g\,d\mu.$$

THEOREM 6.2.8 (Absolute Continuity of the Integral). *If f is a summable* *function on the set E (measurable), then*

$$\int_A f\,\mathrm{d}\mu \to 0 \qquad \textit{if } \mu(A)\to 0,\ A\subset E. \tag{6.2.9}$$

Proof. By Th. 6.1.3,

$$\lambda(A)=\int_A |f|\,\mathrm{d}\mu,$$

is a measure on the algebra of measurable subsets of E, and also $\lambda(E)<\infty$. The sequence of sets $E_\nu=\{x\in E:|f(x)|\geqslant \nu\}$ is decreasing; also $\lim_{\nu\to\infty}E_\nu=\{x\in E:|f(x)|=\infty\}$ and, by Th. 6.2.3, $\mu(\{x\in E:|f(x)|=\infty\})=0$, hence

$$\lim_{\nu\to\infty}\lambda(E_\nu)=0.$$

Let $\varepsilon>0$, then $\lambda(E_m)<\frac{1}{2}\varepsilon$ for some m. Let A be a measurable subset of the set E. Since

$$\left|\int_A f\,\mathrm{d}\mu\right| \leqslant \int_A |f|\mathrm{d}\mu=\lambda(A)=\lambda(A\cap E_m)+\lambda(A\backslash E_m)\leqslant \tfrac{1}{2}\varepsilon+m\mu(A),$$

therefore, if $\mu(A)<\varepsilon/2m$ then $|\int_A f\,\mathrm{d}\mu|<\varepsilon$. ■

Remark. In the case of Lebesgue measure or, more generally, in the case of a measure for which every set of finite measure is a finite union of sets with arbitrarily small measure,[7] the condition (6.2.9),[8] when $\mu(E)<\infty$ implies the summability of the function f over E. This follows from Th. 6.2.2, for then $E=E_1\bigcup\cdots\bigcup E_k$, where $\mu(E_i)<\delta$ where δ corresponds to $\varepsilon=1$ in (6.2.9).

FATOU'S LEMMA. *If f_ν are measurable and non-negative on a measurable set E, then*

$$\int_E \liminf_{\nu\to\infty} f_\nu\,\mathrm{d}\mu \leqslant \liminf_{\nu\to\infty}\int_E f_\nu\,\mathrm{d}\mu. \tag{6.2.10}$$

For $f_\nu(x)\geqslant g_\nu(x)\geqslant 0$ on E where $g_\nu(x)=\inf_{\sigma\geqslant\nu}f_\sigma(x)$ is an increasing sequence on E. Hence, by Th. 6.1.5,

$$\int_E \liminf_{\nu\to\infty} f_\nu\,\mathrm{d}\mu=\int_E\lim_{\nu\to\infty}g_\nu\,\mathrm{d}\mu=\lim_{\nu\to\infty}\int_E g_\nu\,\mathrm{d}\mu\leqslant\liminf_{\nu\to\infty}\int_E f_\nu\,\mathrm{d}\mu.$$

THEOREM 6.2.9 (Lebesgue). *If f_ν are measurable on E and $|f_\nu(x)|\leqslant g(x)$ on E ($\nu=1,2,\ldots$), where g is summable on E, then*

$$\int_E \liminf_{\nu\to\infty} f_\nu\,\mathrm{d}\mu\leqslant\liminf_{\nu\to\infty}\int_E f_\nu\,\mathrm{d}\mu\leqslant\limsup_{\nu\to\infty}\int_E f_\nu\,\mathrm{d}\mu\leqslant\int_E\limsup_{\nu\to\infty}f_\nu\,\mathrm{d}\mu. \tag{6.2.11}$$

Proof. The functions $g+f_\nu$ and $g-f_\nu$ are non-negative and so by Fatou's lemma

$$\int_E \liminf_{\nu\to\infty}(g+f_\nu)\,\mathrm{d}\mu \leqslant \liminf_{\nu\to\infty}\int_E (g+f_\nu)\,\mathrm{d}\mu$$

and

$$\int_E \liminf_{\nu\to\infty}(g-f_\nu)\,\mathrm{d}\mu \leqslant \liminf_{\nu\to\infty}\int_E (g-f_\nu)\,\mathrm{d}\mu.$$

Hence, by Th. 6.2.7 and the properties of lim inf (see (0.2.9) and (0.2.12)) and noting that $|g(x)| < \infty$ a.e. on E, we obtain the inequalities (6.2.11). ■

The inequalities (6.2.11) yield immediately

THEOREM 6.2.10 (Lebesgue). *If f_ν, f are measurable on a measurable set E, $|f_\nu(x)| \leqslant g(x)$ on E ($\nu = 1, 2, \ldots$), where g is summable on E and $f_\nu(x) \to f(x)$ almost everywhere on E, then f summable on E*[9] *and*

$$\int_E f_\nu\,\mathrm{d}\mu \to \int_E f\,\mathrm{d}\mu.$$

From this it follows that if f_ν are measurable on E and $\sum_{\nu=1}^{\infty}\int_E |f_\nu|\,\mathrm{d}\mu < \infty$, then

$$\sum_{\nu=1}^{\infty}\int_E f_\nu\,\mathrm{d}\mu = \int_E \sum_{\nu=1}^{\infty} f_\nu\,\mathrm{d}\mu.$$

THEOREM 6.2.11. *If f_ν and f are measurable on a measurable set E, $|f_\nu(x)| \leqslant g(x)$ on E ($\nu = 1, 2, \ldots$) where g is summable on E and $f_\nu \xrightarrow{(\mu)} f$ on E, then f is summable on E and*

$$\int_E f_\nu\,\mathrm{d}\mu \to \int_E f\,\mathrm{d}\mu$$

Proof. Consider any subsequence J_{α_ν} of the sequence

$$J_\nu = \int_E f_\nu\,\mathrm{d}\mu.$$

By Th. 5.6.5 (Riesz), we can choose a subsequence f_{β_ν} of the sequence f_{α_ν} which converges to f a.e. on E. Hence by Th. 6.2.10, f is summable and $J_{\beta_\nu} \to \int_E f\,\mathrm{d}\mu$. Hence

$$J_\nu \to \int_E f\,\mathrm{d}\mu. \quad ■$$

THEOREM 6.2.12 (Vitali). *Let f, f_ν be summable functions on a set E which is* *measurable and has finite measure and suppose that*

$$\sup_{\nu} \int_A |f_\nu| \mathrm{d}\mu \to 0 \qquad \text{as } \mu(A) \to 0.$$

If $f_\nu \xrightarrow{(\mu)} f$ *on* E *(and hence, in particular, if* $f_\nu \to f$ *almost everywhere on* E*), then*

$$\int_E f_\nu \, \mathrm{d}\mu \to \int_E f \, \mathrm{d}\mu.$$

Proof. Let $\varepsilon > 0$. By hypothesis and by Th. 6.2.8, there exists $\delta > 0$ such that if $\mu(A) < \delta$ and $A \subset E$ then

$$\int_A |f| \mathrm{d}\mu < \tfrac{1}{3}\varepsilon \quad \text{and} \quad \int_A |f_\nu| \mathrm{d}\mu < \tfrac{1}{3}\varepsilon \quad (\nu = 1, 2, \ldots).$$

Let $E_\nu = \{x : |f_\nu(x) - f(x)| \geqslant \varepsilon/3\mu(E)\}$. Then $\mu(E_\nu) \to 0$ and so N exists such that when $\nu \geqslant N$ then $\mu(E_\nu) < \delta$, and so

$$\int_{E_\nu} |f_\nu - f| \mathrm{d}\mu \leqslant \int_{E_\nu} |f_\nu| \mathrm{d}\mu + \int_{E_\nu} |f| \mathrm{d}\mu < 2\varepsilon/3.$$

But since

$$\int_{E \setminus E_\nu} |f_\nu - f| \mathrm{d}\mu \leqslant \frac{\varepsilon}{3\mu(E)} \mu(E \setminus E_\nu) \leqslant \varepsilon/3,$$

therefore

$$\left| \int_E f_\nu \, \mathrm{d}\mu - \int_E f \, \mathrm{d}\mu \right| \leqslant \int_E |f_\nu - f| \mathrm{d}\mu \leqslant \varepsilon \quad \text{for } \nu \geqslant N. \quad \blacksquare$$

Remark. In the case of Lebesgue measure, or, more generally, in the case of a measure such that every set of finite measure is a finite union of sets of arbitrarily small measure, the hypothesis of summability of f and f_ν may be dropped (while obviously maintaining the hypothesis of measurability) because this is then a consequence of the remaining hypotheses.

Indeed, by the remark following Th. 6.2.8, it suffices to show that f satisfies condition (6.2.9). Now, from the sequence f_ν we can select a subsequence f_{α_ν} which converges to f a.e. on E. By hypothesis, given $\varepsilon > 0$ there exists $\delta > 0$ such that if $\mu(A) < \delta$ and $A \subset E$, then $\int_A |f_{\alpha_\nu}| \mathrm{d}\mu \leqslant \varepsilon$, and so by Fatou's lemma $\int_A |f| \mathrm{d}\mu < \varepsilon$.

The necessity of the hypothesis $\mu(E) < \infty$ is shown by the example of the sequence $f_\nu = (1/\nu)\chi_{(0,\nu)}$ when $E = (0, \infty)$ and $\mu = m_1$. This sequence is convergent to zero, $\int_A |f_\nu| \mathrm{d}x \leqslant m_1(A)$ but nonetheless $\int_0^\nu f_\nu \, \mathrm{d}x = 1 \not\to 0$.

The hypothesis of the summability of f cannot be discarded as is shown by the example of the set E consisting of two elements a and b, the measure μ for which $\mu(A)$ equals the number of elements in the set $A \subset E$ and the sequence f_ν given by

$$f_\nu(a) = \nu \quad \text{and} \quad f_\nu(b) = -\nu.$$

Lebesgue Sums

Let f be a measurable function which is finite on a measurable set E of finite measure. Let $\cdots < l_{-1} < l_0 < l_1 < \cdots$, where $l_{-\nu} \to -\infty$, $l_\nu \to \infty$ and let $\varepsilon = \sup_\nu (l_\nu - l_{\nu-1})$.

Let

$$E_\nu = E \cap \{x : l_{\nu-1} \leqslant f(x) < l_\nu\} \quad (\nu = 0, \pm 1, \pm 2, \ldots).$$

If the series

$$s = \sum_{\nu=-\infty}^{\infty} l_{\nu-1}\mu(E_\nu), \quad S = \sum_{\nu=-\infty}^{\infty} l_\nu \mu(E_\nu) \tag{6.2.12}$$

are absolutely convergent then their sums s and S are called respectively *lower* and *upper Lebesgue sums* (for the function f and the set E). Introducing the functions $\varphi(x) = l_{\nu-1}$ and $\psi(x) = l_\nu$ on E_ν ($\nu = 0, \pm 1, \pm 2, \ldots$) we have

$$\varphi(x) \leqslant f(x) \leqslant \psi(x) \quad \text{and} \quad \psi(x) - \varphi(x) \leqslant \varepsilon \text{ on } E. \tag{6.2.13}$$

Summability of φ and ψ is respectively equivalent to the absolute convergence of the first and second series in (6.2.12), and we have

$$s = \int_E \varphi \, d\mu \quad \text{and} \quad S = \int_E \psi \, d\mu.$$

From (6.2.13) we have the inequalities

$$|f(x)| \leqslant |\varphi(x)| + \varepsilon, \quad |\varphi(x)| \leqslant |\psi(x)| + \varepsilon, \quad |\psi(x)| \leqslant |f(x)| + \varepsilon,$$

so that the summabilities of the functions f, φ, ψ are all equivalent. Hence, the summability of f implies the absolute convergence of all Lebesgue sums and the absolute convergence of one Lebesgue sum implies the summability of f. Then, by (6.2.13),

$$s \leqslant \int_E f \, d\mu \leqslant S \quad \text{and} \quad S - s \leqslant \varepsilon\mu(E)$$

from which it follows that

$$s \to \int_E f \, d\mu \quad \text{and} \quad S \to \int_E f \, d\mu \qquad \text{if } \varepsilon \to 0.$$

Relationship of Lebesgue Measure and Integral with the Riemann Integral.

We establish the following

THEOREM 6.2.13. *Let f be a function which is bounded on a closed interval P in $\mathbb{R}^n$. Let $m(x)$, $M(x)$ denote the minimum and the maximum of the function f at the point $x \in P$ respectively. Then, the lower and upper Darboux integrals*[10] *of the* *function f over P are given by*

$$\underline{\int}_P f(x)\,dx = \int_P m(x)\,dx \quad \textit{and} \quad \overline{\int}_P f(x)\,dx = \int_P M(x)\,dx. \tag{6.2.14}$$

Proof. Consider a sequence of subdivisions $P = \bar{P}_{1\nu} \cup \cdots \cup \bar{P}_{k_\nu \nu}$ where $P_{1\nu}, \ldots, P_{k_\nu \nu}$ are disjoint open intervals and $\varepsilon_\nu = \max_i \delta(P_{i\nu}) \to 0$. Let $m_{i\nu} = \inf_{\bar{P}_{i\nu}} f$, $M_{i\nu} = \sup_{\bar{P}_{i\nu}} f$. We have, by definition,

$$\begin{aligned} s_\nu &= \sum_{i=1}^{k_\nu} m_{i\nu} |P_{i\nu}| \to \underline{\int}_P f(x)\,dx, \\ S_\nu &= \sum_{i=1}^{k_\nu} M_{i\nu} |P_{i\nu}| \to \overline{\int}_P f(x)\,dx. \end{aligned} \tag{6.2.15}$$

Let B_ν denote the union of the boundaries of the intervals $P_{1\nu}, \ldots, P_{k_\nu \nu}$. Then

$$P = P_{1\nu} \cup \cdots \cup P_{k_\nu \nu} \cup B_\nu$$

is a disjoint union and $m_n(B_\nu) = 0$. We introduce the functions $\varphi_\nu(x) = m_{i\nu}$, $\Phi_\nu(x) = M_{i\nu}$ in $P_{i\nu}$ $(i = 1, \ldots, k_\nu)$ and $\varphi_\nu(x) = \Phi_\nu(x) = 0$ in B_ν. We then have

$$s_\nu = \int_P \varphi_\nu\,dx \quad \text{and} \quad S_\nu = \int_P \Phi_\nu\,dx. \tag{6.2.16}$$

Let $x \in P \setminus \bigcup_{\nu=1}^{\infty} B_\nu$. For every ν we have $x \in P_{i_\nu \nu}$ for some i_ν, and so

$$\varphi_\nu(x) = m_{i_\nu \nu} = \inf_{\bar{P}_{i_\nu \nu}} f.$$

Hence, since $\delta(\bar{P}_{i_\nu \nu}) \to 0$ we have, by Th. 3.4.1, that $\varphi_\nu(x) \to m(x)$. Similarly $\Phi_\nu(x) \to M(x)$. Hence $\varphi_\nu(x) \to m(x)$ and $\Phi_\nu(x) \to M(x)$ a.e. in P. By Th. 6.2.10 and (6.2.16) we therefore have

$$s_\nu \to \int_P m(x)\,dx \quad \text{and} \quad S_\nu \to \int_P M(x)\,dx.$$

Now, using (6.2.15), we obtain the relations (6.2.14). ■

It follows from this theorem that the Darboux integrals are equal if and only if

$$\int_P (M(x) - m(x))\,dx = 0,$$

or, since $M(x) \geqslant m(x)$, by Th. 6.1.4, if $m(x) = M(x)$ a.e. on P. Then $m(x) = M(x) = f(x)$ a.e. on P and both integrals are equal to $\int_P f(x)\,dx$.

We thus have the following

COROLLARY 6.2.1. *Let f be a function which is bounded on an interval P. A necessary and sufficient condition for f to be Riemann integrable is that the set of points of discontinuity of f (on P) should have Lebesgue-measure zero. In this case the function is summable and the Lebesgue integral is equal to the Riemann integral.*

6.3 CARTESIAN PRODUCT OF MEASURES. FUBINI'S THEOREM

We say that the measure μ on a space X is *σ-finite* if X is the union of a sequence of sets of finite μ-measure.

Let μ_1, μ_2 be σ-finite measures respectively on countably additive algebras $\mathscr{S}_1, \mathscr{S}_2$ in the spaces X_1, X_2. We then have

$$X_1 = \bigcup_{k=1}^{\infty} X_1^{(k)}, \quad X_2 = \bigcup_{k=1}^{\infty} X_2^{(k)}, \quad \mu_1(X_1^{(k)}) < \infty, \quad \mu_2(X_2^{(k)}) < \infty \tag{6.3.1}$$

where we can assume that the sequences $X_1^{(k)}, X_2^{(k)}$ are increasing (by taking $X_1^{(1)} \cup \cdots \cup X_1^{(k)}$ in place of $X_1^{(k)}$ and $X_2^{(1)} \cup \cdots \cup X_2^{(k)}$ in place of $X_2^{(k)}$ if necessary).

Let $X^{(k)} = X_1^{(k)} \times X_2^{(k)}$ and let $\mathscr{H}$ denote the class of all sets E such that $E \subset X^{(k)}$ for some k.

LEMMA 6.3.1. *Let $\mathscr{M}$ be a class of sets such that $\mathscr{M}$ contains $\mathscr{S}_1 \bar{\times} \mathscr{S}_2$ and satisfies the conditions:*

$$\text{if } A, B \in \mathscr{M} \text{ and } A \cap B = \varnothing \text{ then } A \cup B \in \mathscr{M}; \tag{6.3.2}$$

$$\text{the limit of a monotone sequence of sets in } \mathscr{M} \cap \mathscr{H} \text{ belongs to } \mathscr{M}, \tag{6.3.3}$$

then $\mathscr{S}_1 \times \mathscr{S}_2 \subset \mathscr{M}$.

Proof. Let $\mathscr{S}'$ be the class of all sets $A \subset X_1 \times X_2$ satisfying the condition

$$A \cap X^{(k)} \in \mathscr{M} \qquad \text{for all } k.$$

Since $A = \lim_{k\to\infty} A \cap X^{(k)}$, therefore, by (6.3.3), $\mathscr{S}' \subset \mathscr{M}$. Thus, it suffices to check that $\mathscr{S}'$ satisfies the conditions of Th. 4.2.1. We have $\mathscr{S}_1 \bar{\times} \mathscr{S}_2 \subset \mathscr{S}'$, for if $E_1 \in \mathscr{S}_1$, $E_2 \in \mathscr{S}_2$, then

$$(E_1 \times E_2) \cap X^{(k)} = (E_1 \cap X_1^{(k)}) \times (E_2 \cap X_2^{(k)}) \in \mathscr{M}$$

for all k, hence $E_1 \times E_2 \in \mathscr{S}'$. If $A, B \in \mathscr{S}'$, $A \cap B = \varnothing$ then by (6.3.2),

$$(A \cup B) \cap X^{(k)} = (A \cap X^{(k)}) \cup (B \cap X^{(k)}) \in \mathscr{M}$$

for all k, hence $A \cup B \in \mathscr{S}'$. If A_ν is a monotone sequence of sets of $\mathscr{S}'$, then, by (6.3.3)

$$\left(\lim_{\nu\to\infty} A_\nu\right) \cap X^{(k)} = \lim_{\nu\to\infty} (A_\nu \cap X^{(k)}) \in \mathscr{M}$$

for all k, hence $\lim A_\nu \in S'$. ■

For every $E \subset X_1 \times X_2$ we denote by $E_{x_1} = \{x_2 : (x_1, x_2) \in E\}$ the section of E determined by $x_1 \in X_1$, and by $E^{x_2} = \{x_1 : (x_1, x_2) \in E\}$ the section determined by $x_2 \in X_2$. By Th. 4.2.3, if $E \in \mathscr{S}_1 \times \mathscr{S}_2$ then always $E_{x_1} \in \mathscr{S}_1$ and $E^{x_2} \in \mathscr{S}_2$. We have the following.

THEOREM 6.3.1. *Let $E \in \mathscr{S}_1 \times \mathscr{S}_2$, then the function $x_1 \to \mu_2(E_{x_1})$ is μ_1-measurable and the function $x_2 \to \mu_1(E^{x_2})$ is μ_2-measurable.*

Proof. We will only prove the first statement; the proof of the second is analogous. We denote by $\mathscr{M}$ the class of all sets $E \in S_1 \times S_2$ such that

$$x_1 \to \mu_2(E_{x_1}) \text{ is a } \mu_1\text{-measurable function.}$$

The class $\mathscr{M}$ contains all sets of the form $E_1 \times E_2$ where $E_1 \in \mathscr{S}_1$, $E_2 \in \mathscr{S}_2$ for then $x_1 \to \mu_2\ ((E_1 \times E_2)_{x_1}) = \mu_2(E_2)\chi_{E_1}(x_1)$ is μ_1-measurable. By the lemma it therefore suffices to prove that the class $\mathscr{M}$ satisfies conditions (6.3.2), (6.3.3). If $A, B \in \mathscr{M}$ and $A \cap B = \varnothing$, then $(A \cup B)_{x_1} = A_{x_1} \cup B_{x_1}$ and $A_{x_1} \cap B_{x_1} = (A \cap B)_{x_1} = \varnothing$, so that

$$x_1 \to \mu_2((A \cup B)_{x_1}) = \mu_2(A_{x_1}) + \mu_2(B_{x_1})$$

is a μ_1-measurable function, i.e. $A \cup B \in \mathscr{M}$. If A_ν is a monotone sequence of sets in $\mathscr{M} \cap \mathscr{H}$, then $(A_\nu)_{x_1}$ is a monotone sequence of μ_2-measurable sets of finite measure (since each of them is contained in some $X_2^{(k)}$) and $(\lim_{\nu\to\infty} A_\nu)_{x_1} = \lim_{\nu\to\infty} (A_\nu)_{x_1}$, so

$$x_1 \to \mu_2\left(\left(\lim_{\nu\to\infty} A_\nu\right)_{x_1}\right) = \lim_{\nu\to\infty} \mu_2((A_\nu)_{x_1})$$

is a μ_1-measurable function. Hence $\lim_{\nu\to\infty} A_\nu \in \mathscr{M}$. ∎

The function $\mu_1 \times \mu_2$ defined on $S_1 \times S_2$ by

$$(\mu_1 \times \mu_2)(E) = \int_{X_1} \mu_2(E_{x_1})\, d\mu_1 \tag{6.3.4}$$

is a measure satisfying the condition

$$(\mu_1 \times \mu_2)(E_1 \times E_2) = \mu_1(E_1)\mu_2(E_2) \text{ for } E_i \in \mathscr{S}_i, \quad i = 1, 2. \tag{6.3.5}$$

For, if $E = \bigcup_{\nu=1}^{\infty} E_\nu$ is a disjoint union of sets in $\mathscr{S}_1 \times \mathscr{S}_2$, then $E_{x_1} = \bigcup_{\nu=1}^{\infty} (E_\nu)_{x_1}$ is a disjoint union of μ_2-measurable sets, and so

$$\mu_2(E_{x_1}) = \sum_{\nu=1}^{\infty} \mu_2((E_\nu)_{x_1}).$$

Hence, by (6.1.13) and (6.3.4)

$$(\mu_1 \times \mu_2)(E) = \sum_{\nu=1}^{\infty} (\mu_1 \times \mu_2)(E_\nu).$$

Also, if $E = E_1 \times E_2$, $E_1 \in \mathscr{S}_1$, $E_2 \in \mathscr{S}_2$, then

$$\mu_2((E)_{x_1}) = \mu_2(E_2)\chi_{E_1}(x_1)$$

and so (6.3.5) follows from (6.3.4).

The measure defined on $S_1 \times S_2$ by (6.3.4) is called the *cartesian product of the measures* μ_1 *and* μ_2. We then have the following

THEOREM 6.3.2. *The measure $\mu_1 \times \mu_2$ is the unique measure on $\mathscr{S}_1 \times \mathscr{S}_2$ which satisfies* (6.3.5). *Also*

$$(\mu_1 \times \mu_2)(E) = \int_{X_1} \mu_2(E_{x_1})\,\mathrm{d}\mu_1 = \int_{X_2} \mu_1(E^{x_2})\,\mathrm{d}\mu_2 \tag{6.3.6}$$

for $E \in \mathscr{S}_1 \times \mathscr{S}_2$.

Proof. As above, we verify that

$$\int_{X_2} \mu_1(E^{x_2})\,\mathrm{d}\mu_2$$

as a function of the set E is a measure on $\mathscr{S}_1 \times \mathscr{S}_2$ which satisfies condition (6.3.5). Equation (6.3.6) then follows by uniqueness. Let λ and $\bar{\lambda}$ be measures on $\mathscr{S}_1 \times \mathscr{S}_2$ satisfying condition (6.3.5) and let $\mathscr{M}$ denote the class of all sets $E \in \mathscr{S}_1 \times \mathscr{S}_2$ such that $\lambda(E) = \bar{\lambda}(E)$. The class $\mathscr{M}$ contains $\mathscr{S}_1 \bar{\times} \mathscr{S}_2$, and so to show that

$$\lambda(E) = \bar{\lambda}(E) \quad \text{on} \quad \mathscr{S}_1 \times \mathscr{S}_2$$

it suffices, by the lemma, to check that the class $\mathscr{M}$ satisfies conditions (6.3.2), (6.3.3). If A, $B \in \mathscr{M}$ and $A \cap B = \varnothing$ then

$$\lambda(A \cup B) = \lambda(A) + \lambda(B) = \bar{\lambda}(A) + \bar{\lambda}(B) = \bar{\lambda}(A \cup B)$$

and so $A \cup B \in \mathscr{M}$. If A_ν is a monotone sequence of sets in $\mathscr{M} \cap \mathscr{H}$, then

$$\lambda(A_\nu) = \bar{\lambda}(A_\nu) < \infty$$

(for $A_\nu \subset X^{(k_\nu)}$ for some k_ν, and $\lambda(A_\nu) \leqslant \lambda(X^{(k_\nu)}) = \mu_1(X_1^{(k_\nu)})\mu_2(X_2^{(k_\nu)}) < \infty$), hence

$$\lambda\left(\lim_{\nu\to\infty} A_\nu\right) = \lim_{\nu\to\infty} \lambda(A_\nu) = \lim_{\nu\to\infty} \bar{\lambda}(A_\nu) = \bar{\lambda}\left(\lim_{\nu\to\infty} A_\nu\right).$$

Thus $\lim_{\nu\to\infty} A_\nu \in \mathscr{M}$, as required. ∎

From (6.3.6), using Th. 6.1.4, we have

THEOREM 6.3.3. *If $E \in \mathscr{S}_1 \times \mathscr{S}_2$ and $(\mu_1 \times \mu_2)(E) = 0$, then $\mu_2(E_{x_1}) = 0$ almost everywhere on X_1 (relative to the measure μ_1) and $\mu_1(E^{x_2}) = 0$ almost everywhere on X_2 (relative to the measure μ_2).*

If the function $f(x_1, x_2)$ is $(\mu_1 \times \mu_2)$-measurable then, by Th. 4.4.12, $f(x_1, x_2)$ is μ_2-measurable as a function of x_2 for arbitrarily fixed x_1 and is μ_1-measurable as a function of x_1 for arbitrarily fixed x_2.

THEOREM 6.3.4 (Fubini). *Let the function f be defined, $(\mu_1 \times \mu_2)$-measurable and non-negative on the set $E_1 \times E_2$ (E_1 — μ_1-measurable, E_2 — μ_2-measurable). Then, the function*

$$\varphi(x_1) = \int_{E_2} f(x_1, x_2)\,\mathrm{d}\mu_2(x_2)$$

is μ_1-measurable, the function

$$\psi(x_2)=\int_{E_1} f(x_1,x_2)\,\mathrm{d}\mu_1(x_1)$$

is μ_2-measurable and

$$\int_{E_1\times E_2} f\mathrm{d}(\mu_1\times\mu_2)=\int_{E_1}\left[\int_{E_2} f(x_1,x_2)\,\mathrm{d}\mu_2(x_2)\right]\mathrm{d}\mu_1(x_1)$$
$$=\int_{E_2}\left[\int_{E_1} f(x_1,x_2)\,\mathrm{d}\mu_1(x_1)\right]\mathrm{d}\mu_2(x_2). \tag{6.3.7}$$

Proof. We can assume $E_1=X_1$, $E_2=X_2$ for if we extend f over $X_1\times X_2$ by $f(x_1,x_2)=0$ on $\backslash(E_1\times E_2)$ we see that all the integrals (over E_1, E_2, $E_1\times E_2$ respectively) in the statement of the theorem are equal to the integrals over X_1, X_2 and $X_1\times X_2$ respectively.

The theorem is true if $f=\chi_E$ is the characteristic function of a $(\mu_1\times\mu_2)$-measurable set, for then

$$\int_{X_1\times X_2} f\mathrm{d}(\mu_1\times\mu_2)=(\mu_1\times\mu_2)(E)$$

and, since $\chi_E(x_1,x_2)=\chi_{E_{x_1}}(x_2)$, we have

$$\varphi(x_1)=\mu_2(E_{x_1})$$

and similarly

$$\psi(x_2)=\mu_1(E^{x_2}).$$

Thus (6.3.7) is a consequence of (6.3.6)

From Th. 6.1.6, it follows that the theorem is true for linear combinations with non-negative coefficients of characteristic functions of $(\mu_1\times\mu_2)$-measurable sets, that is, for arbitrary simple functions which are $(\mu_1\times\mu_2)$-measurable and non-negative. Now, from Th. 6.1.5, the theorem is true for the limit of an increasing sequence of simple, $(\mu_1\times\mu_2)$-measurable and non-negative functions. Hence by Th. 4.4.10, it is true for any $(\mu_1\times\mu_2)$-measurable, non-negative function. ■

THEOREM 6.3.5 (Fubini). *Let f be a $(\mu_1\times\mu_2)$-summable function on a set $E_1\times E_2$ (E_1 is μ_1-measurable, E_2 is μ_2-measurable). Then $x_2\to f(x_1,x_2)$ is μ_2-summable on E_2 for almost all $x_1\in E_1$ and the function*

$$\varphi(x_1)=\int_{E_2} f(x_1,x_2)\,\mathrm{d}\mu_2(x_2)$$

(defined almost everywhere on E_1) is μ_1-summable on E_1. Similarly $x_1\to f(x_1,x_2)$ is μ_1-summable on E_1 for almost all $x_2\in E_2$ and the function

$$\psi(x_2)=\int_{E_1} f(x_1,x_2)\,\mathrm{d}\mu_1(x_1)$$

(defined almost everywhere on E_2) is μ_2-summable on E_2. Then, formula (6.3.7) *holds.*

Proof. By hypothesis we have $f = f_+ - f_-$, where

$$\int_{E_1 \times E_2} f_+ \mathrm{d}(\mu_1 \times \mu_2) < \infty, \quad \text{and} \quad \int_{E_1 \times E_2} f_- \mathrm{d}(\mu_1 \times \mu_2) < \infty.$$

Thus, for f_+ and f_- equation (6.3.7) holds. Hence, by property (6.1.9),

$$\int_{E_2} f_+(x_1, x_2)\,\mathrm{d}\mu_2(x_2) < \infty \quad \text{and} \quad \int_{E_2} f_-(x_1, x_2)\,\mathrm{d}\mu_2(x_2) < \infty$$

for almost all $x_1 \in E_1$. It follows that $f(x_1, x_2)$ is summable as a function of x_2 for almost all $x_1 \in E_1$ and since,

$$\varphi(x_1) = \int_{E_2} f_+(x_1, x_2)\,\mathrm{d}\mu(x_2) - \int_{E_2} f_-(x_1, x_2)\,\mathrm{d}\mu(x_2),$$

φ is also summable (as the difference of two summable functions). The function ψ can be treated similarly. Equation (6.3.7) for f is obtained by subtracting (6.3.7) for f_- from (6.3.7) for f_+. ■

The Case of Lebesgue Measure and Integrals

Now consider Lebesgue measure on $X_1 = \mathbb{R}^p$ and $X_2 = \mathbb{R}^q$. By Th. 4.3.6, we have

$$\mathscr{B}_p \times \mathscr{B}_q = \mathscr{B}_{p+q}.$$

Then, by Th. 5.4.3, the cartesian product of the restrictions of the measures m_p and m_q to $\mathscr{B}_p$ and $\mathscr{B}_q$ respectively is, by Th. 6.3.2, the restriction of the measure m_{p+q} to $\mathscr{B}_{p+q}$. Thus, in the case of Lebesgue measure, Borel sets and Baire functions we may apply Ths. 6.3.1–6.3.5. By Th. 5.4.3, we have the inclusion

$$\mathscr{L}_p \times \mathscr{L}_q \subset \mathscr{L}_{p+q},$$

(for $\mathscr{L}_p \times \mathscr{L}_q$ is the smallest countably additive algebra containing all sets of the form $E \times F$, where $E \in \mathscr{L}_p$ and $F \in \mathscr{L}_q$). This is a strict inclusion: for example, the set $\{a\} \times F$, where $a \in \mathbb{R}^p$ and F is a set which is not $\mathscr{L}_q$-measurable, belongs only to the right-hand side.[11] Thus $m_p \times m_q$ is not identical to the measure m_{p+q} but is only the restriction of the measure m_{p+q} to $\mathscr{L}_p \times \mathscr{L}_q$, ($m_{p+q}$ is the completion of the measure $m_p \times m_q$). Thus, in the case of sets and functions which are measurable in the sense of Lebesgue, Ths 6.3.1–6.3.5 are not directly applicable.

THEOREM 6.3.6. *If $E \subset \mathbb{R}^{p+q}$ and $m_{p+q}(E) = 0$ then $m_q(E_x) = 0$ for almost all $x \in \mathbb{R}^p$ and $m_p(E^y) = 0$ for almost all $y \in \mathbb{R}^q$.*

For then $E \subset H$ where H is of type G_δ (so $H \in \mathscr{B}_{p+q}$) and $m_{p+q}(H) = 0$. Hence, by Th. 6.3.3, $m_q(H_x)=0$ for almost all $x\in\mathbb{R}^p$ and so, since $E_x\subset H_x$, $m_q(E_x)=0$ for almost all $x\in\mathbb{R}^p$. Similarly for E^y. ■

COROLLARY 6.3.1. *If $f(x,y)=g(x,y)$ almost everywhere on $E\times F$, then for almost all fixed $x_0\in E$, $f(x_0,y)=g(x_0,y)$ almost everywhere on F and for almost all fixed $y_0\in F$, $f(x,y_0)=g(x,y_0)$ almost everywhere on E.*

Let E be an $\mathscr{L}_p$-measurable set, F—an $\mathscr{L}_q$-measurable set, and let $f(x,y)$ be an $\mathscr{L}_{p+q}$-measurable function on $E\times F$. By Th. 5.4.2′, there exist sets $E_0\subset E$, $F_0\subset F$ of type F_σ, so that $E_0\in\mathscr{B}_p$, $F_0\in\mathscr{B}_q$ such that

$$m_p(E\backslash E_0)=0,\quad m_q(F\backslash F_0)=0.$$

Then also

$$m_{p+q}((E\times F)\backslash(E_0\times F_0))=0.$$

By Th. 5.5.4,

$$f(x,y)=g(x,y)\text{ a.e. on }E_0\times F_0$$

where g is a function of the second class of Baire, and so is $\mathscr{L}_{p+q}$-measurable. Hence for almost every fixed $x_0\in E_0$ we have

$$f(x_0,y)=g(x_0,y)\text{ a.e. on }F_0.$$

But since, by Th. 4.4.12, $y\rightarrow g(x_0,y)$ is $\mathscr{L}_q$-measurable as a function of y, therefore

$$y\rightarrow f(x_0,y)\text{ is }\mathscr{L}_q\text{-measurable}$$

for almost all fixed $x_0\in E$.

Similarly

$$f(x,y_0)=g(x,y_0)\text{ a.e. on }E_0$$

and

$$x\rightarrow f(x,y_0)\text{ is }\mathscr{L}_p\text{-measurable}$$

for almost all fixed $y_0\in F$. If $f\geqslant 0$ almost everywhere on $E\times F$ then we may choose the function g so that it is non-negative (if necessary, taking max $(0,g(x,y))$ instead of $g(x,y)$). Since Th. 6.3.4 applies to g, and since the integrals over E_0, F_0, $E_0\times F_0$ have the same values as the integrals over E, F, $E\times F$, we therefore have

THEOREM 6.3.7 (Fubini).[12] *Let f be an $\mathscr{L}_{p+q}$-measurable function which is non-negative almost everywhere on the set $E\times F$ (where E is $\mathscr{L}_p$-measurable and F is $\mathscr{L}_q$-measurable). Then, $y\rightarrow f(x,y)$ is measurable and non-negative for almost all $x\in E$ and the function*

$$x\rightarrow\int_E f(x,y)\,\mathrm{d}y$$

(defined for almost all $x\in E$) is $\mathscr{L}_p$-measurable. Also $x\rightarrow f(x,y)$ is measurable and

non-negative for almost all $y \in F$ *and the function*

$$y \to \int_E f(x, y)\,dx$$

(defined for almost all $y \in F$*) is* $\mathscr{L}_q$*-measurable and*

$$\int_{E \times F} f(x, y)\,dx\,dy = \int_E \left(\int_F f(x, y)\,dy \right) dx = \int_F \left(\int_E f(x, y)\,dx \right) dy. \tag{6.3.8}$$

If f is summable on $E \times F$ then g is summable on $E_0 \times F_0$. Then, since Th. 6.3.5 applies to the function g we have

THEOREM 6.3.8 (Fubini).[12] *Let* f *be a summable function on the set* $E \times F$ *(where* E *is* $\mathscr{L}_p$*-measurable and* F *is* $\mathscr{L}_q$*-measurable). Then* $y \to f(x, y)$ *is summable on* F *for almost all* $x \in E$ *and the function*

$$x \to \int_F f(x, y)\,dy$$

(defined almost everywhere on E*) is summable on* E*. Also* $x \to f(x, y)$ *is summable on* E *for almost all* $y \in F$ *and the function*

$$y \to \int_E f(x, y)\,dx$$

(defined almost everywhere on F*) is summable on* F*. Then relation* (6.3.8) *holds.*

In particular if $f = \chi_A$ is a characteristic function, we obtain from Th. 6.3.7:

THEOREM 6.3.9. *Let* A *be an* $\mathscr{L}_{p+q}$*-measurable set. Then* A_x *is an* $\mathscr{L}_q$*-measurable set for almost all* $x \in \mathbb{R}^p$ *and the function* $x \to m_q(A_x)$ *is measurable. Also* A^y *is an* $\mathscr{L}_p$*-measurable set for almost all* $y \in \mathbb{R}^q$ *and the function* $y \to m_p(A^y)$ *is measurable. Moreover*

$$m_{p+q}(A) = \int_{\mathbb{R}^p} m_q(A_x)\,dx = \int_{\mathbb{R}^q} m_p(A^y)\,dy. \tag{6.3.9}$$

This theorem is easily obtained as a corollary of Th. 6.3.6, Th. 6.3.1 and (6.3.6). Indeed, taking a set $H \subset A$ of type F_σ and such that $m_{p+q}(A \setminus H) = 0$, we have

$$m_q((A \setminus H)_x) = 0$$

for almost all x, so that the set

$$A_x = H_x \cup (A_x \setminus H_x) = H_x \cup (A \setminus H)_x$$

is measurable (H_x is of type F_σ) and $m_q(A_x) = m_q(H_x)$ for almost all x. It follows that the function $m_q(A_x)$ is measurable and hence we obtain the first equation in (6.3.9). The case A^y is similar.

6.4 RIESZ'S THEOREM

Let X be a locally compact metric[13] space. Let $\mathscr{C}_+$ denote the class of all continuous, non-negative functions on X each of which vanishes outside a compact set (which depends on the function). It is easily seen that $\mathscr{C}_+$ is closed under the operations of taking products and of taking linear combinations with non-negative coefficients. Similarly $\psi-\varphi\in\mathscr{C}_+$ if $\varphi, \psi\in\mathscr{C}_+$ and $\varphi\leqslant\psi$.

Let μ be a measure on $\mathscr{B}(X)$ which is finite on compact sets, then

$$I(\varphi)=\int_X \varphi\,\mathrm{d}\mu$$

is a functional on $\mathscr{C}_+$ which is finite, non-negative and additive:

$$I(\varphi+\psi)=I(\varphi)+I(\psi).$$

Conversely we have:

RIESZ'S THEOREM. *If $I(\varphi)$ is a functional on $\mathscr{C}_+$ which is finite, non-negative and additive,*

$$I(\varphi+\psi)=I(\varphi)+I(\psi),$$

then there exists a measure μ on $\mathscr{B}(X)$ which is finite on compact sets and is such that

$$I(\varphi)=\int_X \varphi\,\mathrm{d}\mu \qquad \textit{for } \varphi\in\mathscr{C}_+.^{13} \tag{6.4.1}$$

For any open set G we denote by $\Gamma(G)$ the class of all functions $\varphi\in\mathscr{C}_+$ such that

$$\varphi\leqslant 1 \quad \text{and} \quad \overline{\{x:\varphi(x)\neq 0\}}\subset G.$$

LEMMA. *If $\varphi\in\Gamma(G_1\cup\cdots\cup G_k)$, where the sets G_i are open, then there exist functions $\varphi_1\in\Gamma(G_1),\ldots,\varphi_k\in\Gamma(G_k)$ such that $\varphi=\varphi_1+\cdots+\varphi_k$.*

Proof. The set $F=\overline{\{x:\varphi(x)\neq 0\}}$ is compact. By the local compactness of X there exists a compact set E such that

$$F\subset \operatorname{Int} E \quad \text{and} \quad E\subset G_1\cup\cdots\cup G_k.$$

The sets $H_i=G_i\cap E$ $(i=1,\ldots,k)$ form an open covering of the compact space E.

It suffices to find functions $\psi_1,\ldots,\psi_k$ which are non-negative and continuous on E and are such that

$$\psi=\psi_1+\cdots+\psi_k>0 \text{ on } E \quad \text{and} \quad \overline{\{x:\psi_i(x)\neq 0\}}\subset H_i$$

for then the functions

$$\varphi_i(x)=\begin{cases}\varphi(x)\psi_i(x)/\psi(x) & \text{on } \operatorname{Int} E\\ 0 & \text{on } X\backslash F\end{cases}$$

satisfy the required conditions. We note that in the sequence $H_1,\ldots,H_k$ we can replace any H_s by H'_s which are open and such that

$$\bar{H}'_s \subset H_s,$$

without losing the covering property (we can choose H'_s such that $E \backslash \bigcup_{i \neq s} H_i \subset H'_s$). It follows that there exists an open covering $\{D_i\}$ such that $\bar{D}_i \subset H_i$. It then suffices to take $\psi_i(x) = \rho(x, E \backslash D_i)$.[14] ■

Proof of Riesz's Theorem

Clearly

$$I(0) = 0 \quad \text{and} \quad I(\varphi) \leqslant I(\psi) \quad \text{if} \quad \varphi \leqslant \psi$$

(for then $\psi - \varphi \in \mathscr{C}_+$). Next we have

$$I(c\varphi) = cI(\varphi) \qquad \text{for } c \geqslant 0 \tag{6.4.2}$$

for the function $t \to I(t\varphi)$ is increasing and additive (see Note 5.15). Let us define a measure μ on $\mathscr{B}(X)$, which will satisfy the required condition, by

$$\mu(E) = \inf_{E \subset G \text{open}} \left(\sup_{\Gamma(G)} I(\varphi) \right).$$

To prove that μ is a measure, it suffices, by Ths 5.2.3–5.2.4, to show that the open set function

$$\lambda(G) = \sup_{\Gamma(G)} I(\varphi)$$

satisfies conditions (1)–(5) of §5.2.[15] Properties (1)–(3) are obvious.

Now let G_n be a sequence of open sets. Let $L < \lambda(\bigcup_1^\infty G_n)$. Then there exists $\varphi \in \Gamma(\bigcup_1^\infty G_n)$ such that $L < I(\varphi)$. Since the set $\overline{\{x : \varphi(x) \neq 0\}}$ is compact we must have $\varphi \in \Gamma(G_1 \cup \cdots \cup G_N)$ for some N and so, by the lemma $\varphi = \varphi_1 + \cdots + \varphi_N$ for some $\varphi_i \in \Gamma(G_i)$. Thus

$$L < I(\varphi) = I(\varphi_1) + \cdots + I(\varphi_N) \leqslant \sum_1^\infty \lambda(G_n)$$

and we have shown that

$$\lambda\left(\bigcup_1^\infty G_n \right) \leqslant \sum_1^\infty \lambda(G_n)$$

so that λ satisfies condition (4).

Finally, we prove property (5a) (stronger than property (5)):

$$\lambda(G \cup H) = \lambda(G) + \lambda(H)$$

for G, H open and disjoint. Now, if $\varphi \in \Gamma(G)$ and $\psi \in \Gamma(H)$ then

$$\varphi + \psi \in \Gamma(G \cup H)$$

(for we cannot have simultaneously $\varphi(x) \neq 0$ and $\psi(x) \neq 0$, so that $\varphi + \psi \leqslant 1$).

Therefore

$$I(\varphi)+I(\psi)=I(\varphi+\psi)\leqslant\lambda(G\cup H).$$

Taking suprema we have

$$\lambda(G)+\lambda(H)\leqslant\lambda(G\cup H)$$

which, together with property (4), implies equality.

Thus μ is a measure on $\mathscr{B}(X)$ and

$$\begin{aligned}\mu(G)&=\sup_{\Gamma(G)} I &&\text{for } G \text{ open},\\ \mu(E)&=\inf_{E\subset G\text{open}}\mu(G) &&\text{for } E\in\mathscr{B}(X)\end{aligned}\tag{6.4.3}$$

(since if D is open then $\mu(D)=\lambda(D)$).

Let $f\in\mathscr{C}_+$. Applying the preceding part of the proof to the functional $\varphi\to I(f\varphi)$ we see that the set function μ_f given by

$$\begin{aligned}\mu_f(G)&=\sup_{\varphi\in\Gamma(G)} I(f\varphi) &&\text{for } G \text{ open}\\ \mu_f(E)&=\inf_{E\subset G\text{open}}\mu_f(G) &&\text{for } E\in\mathscr{B}(X)\end{aligned}\tag{6.4.4}$$

is a measure on $\mathscr{B}(X)$. We show that

$$\mu_f(E)=\int_E f\,\mathrm{d}\mu \qquad \text{for } E\in\mathscr{B}(X).$$

By Th. 6.1.7, it suffices to show that μ_f satisfies the condition

$$\left(\inf_E f\right)\mu(E)\leqslant\mu_f(E)\leqslant\left(\sup_E f\right)\mu(E)\qquad\text{for } E\in\mathscr{B}(X).\tag{6.4.5}$$

But by (6.4.2),

$$\left(\inf_H f\right)I(\varphi)\leqslant I(f\varphi)\leqslant\left(\sup_H f\right)I(\varphi)$$

provided $\overline{\{x:\varphi(x)\neq 0\}}\subset H$. Let $E\subset H$, H open. Taking suprema and then infima we obtain, first for E open and then for $E\in\mathscr{B}(X)$, the inequality

$$\left(\inf_H f\right)\mu(E)\leqslant\mu_f(E)\leqslant\left(\sup_H f\right)\mu(E).$$

This implies (6.4.5), since

$$\sup_{E\subset H}\inf_H f=\inf_E f \quad\text{and}\quad \inf_{E\subset H}\sup_H f=\sup_E f$$

(in the first of these formulae the inequality $\leqslant$ is obvious; the inequality $\geqslant$ follows from the fact that if $a<\inf_E f$, then $\inf_H f\geqslant a$ for the set $H=\{x:f(x)>a\}$ which is open and contains E; similarly for the second formula).

The equality (6.4.1) in Riesz's theorem is therefore a consequence of the relation

$$I(f) = \mu_f(X).$$

This follows from (6.4.4) since $I(f\varphi) \leqslant I(f)$ for $\varphi \in \Gamma(X)$ and equality occurs for a function $\varphi \in \Gamma(X)$ which equals 1 on the compact set $\overline{\{x : f(x) \neq 0\}}$.

The measure μ is finite on compact sets. Indeed, if F is a compact set, then taking a function $\varphi \in \mathscr{C}_+$ such that $\varphi = 1$ on F, we have

$$\mu(F) = \int_X \chi_F \, \mathrm{d}\mu \leqslant \int_X \varphi \, \mathrm{d}\mu = I(\varphi) < \infty .$$

This completes the proof of Riesz's theorem. ■

Uniqueness of Measure in Riesz's Theorem[16]

The measure μ constructed in the theorem on $\mathscr{B}(X)$ is the *unique regular measure*, i.e. such that

(1) $\mu(G) = \sup\limits_{G \supset F \text{compact}} \mu(F)$ for G open,

(2) $\mu(E) = \inf\limits_{E \subset G \text{open}} \mu(G)$ for $E \in \mathscr{B}(X)$,[17]

which satisfies Riesz's theorem.[18]

These conditions follow from conditions (6.4.3) which are equivalent to conditions (1), (2) since

$$\sup_{\Gamma(G)} I = \sup_{G \supset F \text{compact}} \mu(F).$$

This holds because for $\varphi \in \Gamma(G)$ we have $I(\varphi) \leqslant \mu(\overline{\{x : \varphi(x) \neq 0\}})$ and for compact $F \subset G$ there exists $\varphi \in \Gamma(G)$ such that $\varphi = 1$ on F and so $\mu(F) \leqslant I(\varphi)$.

The class $\mathscr{B}^0(X)$ of unions of sequences of relatively compact Borel sets is a countably additive algebra which is generated (as is easily shown) both by the class of open relatively compact sets and equally by the class of compact sets. If X is σ-compact then $\mathscr{B}^0(X) = \mathscr{B}(X)$.

If μ is finite on compact sets then equality (2) holds on $\mathscr{B}^0(X)$ (by Th. 5.2.5, since $\mathscr{B}^0(X) \subset \mathscr{B}_\mu$). On the other hand, equality (1) is always satisfied if G is σ-compact. In particular (1) holds for G open and relatively compact.[19] Thus in Riesz's theorem the measure is uniquely defined on $\mathscr{B}^0(X)$.[20]

In the case where X is σ-compact, every measure μ on $\mathscr{B}(X)$ which is finite on compact sets is regular (for then every open set is σ-compact).[21] Hence, if we add the supplementary hypothesis that the space X is σ-compact, we have the uniqueness of the measure in Riesz's theorem.[22]

A measure μ on $\mathscr{B}^0(X)$ which is finite on compact sets always has a unique (by (1) and (2)) regular extension $\tilde{\mu}$ on $\mathscr{B}(X)$, and

$$\int \varphi \, \mathrm{d}\mu = \int \varphi \, \mathrm{d}\tilde{\mu} \qquad \text{for } \varphi \in \mathscr{C}_+$$

(the left-hand term represents an integral over any compact set on whose complement $\varphi = 0$). Indeed, the functional

$$I(\varphi) = \int \varphi \, d\mu \qquad \text{for } \varphi \in \mathscr{C}_+$$

satisfies the hypotheses of Riesz's theorem, and so

$$I(\varphi) = \int \varphi \, d\tilde{\mu} \qquad \text{on } \mathscr{C}_+$$

for some regular measure $\tilde{\mu}$ on $\mathscr{B}(X)$. For a set G open and relatively compact (and hence σ-compact), we have

$$\mu(G) = \sup_{\Gamma(G)} I = \sup_{G \supset F \text{ compact}} \tilde{\mu}(F) = \tilde{\mu}(G)$$

from which, using (2) for $\tilde{\mu}$, we obtain $\mu(E) \leqslant \tilde{\mu}(E)$ if E is Borel and relatively compact. But then E is contained in some open relatively compact set G and so $\mu(E) \geqslant \tilde{\mu}(E)$ since $\mu(G \backslash E) \leqslant \tilde{\mu}(G \backslash E)$. Therefore $\mu = \tilde{\mu}$ on Borel relatively compact sets and therefore also on $\mathscr{B}^0(X)$.

Hence, in Riesz's theorem

$$I(\varphi) = \int \varphi \, d\mu \qquad \text{on } \mathscr{C}_+$$

for a unique measure μ on $\mathscr{B}^0(X)$ which is finite on compact sets.[23]

NOTES

1. For, from inequalities (6.1.14) it follows that $\lambda(D) = \infty \cdot \mu(D)$.
2. Here $\#E$ denotes the number of elements in the set E (taking $\#E = \infty$ if E is not finite).
3. Using the inequality $\chi_{E_1 \cup \cdots \cup E_r} \leqslant \chi_{E_1} + \cdots + \chi_{E_r}$.
4. Using (6.1.6), Th. 4.4.8 and Th. 5.5.1.
5. For (6.2.7) to hold, it suffices to assume that the left-hand integral exists or that the right-hand series is absolutely convergent.
6. For (6.2.8) to hold, it suffices to assume that the right-hand integrals exist and that the right-hand side is not of the form $\pm\infty \mp \infty$.
7. Lebesgue measure satisfies this condition; indeed, if $m_n(E) < \infty$ and $\varepsilon > 0$, then taking a decomposition $\mathbb{R}^n = \bigcup_{i=1}^{\infty} P_i$ into disjoint intervals such that $m_n(P_i) < \varepsilon$, we have $m_n(\bigcup_{i=k}^{\infty} (E \cap P_i)) < \varepsilon$ for k sufficiently large, so that $E = (E \cap P_1) \cup \cdots \cup (E \cap P_{k-1}) \cup \bigcup_{i=k}^{\infty} (E \cap P_i)$ is a finite union of sets of measure $< \varepsilon$. On the other hand, this condition is not satisfied by any measure for which there is a singleton having positive measure.
8. This means that given $\varepsilon > 0$ there exists $\delta > 0$ such that if $\mu(A) < \delta$ then the integral in (6.2.9) exists and has absolute value $< \varepsilon$.
9. Because $|f(x)| \leqslant g(x)$ a.e. on E.
10. These integrals are defined by (6.2.15) (for any function f bounded at P, the sequence s_ν and S_ν always tend to a limit which is independent of the sequence of subdivisions with the given properties; this is also a consequence of the remainder of the proof); the equality of the integrals is equivalent to the existence of the Riemann integral, in which case they all have the same value.

11. Indeed, the set $\{a\} \times F$ is contained in the set $\{a\} \times \mathbb{R}^q$ of measure zero, so it is $\mathscr{L}_{p+q}$-measurable. Also, by Th. 4.2.3, since $(\{a\} \times F)_a = F$, we cannot have $\{a\} \times F \in \mathscr{L}_p \times \mathscr{L}_q$.
12. A direct proof of this theorem is given in Appendix 2.
13. It is not essential that the space be metric: the proof extends without change to the case of a locally compact Hausdorff topological space.
14. In the case of a topological space we take an open covering $\{C_i\}$ such that $\bar{C}_i \subset D_i$ and, by Urysohn's lemma, functions ψ_i equal to 1 on $\bar{C}_i$ and 0 on $E \backslash D_i$.
15. In the case of a topological space we make use of the remark following Th. 5.2.4; for condition (5b) is satisfied (since $\Gamma(G) = \bigcup_{\bar{D} \subset G} \Gamma(D)$) and so is condition (5a) (verified subsequently).
16. The details of the proof have been left to the reader and are not difficult to supply.
17. In the case in which μ is finite, (1) is equivalent to the condition that G is the union of a σ-compact set and a set of μ-measure zero, and (2) is equivalent to the condition that E is the difference between a set of type G_δ and a set of measure zero.
18. See, W. Rudin, *Real and Complex Analysis.*
19. In the case of a topological space, condition (1) has to be assumed.
20. In the case of a topological space one must either assume condition (1) (see Note 5.3) or replace $\mathscr{B}^0(X)$ by the countably additive algebra $\mathscr{B}^\sigma(X)$ generated by the class $\mathscr{U}^\sigma$ of σ-compact open sets, for (2) holds in $\mathscr{B}^\sigma(X)$ with $G \in \mathscr{U}^\sigma$. (Equivalently, the class $\mathscr{C}^\delta$ of compact sets of type G_δ generates $\mathscr{B}^\sigma(X)$; we have $\mathscr{B}^\sigma(X) \subset \mathscr{B}^0(X)$ where equality holds if X is metrizable; we note further that if F is compact and is contained in G open, then $F \subset E \subset G$ for some $E \in \mathscr{U}^\sigma$ and for some $E \in \mathscr{C}^\delta$.) Indeed, the class $\mathscr{K}$ of sets satisfying condition (2) is countably additive and so contains the class

$$\mathscr{K}_0 = \{E \in \mathscr{B}^0(X) : E \cap G \in \mathscr{K} \text{ for } G \text{ relatively compact in } \mathscr{U}^\sigma\};$$

now, the class $\mathscr{K}_0$ is monotone and contains the class $\mathscr{J}$ of intersections of decreasing sequences of sets in $\mathscr{U}^\sigma$, which is finitely additive (unions) and multiplicative and contains $\mathscr{U}^\sigma \cup \mathscr{C}^\delta$, and hence also finite unions of finite intersections of sets in $\mathscr{U}^\sigma \cup \mathscr{C}^\delta$, which form a finitely additive algebra $\mathscr{R}$ generating $\mathscr{B}^\sigma(X)$; hence $\mathscr{K}_0 \supset \mathscr{R}$, so that by Th. 4.1.8, $\mathscr{B}^\sigma(X) \subset \mathscr{K}_0$.
21. In the case of a topological space, condition (1) must be assumed.
22. In the case of a topological space we find only that regularity of a measure which is finite on compact sets is equivalent to condition (1).
23. In the case of a topological space one must replace $\mathscr{B}^0(X)$ by $\mathscr{B}^\sigma(X)$ (see Note 20) and assume that μ is finite on compact sets (in $\mathscr{B}^\sigma(X)$). In this case φ (after restriction to a set in $\mathscr{C}^\delta$, outside which it vanishes) is measurable, for $\{\varphi \geqslant a\} = \bigcap_1^\infty \{\varphi > a - 1/n\} \in \mathscr{C}^\delta$. Similarly, we obtain $\mu(G) = \tilde{\mu}(G)$ for $G \in \mathscr{U}^\sigma$ and $\mu(E) \leqslant \tilde{\mu}(E)$ for $E \in \mathscr{B}^\sigma(X)$ noting that condition (2) holds in $\mathscr{B}^\sigma(X)$ with $G \in U^\sigma$ (see Note 20).

CHAPTER 7

DIFFERENTIATION

7.1 DIFFERENTIABILITY ALMOST EVERYWHERE

Let $E \subset \mathbb{R}^n$ and let $\mathscr{R}$ be a family of closed, n-dimensional cubes. We say that *the family $\mathscr{R}$ is a Vitali covering of the set E* if given $x \in E$ and $\varepsilon > 0$ there exists a cube $Q \in \mathscr{R}$ such that $x \in Q$ and $\delta(Q) < \varepsilon$.

It follows from this definition that if G is an open set and $x \in E \cap G$, then there exists a cube $Q \in \mathscr{R}$ of sufficiently small diameter $\delta(Q)$ such that $x \in Q \subset G$ (it is sufficient that $\delta(Q) < \rho(x, \backslash G)$). Hence, if G is an open set containing E then the family $\mathscr{R}_1$ of all cubes in $\mathscr{R}$ contained in G is also a Vitali covering of E.

We have the following

THEOREM 7.1.1 (Vitali Covering Theorem). *Let the family $\mathscr{R}$ of closed cubes be a Vitali covering of a bounded set*[1] *E. Then there is a finite or countable sequence of disjoint cubes $Q_1, Q_2, \ldots$ in $\mathscr{R}$ such that*

$$m_n^*\left(E \backslash \bigcup_\nu Q_\nu\right) = 0. \tag{7.1.1}$$

The family $\mathscr{R}$ of closed cubes is called *scattered* if, given any finite system of disjoint cubes $Q_1, Q_2, \ldots, Q_k$ from the family, there exists a cube $Q \in \mathscr{R}$ which is disjoint from $Q_1 \cup \cdots \cup Q_k$.

LEMMA 7.1.1. *Let $\mathscr{R}$ be a non-empty, scattered family of closed cubes whose union is bounded. Then, from the family $\mathscr{R}$ we can select a sequence Q_ν of disjoint cubes with the property that for every cube $Q \in \mathscr{R}$ there exists p such that*

$$Q_p \cap Q \neq \varnothing \quad \textit{and} \quad \delta(Q_p) > \tfrac{1}{2}\delta(Q). \tag{7.1.2}$$

Proof. We define the sequence Q_ν by induction. Since $0 < \sup_{\mathscr{R}} \delta(Q) < \infty$ therefore there exists a cube $Q_1 \in \mathscr{R}$ such that

$$\delta(Q_1) > \tfrac{1}{2} \sup_{\mathscr{R}} \delta(Q).$$

If now we have already chosen disjoint cubes $Q_1, \ldots, Q_k$ from $\mathscr{R}$ then, since the family $\mathscr{R}_k$ of cubes which belong to $\mathscr{R}$ and are disjoint with $Q_1 \cup \cdots \cup Q_k$ is non-empty and $0<\sup_{\mathscr{R}_k}\delta(Q)<\infty$, therefore there exists a cube $Q_{k+1}\in\mathscr{R}_k$ such that

$$\delta(Q_{k+1})>\tfrac{1}{2}\sup_{\mathscr{R}_k}\delta(Q).$$

In this way we have defined a sequence Q_ν of disjoint cubes from $\mathscr{R}$. We must have $\sum_{\nu=1}^{\infty}|Q_\nu|<\infty$, and so $|Q_\nu|\to 0$, or equivalently

$$\delta(Q_\nu)\to 0.$$

Let $Q\in\mathscr{R}$. There must exist indices s such that $Q_s\cap Q\neq\varnothing$, for otherwise for all s we would have $Q\in\mathscr{R}_s$ and

$$\delta(Q_{s+1})>\tfrac{1}{2}\delta(Q)$$

which is impossible. Let p be the smallest of these indices. We then have

$$Q_p\cap Q\neq\varnothing.$$

If $p=1$, then

$$\delta(Q_1)>\tfrac{1}{2}\delta(Q).$$

If $p>1$, then $(Q_1\cup\cdots\cup Q_{p-1})\cap Q=\varnothing$ and so $Q\in\mathscr{R}_{p-1}$, whence

$$\delta(Q_p)>\tfrac{1}{2}\delta(Q). \quad \blacksquare$$

Proof of theorem. Let G be an open set which is bounded and contains E. Then the family $\mathscr{R}_1$ of all the cubes of $\mathscr{R}$ which are contained in G is also a Vitali covering of the set E. The theorem is clearly true if there exist disjoint cubes $Q_1,\ldots,Q_k$ from $\mathscr{R}_1$ which together cover E. Otherwise, the family $\mathscr{R}_1$ is scattered; for if $Q_1,\ldots,Q_k$ are disjoint cubes in $\mathscr{R}_1$ then there exist $x\in E\backslash(Q_1\cup\cdots\cup Q_k)$ and a cube $Q\in\mathscr{R}_1$ such that $x\in Q\subset\backslash(Q_1\cup\cdots\cup Q_k)$. Thus, there exists a sequence $\{Q_\nu\}$ of disjoint cubes in $\mathscr{R}_1$ such that the properties in the lemma are satisfied. Let Q_ν^* be a cube with the same centre as Q_ν but whose sides are five times longer. We assert that for any q

$$E\backslash\bigcup_{\nu=1}^{\infty}Q_\nu\subset\bigcup_{\nu=q}^{\infty}Q_\nu^*. \tag{7.1.3}$$

For let

$$x\in E\backslash\bigcup_{\nu=1}^{\infty}Q_\nu.$$

Since $x\in\backslash(Q_1\cup\cdots\cup Q_{q-1})$, therefore there exists a cube $Q\in\mathscr{R}_1$ such that $x\in Q$ and $Q\subset\backslash(Q_1\cup\cdots\cup Q_{q-1})$. Then there exists p such that (7.1.2) holds and since $Q\cap Q_1=\cdots=Q\cap Q_{q-1}=\varnothing$, we must have $p\geqslant q$. But also by (7.1.2) it follows that $Q\subset Q_p^*$. Hence $x\in Q_p^*$ and

$$x\in\bigcup_{\nu=q}^{\infty}Q_\nu^*.$$

From the inclusion (7.1.3) we have

$$m^*\left(E\setminus\bigcup_{\nu=1}^{\infty} Q_\nu\right)\leqslant \sum_{\nu=q}^{\infty}|Q_\nu^*| = 5^n\sum_{\nu=q}^{\infty}|Q_\nu|.$$

From this, since q is arbitrary and since

$$\sum_{\nu=1}^{\infty}|Q_\nu|\leqslant m_n(G)<\infty,$$

we obtain (7.1.1). ■

Let $E\subset\mathbb{R}^n$. A point x is called a *density point of the set* E if

$$\lim_{\delta(Q)\to 0,\, x\in Q}\frac{m_n^*(E\cap Q)}{|Q|}=1 \qquad (Q \text{ is an } n\text{-cube}). \tag{7.1.4}$$

Then, if the set E is measurable, since

$$\frac{m_n(E\cap Q)+m_n(Q\setminus E)}{|Q|}=1,$$

we have

$$\lim_{\delta(Q)\to 0,\, x\in Q}\frac{m_n(Q\setminus E)}{|Q|}=0. \tag{7.1.5}$$

Clearly, if G is an open set and $x\in G$, then x is a density point of a set E if and only if it is a density point of $G\cap E$. We have the following

THEOREM 7.1.2 (Lebesgue's Theorem on Density Points). *Almost every point of an arbitrary set $E\subset\mathbb{R}^n$ is a density point of E.*

Proof.[2] Let Z be the set of points of E which are not density points of E. We can assume that the set E is bounded (for, taking a sequence G_n of open sets which are bounded and whose union is $\mathbb{R}^n$, it suffices to prove the theorem for $E\cap G_n$). Since

$$\limsup_{\delta(Q)\to 0,\, x\in Q}\frac{m_n^*(E\cap Q)}{|Q|}\leqslant 1$$

therefore

$$Z=\bigcup_{k=1}^{\infty} H\left(1-\frac{1}{k}\right)$$

where

$$H=H(\alpha)=E\cap\left\{x\colon \liminf_{\delta(Q)\to 0,\, x\in Q}\frac{m_n^*(E\cap Q)}{|Q|}<\alpha\right\}.$$

It therefore suffices to show that if $0<\alpha<1$ then $m_n^*(H)=0$. Let $\varepsilon>0$ and let G be an open set containing H such that

$$m_n(G)\leqslant m_n^*(H)+\varepsilon.$$

From the definition of the set H it follows that the family of cubes Q contained in G and such that

$$\frac{m_n^*(E\cap Q)}{|Q|}<\alpha$$

is a Vitali covering of H. Hence, by Th. 7.1.1, there exists a sequence Q_ν of disjoint sets from this family such that

$$m_n^*\left(H\setminus\bigcup_\nu Q_\nu\right)=0.$$

Since

$$H=\bigcup_\nu(H\cap Q_\nu)\cup\left(H\setminus\bigcup_\nu Q_\nu\right)\subset\bigcup_\nu(E\cap Q_\nu)\cup\left(H\setminus\bigcup_\nu Q_\nu\right),$$

we have

$$m_n^*(H)\leqslant\sum_\nu m_n^*(E\cap Q_\nu)<\alpha\sum_\nu|Q_\nu|\leqslant\alpha m_n(G)\leqslant\alpha(m_n^*(H)+\varepsilon).$$

Hence, in the limit as $\varepsilon\to 0$,

$$m_n^*(H)\leqslant\alpha m_n^*(H)$$

and since $0<\alpha<1$, we must have $m_n^*(H)=0$. ■

Let f be a function defined on a set $E\subset\mathbb{R}^n$. We say that f is *approximately continuous* at $x\in E$ if there exists a measurable set $F\subset E$ for which x is a density point and such that the restriction f_F is continuous. From Th. 5.5.2 (Luzin) and Th. 7.1.2, there follows:

THEOREM 7.1.3. *A measurable function is approximately continuous almost everywhere.*

Now let F be a function defined on n-dimensional intervals and let $x\in\mathbb{R}^n$. The limits

$$\bar{D}F(x)=\limsup_{\delta(Q)\to 0,x\in Q}\frac{F(Q)}{|Q|},\quad \underline{D}F(x)=\liminf_{\delta(Q)\to 0,x\in Q}\frac{F(Q)}{|Q|},\tag{7.1.6}$$

where Q is an n-dimensional cube, are called the *upper derivative and the lower derivative of the function F at the point x*,[3] respectively. If both these derivatives are equal, then their common value

$$DF(x)=\lim_{\delta(Q)\to 0,x\in Q}\frac{F(Q)}{|Q|}\tag{7.1.7}$$

is called the *derivative of the function F at the point x*. Clearly

$$D(\alpha F+\beta G)(x)=\alpha DF(x)+\beta DG(x)\tag{7.1.8}$$

if $DF(x)$ and $DG(x)$ exist.[4]

In the one-dimensional case let f be a finite function of one variable defined on an interval Δ. The function F whose value at the interval $[\xi, \eta] \subset \Delta$ is defined by

$$F([\xi, \eta]) = f(\xi) - f(\eta) \tag{7.1.9}$$

is additive. We call F the *interval function associated with f*. Conversely, every additive interval function F on intervals $[\xi, \eta] \subset \Delta$ is associated with the function

$$f(x) = F([a, x]) \quad \text{if} \quad x > a, \; f(a) = 0, \; f(x) = -F([x, a]) \quad \text{if} \quad x < a.$$

We always have $DF(x) = f'(x)$, where both sides exist or fail to exist together. For the existence of the limit

$$DF(x) = \lim_{\eta - \xi \to 0+, \xi \leqslant x \leqslant \eta} \frac{f(\eta) - f(\xi)}{\eta - \xi} \tag{7.1.10}$$

implies the existence of $f'(x)$. Conversely, from the identity

$$\frac{f(\eta) - f(\xi)}{\eta - \xi} = \alpha \frac{f(\xi) - f(x)}{\xi - x} + \beta \frac{f(\eta) - f(x)}{\eta - x}$$

(for $\xi < x < \eta$, where $\alpha = (x - \xi)/(\eta - \xi) > 0$, $\beta = (\eta - x)/(\eta - \xi) > 0$ and $\alpha + \beta = 1$) it follows that if $f'(x)$ exists then it is the unique limit in (7.1.10).[5]

LEMMA 7.1.2. *Let F be an additive interval function*[6] *which is non-negative and defined for intervals contained in the interval P. If* $\bar{D}F(x) > M$ *in a set* $E \subset P$, *then*

$$F(P) \geqslant M m_n^*(E). \tag{7.1.11}$$

Proof. Let $M > 0$. The family of cubes Q which are contained in P and are such that

$$\frac{F(Q)}{|Q|} > M$$

constitutes a Vitali covering for the set E. Hence, by Th. 7.1.1, there exists a sequence Q_ν of disjoint cubes in this family for which

$$m_n^*\left(E \backslash \bigcup_\nu Q_\nu\right) = 0.$$

Thus, since $E \subset \bigcup_\nu Q_\nu \cup (E \backslash \bigcup_\nu Q_\nu)$,

$$m_n^*(E) \leqslant \sum_\nu |Q_\nu|.$$

Now since, by Th. 5.3.2, proceeding to the limit if necessary,

$$\sum_\nu F(Q_\nu) \leqslant F(P)$$

therefore

$$M m_n^*(E) \leqslant \sum_\nu M |Q_\nu| \leqslant \sum_\nu F(Q_\nu) \leqslant F(P). \quad \blacksquare$$

LEMMA 7.1.3. *Under the hypotheses of lemma 7.1.2, if $\bar{D}F(x) > M$ on some set $E \subset P$, then $\underline{D}F(x) \geqslant M$ on density points of E.*

Proof. Let x be a density point of the set E. If $x \in Q \subset P$, then applying lemma 7.1.2 to the cube Q and the set $E \cap Q$, we have $F(Q) \geqslant Mm_n^*(E \cap Q)$, so that

$$\frac{F(Q)}{|Q|} \geqslant M \frac{m_n^*(E \cap Q)}{|Q|}.$$

In the limit as $\delta(Q) \to 0$ we obtain $\underline{D}F(x) \geqslant M$. ■

THEOREM 7.1.4 (Lebesgue). *An additive, non-negative interval function which is defined for intervals contained in an interval P has a finite derivative almost everywhere in P.*

Proof. Let F be an interval function which satisfies the assumptions of the theorem. Let $Z_\infty = \{x : \bar{D}F(x) = \infty\}$. Then for $x \in Z_\infty$, $\bar{D}F(x) > M$ for any $M < \infty$. Hence, by lemma 7.1.2, $F(P) \geqslant Mm_n^*(Z_\infty)$. Therefore

$$m_n(Z_\infty) = 0.$$

Let w_ν be an enumeration of all the rational numbers and let $E_\nu = \{x : \bar{D}F(x) > w_\nu\}$. By lemma 7.1.3, $\underline{D}F(x) \geqslant w_\nu$ on $E_\nu \backslash Z_\nu$, where Z_ν is the set of points of E_ν which are not density points of E_ν. But by Th. 7.1.2,

$$m_n(Z_\nu) = 0.$$

The set

$$Z = Z_\infty \cup \bigcup_{\nu=1}^{\infty} Z_\nu$$

has measure zero. We assert that $DF(x)$ exists and is bounded on $P \backslash Z$. For let $x \in P \backslash Z$. Then $\bar{D}F(x) < \infty$. There exists a sequence w_{α_ν} which converges to $\bar{D}F(x)$ and is such that $w_{\alpha_\nu} < \bar{D}F(x)$. Then $x \in E_{\alpha_\nu} \backslash Z_{\alpha_\nu}$ and so $\underline{D}F(x) \geqslant w_{\alpha_\nu}$. Thus in the limit $\underline{D}F(x) \geqslant \bar{D}F(x)$. Hence

$$\underline{D}F(x) = \bar{D}F(x) < \infty. \quad ■$$

If f is a finite, increasing function on the interval $[a, b]$, then, by (7.1.9), the interval function F associated with f is non-negative.

From Th. 1.4.1, we therefore have

THEOREM 7.1.5 (Lebesgue). *Every function f of bounded variation on an interval $[a, b]$ has a finite derivative almost everywhere in this interval.*

In particular, every function which satisfies a Lipschitz condition has a finite derivative almost everywhere.

Let $x_0 \in \mathbb{R}^n$ and let f be a function which is measurable in a neighbourhood of the point x_0. If $|f(x_0)| < \infty$ and

$$\lim_{\delta(Q)\to 0, x_0\in Q} \frac{1}{|Q|}\int_Q |f(x)-f(x_0)|\,dx = 0 \tag{7.1.12}$$

then we say that the point x_0 is a *Lebesgue point of the function f*. It is easily seen that if x_0 is a point at which f is continuous[7] then x_0 is a Lebesgue point of f.

THEOREM 7.1.6. *Let f be summable on the interval P. Then, almost all points of P are Lebesgue points of the function f.*

Proof. We have to show that

$$h(x) = \limsup_{\delta(Q)\to 0, x\in Q} \frac{1}{|Q|}\int_Q |f(u)-f(x)|\,du = 0$$

almost everywhere in P. To do this it is clearly sufficient to show that for all $c > 0$ the set

$$Z = \{x : h(x) > c\}$$

has measure zero. By Th. 5.5.2 (Luzin), there is an increasing sequence of closed sets $F_k \subset P$ such that f_{F_k} is continuous and $m_n(P\setminus F_k)\to 0$.

Let H_k be the set of density points of the set F_k and let

$$H = \bigcup_1^\infty H_k.$$

Since $m_n(P\setminus H) = 0$ it therefore suffices to show that for all k

$$m_n(H_k \cap Z) = 0.$$

Now, the interval function

$$G(R) = \int_{R\setminus F_k} |f(u)|\,du$$

is finite, additive and non-negative and obeys the inequality

$$h(x) \leqslant \bar{D}G(x) \qquad \text{for } x\in H_k. \tag{7.1.13}$$

This follows by taking the limit as $\delta(Q)\to 0$ and $x\in Q$ in the inequality

$$\frac{1}{|Q|}\int_Q |f(u)-f(x)|\,du \leqslant \frac{1}{|Q|}\int_{Q\cap F_k} |f(u)-f(x)|\,du$$
$$+\frac{1}{|Q|}\int_{Q\setminus F_k} |f(u)|\,du + |f(x)|\frac{m_n(Q\setminus F_k)}{|Q|}$$

for the first and third terms on the right-hand side tend to zero (by the continuity of f_{F_k} and the fact that x is a density point of F_k). By lemma 7.1.2, the inequality (7.1.13) implies that

$$cm_n^*(H_k\cap Z) \leqslant \int_{P\setminus F_k} |f(u)|\,du.$$

By Th. 6.2.8 on the absolute continuity of the integral, the right-hand side tends to zero as k increases. But the left-hand side can only increase with k so that we must have

$$m_n^*(H_k \cap Z) = 0. \quad \blacksquare$$

Let f be a summable function on the interval P and let

$$F(R) = \int_R f(x)\,\mathrm{d}x.$$

If x_0 is a Lebesgue point and $\delta(Q) \to 0$, where Q is a cube contained in P and containing x_0, then

$$\left|\frac{F(Q)}{Q} - f(x_0)\right| = \frac{1}{|Q|}\left|\int_Q (f(x) - f(x_0))\,\mathrm{d}x\right| \leqslant \frac{1}{|Q|}\int_Q |f(x) - f(x_0)|\,\mathrm{d}x \to 0.$$

Th. 7.1.6 therefore implies

THEOREM 7.1.7 (Lebesgue's Theorem on Differentiation of Integrals).[8] *Let f be a summable function on the interval P and let*

$$F(R) = \int_R f(x)\,\mathrm{d}x.$$

Then $DF(x) = f(x)$ almost everywhere on P.

In particular, in the case of a function f of a single variable which is summable on the interval $[a, b]$, the interval function F is associated with the function

$$\int_a^x f(t)\,\mathrm{d}t.$$

Therefore

$$\frac{\mathrm{d}}{\mathrm{d}x}\int_a^x f(t)\,\mathrm{d}t = f(x) \qquad \text{a.e. in } [a, b]. \tag{7.1.14}$$

Rademacher's Theorem

We now prove Rademacher's theorem on the almost everywhere differentiability of functions of several variables which satisfy a Lipschitz condition. For simplicity we confine ourselves to the case of two variables; the same method is applicable in the general case.

THEOREM 7.1.8 (Rademacher). *Let the function f satisfy the Lipschitz condition*

$$|f(\bar{x}, \bar{y}) - f(x, y)| \leqslant M(|\bar{x} - x| + |\bar{y} - y|), ((x, y), (\bar{x}, \bar{y}) \in P) \tag{7.1.15}$$

in the interval P. Then f has a differential almost everywhere in P.

LEMMA 7.1.4. *If the function φ is increasing in the interval $[a,b]$ and $\varphi'(x) > A$ on the set $E \subset [a,b]$ then $\varphi(b) - \varphi(a) \geqslant Am^*(E)$.*

This is a direct consequence of lemma 7.1.2 applied to the interval function associated with φ.

LEMMA 7.1.5. *Let the function g be increasing with respect to each variable in the interval $R = [a, a+h] \times [b, b+k] (h, k > 0)$. If*

$$\frac{\partial g}{\partial x}(x,y) > A \quad \textit{and} \quad \frac{\partial g}{\partial y}(x,y) > B$$

on a measurable set $E \subset R$, where $m_2(R \setminus E) < \varepsilon^2$ with $0 < \varepsilon < \min(h,k)$ then

$$g(a+h, b+k) - g(a,b) \geqslant A(h - 2\varepsilon) + B(k - 2\varepsilon).$$

Proof. There exists $y_0 \in [b, b+\varepsilon]$ such that the set E^{y_0} is measurable and

$$m_1(E^{y_0} \cap [a, a+h-\varepsilon]) > h - 2\varepsilon.$$

For otherwise, by Th. 6.3.9, we would have

$$m_2(E \cap ([a, a+h-\varepsilon] \times [b, b+\varepsilon]) \leqslant (h - 2\varepsilon)\varepsilon$$

from which

$$m_2(R \setminus E) \geqslant m_2(([a, a+h-\varepsilon] \times [b, b+\varepsilon]) \setminus E) \geqslant \varepsilon^2$$

contrary to hypothesis. Thus, by lemma 7.1.4,

$$g(x, y_0) - g(a, y_0) \geqslant A(h - 2\varepsilon) \quad \text{for} \quad a + h - \varepsilon \leqslant x \leqslant a + h. \tag{7.1.16}$$

Similarly, there exists $x_0 \in [a+h-\varepsilon, a+h]$ such that the set E_{x_0} is measurable and $m_1(E_{x_0} \cap [b+\varepsilon, b+k]) > k - 2\varepsilon$. Hence

$$g(x_0, b+k) - g(x_0, y) \geqslant B(k - 2\varepsilon) \quad \text{for} \quad b \leqslant y \leqslant b + \varepsilon. \tag{7.1.17}$$

In particular, the inequality (7.1.16) holds for $x = x_0$ and (7.1.17) holds for $y = y_0$. Thus

$$g(a+h, b+k) - g(a,b) \geqslant g(x_0, b+k) - g(a, y_0) \geqslant A(h - 2\varepsilon) + B(k - 2\varepsilon). \quad \blacksquare$$

LEMMA 7.1.6. *Let the function g be increasing with respect to both of its variables and continuous[9] in the interval P. Then, for almost all $(x,y) \in P$ the derivatives $\partial g/\partial x, \partial g/\partial y$ exist are finite and*

$$\liminf_{\lambda \to 0+} \frac{g(x + \lambda p, y + \lambda q) - g(x,y)}{\lambda} \geqslant p \frac{\partial g}{\partial x}(x,y) + q \frac{\partial g}{\partial y}(x,y),$$

where $p, q > 0$.

Proof. We claim that the derivatives $\partial g/\partial x, \partial g/\partial y$ exist and are finite a.e. in P

and are measurable functions. Indeed, the functions

$$\varphi_\nu(x, y) = \inf_{0<|h|<1/\nu} \frac{1}{h}(g(x+h, y) - g(x, y))$$

$$\psi_\nu(x, y) = \sup_{0<|h|<1/\nu} \frac{1}{h}(g(x+h, y) - g(x, y))$$

are semicontinuous on P and $(\partial g/\partial x)(x, y)$ exists and equals $\lim_{\nu\to\infty} \varphi_\nu(x, y)$ on the set $\{(x, y)\colon \lim_{\nu\to\infty} \varphi_\nu(x, y) = \lim_{\nu\to\infty} \psi_\nu(x, y)\}$. It follows that $\partial g/\partial x$ is measurable and that the set N of all points of P at which $\partial g/\partial x$ does not exist or is infinite is a measurable set. By Th. 7.1.5, $m_1(N^y) = 0$ for all y, and hence, by Th. 6.3.9, $m_2(N) = 0$. Similarly for $\partial g/\partial y$.

We now show that at every density point of the set

$$E = E(A, B) = \left\{(x, y)\colon \frac{\partial g}{\partial x}(x, y) > A, \frac{\partial g}{\partial y}(x, y) > B\right\}$$

we have the inequality

$$\liminf_{\lambda\to 0+} \frac{g(x+\lambda p, y+\lambda q) - g(x, y)}{\lambda} \geqslant pA + qB \text{ for } p, q > 0. \tag{7.1.18}$$

Let (x, y) be a density point of E. Let $r = \max(p, q)$. Choose $\varepsilon > 0$. Then, for $\lambda > 0$ sufficiently small we have

$$\frac{m_2(Q\backslash E)}{m_2(Q)} < \varepsilon^2,$$

where $Q = [x, x+\lambda r] \times [y, y+\lambda r]$. The interval $R = [x, x+\lambda p] \times [y, y+\lambda q]$ is contained in Q and we have

$$m_2(R\backslash E) \leqslant m_2(Q\backslash E) < (\lambda r\varepsilon)^2.$$

so that, by lemma 7.1.5,

$$g(x+\lambda p, y+\lambda q) - g(x, y) \geqslant A(\lambda p - 2\lambda r\varepsilon) + B(\lambda q - 2\lambda r\varepsilon).$$

Dividing this inequality on both sides by $\lambda > 0$ and taking the limit (first as $\lambda \to 0+$ and then as $\varepsilon \to 0+$) we obtain the inequality (7.1.18).

Now let $\{w_\nu\}$ be an enumeration of all the rational numbers. Except for a set $Z_{\nu\sigma}$ of measure zero, every point of the set $E(w_\nu, w_\sigma)$ is a density point of this set. Let Z_0 be the set of all points of the interval P at which one of the derivatives $\partial g/\partial x$, $\partial g/\partial y$ does not exist or is not finite. Then the set

$$Z = Z_0 \cup \bigcup_{\nu,\sigma} Z_{\nu\sigma}$$

also has measure zero. Let $(x, y) \in P\backslash Z$. Then, there exist sequences w_{α_ν}, w_{β_ν} such that

$$w_{\alpha_\nu} \to \frac{\partial g}{\partial x}(x, y), \quad w_{\beta_\nu} \to \frac{\partial g}{\partial y}(x, y)$$

$$w_{\alpha_\nu} < \frac{\partial g}{\partial x}(x, y), \quad w_{\beta_\nu} < \frac{\partial g}{\partial y}(x, y).$$

Then $(x, y) \in E(w_{\alpha_\nu}, w_{\beta_\nu}) \backslash Z_{\alpha_\nu \beta_\nu}$, so that

$$\liminf_{\lambda \to 0+} \frac{g(x + \lambda p, y + \lambda q) - g(x, y)}{\lambda} \geqslant p w_{\alpha_\nu} + q w_{\beta_\nu}.$$

Hence, in the limit as $\nu \to \infty$ we obtain the inequality of lemma 7.1.6. ■

Proof of the theorem. By (7.1.15) the functions $M(x + y) + f(x, y)$ and $M(x + y) - f(x, y)$ are increasing with respect to both variables and are continuous in P. Applying lemma 7.1.6 to both of these functions we obtain (after subtracting $M(p + q)$ from both sides and after changing signs in the case of the second function) the inequalities

$$\liminf_{\lambda \to 0+} \frac{f(x + \lambda p, y + \lambda q) - f(x, y)}{\lambda} \geqslant p \frac{\partial f}{\partial x}(x, y) + q \frac{\partial f}{\partial y}(x, y)$$

and

$$\limsup_{\lambda \to 0+} \frac{f(x + \lambda p, y + \lambda q) - f(x, y)}{\lambda} \leqslant p \frac{\partial f}{\partial x}(x, y) + q \frac{\partial f}{\partial y}(x, y)$$

so that

$$\lim_{\lambda \to 0+} \frac{f(x + \lambda p, y + \lambda q) - f(x, y)}{\lambda} = p \frac{\partial f}{\partial x}(x, y) + q \frac{\partial f}{\partial y}(x, y), \qquad (7.1.19)$$

where $p, q > 0$, for almost all (x, y). If, in place of f, we use the functions

$$(x, y) \to f(-x, y), \quad (x, y) \to f(x, -y), \quad (x, y) \to f(-x, -y)$$

we obtain (7.1.19) when $p, q \neq 0$ and $(x, y) \in P \backslash Z'$ where $m_2(Z') = 0$. Let $(x, y) \in P \backslash Z'$ and let $h_\nu \to 0$, $k_\nu \to 0$ with $\lambda_\nu = \sqrt{(h_\nu^2 + k_\nu^2)} > 0$. The functions

$$\varphi_\nu(p, q) = \frac{f(x + \lambda_\nu p, y + \lambda_\nu q) - f(x, y)}{\lambda_\nu}$$

defined on the set $\{(p, q) : p^2 + q^2 = 1\}$ are uniformly bounded (by $2M$) and satisfy a Lipschitz condition with common constant M. Thus they are equicontinuous. Since

$$\varphi_\nu(p, q) \to p \frac{\partial f}{\partial x}(x, y) + q \frac{\partial f}{\partial y}(x, y) \qquad \text{for } p, q \neq 0,$$

therefore, by Th. 3.2.4,

$$\varphi_\nu(p, q) \to p \frac{\partial f}{\partial x}(x, y) + q \frac{\partial f}{\partial y}(x, y) \quad \text{uniformly on} \quad \{(p, q) : p^2 + q^2 = 1\}$$

Putting $p_\nu = h_\nu/\lambda_\nu$ and $q_\nu = k_\nu/\lambda_\nu$ we have $p_\nu^2 + q_\nu^2 = 1$ and hence

$$\frac{f(x+h_\nu, y+k_\nu) - f(x,y) - \dfrac{\partial f}{\partial x}(x,y)h_\nu - \dfrac{\partial f}{\partial y}(x,y)k_\nu}{\sqrt{(h_\nu^2 + k_\nu^2)}}$$

$$= \frac{f(x+\lambda_\nu p_\nu, y+\lambda_\nu q_\nu) - f(x,y)}{\lambda_\nu} - \frac{\partial f}{\partial x}(x,y)p_\nu - \frac{\partial f}{\partial y}(x,y)q_\nu$$

$$= \varphi_\nu(p_\nu, q_\nu) - \frac{\partial f}{\partial x}(x,y)p_\nu - \frac{\partial f}{\partial y}(x,y)q_\nu \to 0. \quad \blacksquare$$

More generally, from Rademacher's theorem we obtain

THEOREM 7.1.9 (Stepanov). *A function f defined on an open set $G \subset \mathbb{R}^n$ and satisfying, at every point a of the set $E \subset G$, the condition*

$$\limsup_{x \to a} \frac{|f(x) - f(a)|}{|x - a|} < \infty$$

has a differential at almost every point of E.[10]

7.2 INTERVAL FUNCTIONS OF BOUNDED VARIATION

Let F be an additive function of n-dimensional intervals which is defined for intervals contained in an interval P. The quantity

$$|F|(P) = \sup\left\{\sum_{i=1}^{k} |F(P_i)| : P = \bigcup_{i=1}^{k} P_i\right\} \tag{7.2.1}$$

where P_i are non-overlapping intervals, is called the *variation of the function* F on the interval P. We say that F is a *function of bounded variation* (on the interval P) if $|F|(P) < \infty$. In the case of a function f of one variable, this definition, applied to the associated function

$$F([c,d]) = f(d) - f(c)$$

is compatible with the definitions in §1.3.

Interval functions of bounded variation have a number of properties which are analogous to those which were established in Chapter 1 for functions of a single variable. It follows from the definition that an additive and non-negative (or non-positive) function F is a function of bounded variation and that $|F|(P) = F(P)$ (or $|F|(P) = -F(P)$, respectively). It follows from (7.2.1) that

$$|F(P)| \leqslant |F|(P). \tag{7.2.2}$$

Variation is an additive, non-negative interval function: if Q and R are adjacent intervals, then

$$|F|(Q \cup R) = |F|(Q) + |F|(R). \tag{7.2.3}$$

Indeed, if $Q \cup R = \bigcup_{i=1}^{k} P_i$, where P_i are non-overlapping intervals, then

$$\sum |F(P_i)| \leqslant \sum |F(P_i \cap Q)| + \sum |F(P_i \cap R)| \leqslant |F|(Q) + |F|(R)$$

(we put $F(P' \cap Q') = 0$ if P' and Q' do not overlap). Now, taking the supremum of the left-hand side we have

$$|F|(Q \cup R) \leqslant |F|(Q) + |F|(R).$$

Again, if $Q = \bigcup_{i=1}^{k} Q_i$ and $R = \bigcup_{j=1}^{l} R_j$, where the Q_i and the R_j do not overlap, then

$$\sum |F(Q_i)| + \sum |F(R_j)| \leqslant |F|(Q \cup R),$$

so that, taking two suprema on the left,

$$|F|(Q) + |F|(R) \leqslant |F|(Q \cup R).$$

It follows from (7.2.3) that

$$|F|(R) = |F|(R_1) + \cdots + |F|(R_k),$$

where the R_i do not overlap and $R = R_1 \cup \cdots \cup R_k$. Thus, if F is a function of bounded variation on each of the intervals $R_1, \ldots, R_k$, then F is of bounded variation on R.

A linear combination of functions F, G of bounded variation is a function of bounded variation, for

$$\sum |\alpha F(P_i) + \beta G(P_i)| \leqslant |\alpha| \sum |F(P_i)| + |\beta| \sum |G(P_i)| \leqslant |\alpha| |F|(P) + |\beta| |G|(P)$$

and so

$$|\alpha F + \beta G|(P) \leqslant |\alpha| |F|(P) + |\beta| |G|(P).$$

As in §1.5, we define the *Riemann–Stieltjes integral*

$$\int_P f(x) \, \mathrm{d}G^{11}$$

for a function f bounded on an interval P and an interval function G of bounded variation, as the limit of approximating sums

$$\sum_{i=1}^{k} f(\xi_i) G(P_i)$$

(where $\xi_i \in P_i$, $P = \bigcup_{i=1}^{k} P_i$, P_i do not overlap) as $\max_i \delta(P_i) \to 0$. A series of analogous properties can be proved, such as existence of the integral when the function f is continuous.

The quantities

$$\begin{aligned} F^+(P) &= \sup \{\textstyle\sum F(P_i) : \bigcup P_i \subset P\} \\ F^-(P) &= -\inf \{\textstyle\sum F(P_i) : \bigcup P_i \subset P\} = \sup \{\textstyle\sum -F(P_i) : \bigcup P_i \subset P\} \end{aligned} \tag{7.2.4}$$

where $\{P_i\}$ is a system of non-overlapping intervals which are finite or empty (in which case $\bigcup P_i = \varnothing$ and $\sum F(P_i) = 0$), are called the *upper variation* and the *lower variation of the function* F on the interval P.

THEOREM 7.2.1 (Jordan Canonical Decomposition). *Let F be an additive function of bounded variation on an interval P. Then the variations*

$$F^+(R),\ F^-(R),\ |F|(R)$$

are additive, non-negative functions defined for intervals $R \subset P$ and

$$F = F^+ - F^- \quad \textit{and} \quad |F| = F^+ + F^-.\text{[12]} \tag{7.2.5}$$

Proof. It follows from definition (7.2.4) that F^+ and F^- are non-negative. Let $R = \bigcup_{i=1}^k R_i$ where the R_i do not overlap. We have

$$\sum_i |F(R_i)| = \sum_{i'} F(R_{i'}) + \sum_{i''} (-F(R_{i''})) \leqslant F^+(R) + F^-(R)$$

where $F(R_{i'}) \geqslant 0$ and $F(R_{i''}) < 0$. Hence, taking the supremum on the left-hand side

$$|F|(R) \leqslant F^+(R) + F^-(R).$$

Let $\{P_i\}$ be a finite or empty system of non-overlapping intervals contained in R and let $\{Q_j\}$ be another such system. Let $\{R_\nu\}$ be a normal system (see §5.3) for R, P_i, Q_j and let $R_{\nu'}$ be all the intervals contained in $\bigcup P_i$ and such that $F(R_{\nu'}) \geqslant 0$ while $R_{\nu''}$ are all intervals contained in $\bigcup Q_j$ and such that $F(R_{\nu''}) < 0$. Then

$$\sum F(P_i) = \sum_i \sum_{R_\nu \subset P_i} F(R_\nu) \leqslant \sum_{\nu'} F(R_{\nu'})$$

and

$$\sum F(Q_j) = \sum_j \sum_{R_\nu \subset Q_j} F(R_\nu) \geqslant \sum_{\nu''} F(R_{\nu''}).$$

Therefore

$$\sum F(P_i) + \sum -F(Q_j) \leqslant \sum_{\nu'} F(R_{\nu'}) - \sum_{\nu''} F(R_{\nu''}) \leqslant \sum_{R_\nu \subset R} |F(R_\nu)| \leqslant |F|(R)$$

and so, by taking suprema twice on the left-hand side, we have

$$F^+(R) + F^-(R) \leqslant |F|(R).$$

Thus we have proved the second equality in (7.2.5). This immediately implies that F^+ and F^- are finite.

Let $\{P_i\}$ be a finite or empty system of non-overlapping intervals contained in R. There exists a finite or empty system $\{Q_j\}$ of non-overlapping intervals such that P_i, Q_j do not overlap and

$$R = \bigcup P_i \cup \bigcup Q_j$$

(it suffices to take a normal system for R, P_i and to let $\{Q_j\}$ consist of all those intervals which do not overlap with any of the P_i). Then

$$\sum F(P_i) = F(R) + \sum -F(Q_j) \leqslant F(R) + F^-(R)$$

and so, taking the supremum on the left-hand side, $F^+(R) \leqslant F(R) + F^-(R)$, that is,

$$F^+(R) - F^-(R) \leqslant F(R).$$

Similarly

$$\sum -F(P_i) = -F(R) + \sum F(Q_j) \leqslant -F(R) + F^+(R)$$

so that $F^-(R) \leqslant -F(R) + F^+(R)$, or

$$F(R) \leqslant F^+(R) - F^-(R).$$

This establishes the first of the equalities (7.2.5).

The additivity of the functions F^+ and F^- follows from the additivity of the functions F and $|F|$, by (7.2.5). ■

We see, therefore, that an additive interval function is a function of bounded variation if and only if it is the difference of two additive, non-negative interval functions.

By Th. 7.1.4 and the Jordan decomposition we have

THEOREM 7.2.2 (Lebesgue). *An interval function of bounded variation on an interval P has a finite derivative almost everywhere on P.*

THEOREM 7.2.3. *If F is an interval function of bounded variation on an interval P, then DF is summable on P. If also F is non-negative, then*

$$\int_P DF(x)\,\mathrm{d}x \leqslant F(P). \tag{7.2.6}$$

In particular, if the function f is increasing in the interval $[a, b]$ then

$$\int_a^b f'(t)\,\mathrm{d}t \leqslant f(b) - f(a). \tag{7.2.7}$$

Proof. Let $P_1^{(\nu)}, \ldots, P_{k_\nu}^{(\nu)}$ be a sequence of systems of non-overlapping intervals such that

$$P = P_1^{(\nu)} \cup \cdots \cup P_{k_\nu}^{(\nu)} \quad \text{and} \quad \lim_{\nu \to \infty} \left(\max_i \delta(P_i^{(\nu)}) \right) = 0.$$

We can require that all those $P_i^{(\nu)}$ which, together with their boundaries, are contained in P, should be cubes. Let $\varphi_\nu(x)$ be the function defined in P by:

$$\varphi_\nu(x) = \frac{F(P_i^{(\nu)})}{|P_i^{(\nu)}|}$$

when x lies in the interior of $P_i^{(\nu)}$, but which is defined arbitrarily on the boundaries of the intervals $P_1^{(\nu)}, \ldots, P_{k_\nu}^{(\nu)}$. Then

$$\varphi_\nu(x) \to DF(x)$$

almost everywhere in P, namely, where $DF(x)$ exists and is finite and not including the union of the boundaries of the intervals $P_i^{(\nu)}$. Moreover

$$\int_P \varphi_\nu(x)\,\mathrm{d}x = F(P).$$

It follows that DF is measurable and, by Fatou's lemma (§6.2) in the case when F is non-negative, that

$$\int_P DF(x)\,\mathrm{d}x \leqslant \liminf_{\nu\to\infty} \int_P \varphi_\nu(x)\,\mathrm{d}x = F(P).$$

In the general case the summability of DF follows from the summability of DF^+ and DF^-.

7.3 ABSOLUTELY CONTINUOUS INTERVAL FUNCTIONS

Let F be an additive function of n-dimensional intervals which is defined for intervals contained in P. We say that the function is *absolutely continuous* on the interval P if for every $\varepsilon > 0$ there exists $\delta > 0$ such that if $P_1, \ldots, P_k$ is a system of non-overlapping intervals which are contained in P and are such that $|P_1| + \cdots + |P_k| \leqslant \delta$, then $|F(P_1)| + \cdots + |F(P_k)| \leqslant \varepsilon$. It follows from the definition that a linear combination of absolutely continuous functions on P is also an absolutely continuous function on P. It is easy to prove that if P is a (finite) union of non-overlapping intervals and if F is absolutely continuous on each of these intervals then F is absolutely continuous on P.

THEOREM 7.3.1. *An interval function F which is absolutely continuous on an interval P is a function of bounded variation and the variations F^+, F^-, $|F|$ are absolutely continuous on P. If F is of bounded variation and $|F|$ is absolutely continuous then F is absolutely continuous.*

Proof. Suppose that F is absolutely continuous on P. Let $\delta > 0$ correspond to $\varepsilon > 0$ in the definition of absolute continuity. Let $P_1, \ldots, P_k$ be a system of intervals which are non-overlapping and are contained in P, such that

$$|P_1| + \cdots + |P_k| \leqslant \delta.$$

Let $P_i = \bigcup_{\nu=1}^{k_i} R_{i\nu}$ $(i = 1, \ldots, k)$, where the $R_{i\nu}$ do not overlap. Since $\sum_{i=1}^{k} \sum_\nu |R_{i\nu}| \leqslant \delta$, therefore $\sum_{i=1}^{k} \sum_\nu |F(R_{i\nu})| \leqslant \varepsilon$. Now, taking a k-fold supremum on the left-hand side, we obtain

$$\sum_{i=1}^{k} |F|(P_i) \leqslant \varepsilon.$$

It follows that

(i) $|F|(P) < \infty$ (for, if we decompose P into intervals $R_1, \ldots, R_k$ of content $\leqslant \delta$ we have $|F|(R_k) \leqslant \varepsilon$, so that $|F|(P) \leqslant k\varepsilon$);

(ii) $|F|$ is absolutely continuous, and hence, by (7.2.5), so are F^+ and F^-.

If F is a function of bounded variation and $|F|$ is absolutely continuous, then the absolute continuity of F follows from inequality (7.2.2). ■

From Th. 7.2.3 we have the following

COROLLARY. *An absolutely continuous function on an interval P has a finite derivative almost everywhere on P which is summable on P.*

THEOREM 7.3.2. *If F is an absolutely continuous interval function and $DF(x) = 0$ almost everywhere on P, then F is identically equal to zero.*

Proof. Let $P_0 \subset P$ and let $\varepsilon > 0$. Let $\delta > 0$ correspond to ε in the definition of absolute continuity. The family of cubes Q contained in P_0 and such that

$$|F(Q)| < \varepsilon|Q|,$$

covers the set $E = \{x : DF(x) = 0\} \cap P_0$ in the sense of Vitali. Hence, by Th. 7.1.1, there is a sequence $Q_1, Q_2, \ldots$ of disjoint cubes such that

$$m_n\left(E \setminus \bigcup_i Q_i\right) = 0.$$

Hence

$$m_n\left(P_0 \setminus \bigcup_i Q_i\right) = 0$$

(for P_0 and E differ by a set of measure zero). If the sequence $Q_1, Q_2, \ldots$ is finite then $P_0 = \bigcup_i Q_i$ so that

$$|F(P_0)| \leqslant \sum_i |F(Q_i)| < \sum_i \varepsilon|Q_i| = \varepsilon|P_0|.$$

If the sequence is infinite then N exists such that

$$m_n\left(P_0 \setminus \bigcup_{i=1}^{N} Q_i\right) < \delta.$$

Now taking a system of intervals $R_1, \ldots, R_k$ which do not overlap among themselves or with $Q_1, \ldots, Q_N$ and are such that

$$P_0 = \bigcup_{i=1}^{N} Q_i \cup \bigcup_{j=1}^{k} R_j,$$

we have

$$\sum_{j=1}^{k} |R_j| = m_n\left(P_0 \setminus \bigcup_{i=1}^{N} Q_i\right) < \delta$$

so that $\sum_{j=1}^{k} F(R_j) < \varepsilon$. But since

$$\sum_{i=1}^{N} |F(Q_i)| \leqslant \sum_{i=1}^{N} \varepsilon|Q_i| \leqslant \varepsilon|P_0|$$

therefore $|F(P_0)| \leqslant \varepsilon + \varepsilon |P_0|$. This inequality must hold in both cases and for any $\varepsilon > 0$. Hence $F(P_0) = 0$ for any interval $P_0 \subset P$. ■

It follows from Th. 6.2.8, that if f is a summable function on an interval P, then

$$F(R) = \int_R f(x)\,dx$$

is an absolutely continuous function on P. Conversely, every absolutely continuous function is of this form, namely, it is the integral of its derivative.

THEOREM 7.3.3. *If F is an absolutely continuous interval function on an interval P, then*

$$F(R) = \int_R DF(x)\,dx \qquad \textit{for } R \subset P. \tag{7.3.1}$$

Proof. The function

$$G(R) = \int_R DF(x)\,dx$$

is absolutely continuous, and, by Th. 7.1.7,

$$D(G - F)(x) = DG(x) - DF(x) = 0$$

a.e. in P. Hence, by Th. 7.3.2, we have that $G(R) = F(R)$ for $R \subset P$. ■

Remark. It follows that formula (7.3.1) holds (for all $R \subset P$) only if F is absolutely continuous on P.

Now let F be any interval function of bounded variation on the interval P. By Th. 7.2.3, DF is summable on P and so

$$F(R) = \int_R DF(x)\,dx + H(R) \tag{7.3.2}$$

where H is a function of bounded variation such that, by Th. 7.1.7,

$$DH(x) = 0$$

a.e. on P. A function of bounded variation whose derivative is zero almost everywhere is called a *singular function.*[13] Thus the formula (7.3.2) represents the decomposition of an arbitrary function of bounded variation into the sum of an absolutely continuous function and a singular function. This is called a *canonical Lebesgue decomposition.* There is precisely one such decomposition:

THEOREM 7.3.4 (Canonical Lebesgue Decomposition). *Every function F of bounded variation on an interval P may be represented uniquely in the form of a sum $F = G + H$, where G is an absolutely continuous function and H is a singular function.*

Proof. If $F=\bar{G}+\bar{H}$ is another such decomposition then $G-\bar{G}=\bar{H}-H$ so that

$$D(G-\bar{G})(x)=D\bar{H}(x)-DH(x)=0$$

a.e. on P. But since $G-\bar{G}$ is absolutely continuous on P we have by Th. 7.3.2, that $G-\bar{G}=0$. Thus $G=\bar{G}$ and $H=\bar{H}$. ■

Remark. If the function F is non-negative then, by Th. 7.2.3, the 'absolutely continuous part' G and the 'singular part' H are both non-negative.

THEOREM 7.3.5. *If f is a finite and continuous function on a closed interval P, and if G is an absolutely continuous interval function on P, then the Riemann–Stieltjes integral exists and is given by*

$$\int_P f\,\mathrm{d}G=\int_P f(x)DG(x)\,\mathrm{d}x. \tag{7.3.3}$$

Proof. Let $P=P_1^{(\nu)}\cup\cdots\cup P_{k_\nu}^{(\nu)}$ be a sequence of subdivisions of the interval P into non-overlapping intervals such that

$$\lim_{\nu\to\infty}\left(\max_i \delta(P_i^{(\nu)})\right)=0$$

and let $\xi_i^{(\nu)}\in P_i^{(\nu)}$. We define the function $\psi_\nu(x)$ on P by $\psi_\nu(x)=f(\xi_i^{(\nu)})$ on $\tilde{P}_i^{(\nu)}$, where $\tilde{P}_1^{(\nu)}=P_1^{(\nu)}$, $\tilde{P}_i^{(\nu)}=P_i^{(\nu)}\backslash(P_1^{(\nu)}\cup\cdots\cup P_{i-1}^{(\nu)})$. Then

$$\psi_\nu(x)\to f(x)$$

(uniformly) on P. Also, by Th. 7.3.3,

$$\begin{aligned}\int_P \psi_\nu(x)DG(x)\,\mathrm{d}x&=\sum_{i=1}^{k_\nu}\int_{P_i^{(\nu)}} f(\xi_i^{(\nu)})DG(x)\,\mathrm{d}x\\&=\sum_{i=1}^{k_\nu} f(\xi_i^{(\nu)})G(P_i^{(\nu)})\end{aligned}$$

noting that $\tilde{P}_i^{(\nu)}$ and $P_i^{(\nu)}$ differ by a set of measure zero. Since

$$|\psi_\nu(x)DG(x)|\leqslant M|DG(x)|,$$

where M is a constant bounding $|f|$ on P, therefore, by Th. 6.2.10,

$$\sum_{i=1}^{k_\nu} f(\xi_i^{(\nu)})G(P_i^{(\nu)})\to\int_P f(x)DG(x)\,\mathrm{d}x. \quad ■$$

7.4 THE CASE OF FUNCTIONS OF A SINGLE VARIABLE

We say that the function f of a single variable, defined on the interval $[a,b]$, is *absolutely continuous* on this interval if for any $\varepsilon>0$ there exists $\delta>0$ such that if $(\alpha_1,\beta_1),\ldots,(\alpha_k,\beta_k)$ are disjoint intervals contained in $[a,b]$ and such that

$\sum_{i=1}^{k}(\beta_i-\alpha_i)\leqslant\delta$ then

$$\sum_{i=1}^{k}|f(\beta_i)-f(\alpha_i)|\leqslant\varepsilon.$$

This definition is consistent with the definition of absolute continuity for the interval function associated with f. Hence, all the theorems in the previous section have their counterparts in terms of f.

It follows from the definition that an absolutely continuous function is continuous. A function which is absolutely continuous on the intervals $[a,b]$ and $[b,c]$ is absolutely continuous on $[a,c]$. A function which is absolutely continuous on $[a,b]$ is absolutely continuous on any sub-interval of $[a,b]$. Linear combinations and products of absolutely continuous functions are also absolutely continuous (for example, in the case of products this follows from the inequalities

$$\sum_{i=1}^{k}|f(\beta_i)g(\beta_i)-f(\alpha_i)g(\alpha_i)|\leqslant M\left(\sum_{i=1}^{k}|f(\beta_i)-f(\alpha_i)|+\sum_{i=1}^{k}|g(\beta_i)-g(\alpha_i)|\right)$$

where M is a constant which bounds $|f|$ and $|g|$[14]). A function which satisfies a Lipschitz condition is absolutely continuous.

THEOREM 7.4.1. *If f is an absolutely continuous function on the interval $[a,b]$, then*

$$m_1^*(f(E))\to 0 \qquad \text{if } m_1^*(E)\to 0,\quad E\subset[a,b]. \tag{7.4.1}$$

The image under f of a set of measure zero is a set of measure zero. The image of a measurable set is measurable.

Proof. It suffices to demonstrate property (7.4.1) for this implies the second assertion and also the third, by Th. 5.4.6. Thus, let $\varepsilon>0$ and let $\delta>0$ correspond to ε in the definition of absolute continuity. Let $m_1^*(E)<\delta$. By lemma 5.4.3, there is an open set G containing E such that $m_1(G)<\delta$. But G is a finite or countable union of open intervals (α_i,β_i) such that $\sum_i(\beta_i-\alpha_i)<\delta$. Let $\gamma_i<\delta_i$ be points of the interval $[\alpha_i,\beta_i]$ such that the function f takes its maximum value at one of these points and its minimum value at the other (on the interval $[\alpha_i,\beta_i]$). Then $f((\alpha_i,\beta_i))\subset\theta_i$, where θ_i is a closed interval with endpoints $f(\gamma_i)$, $f(\delta_i)$ (possibly degenerating to a point). Also $\sum_{i=1}^{k}|\delta_i-\gamma_i|<\delta$, so that

$$\sum_{i=1}^{k}|\theta_i|=\sum_{i=1}^{k}|f(\delta_i)-f(\gamma_i)|\leqslant\varepsilon,$$

and hence, proceeding to the limit if necessary, $\sum_i|\theta_i|\leqslant\varepsilon$. But since

$$f(E)\subset f(G)=f\left(\bigcup_i(\alpha_i,\beta_i)\right)=\bigcup_i f((\alpha_i,\beta_i))\subset\bigcup_i\theta_i,$$

therefore $m_1^*(f(E))\leqslant\varepsilon$. ■

We note that the variation, regarded as an interval function, is associated with the function $x \to W_a^x(f)$ (see also Th. 1.4.1). Hence, by Th. 7.3.1 and its Corollary, we have

THEOREM 7.4.2. *An absolutely continuous function on $[a, b]$ is a function of bounded variation on $[a, b]$. A function f of bounded variation on $[a, b]$ is absolutely continuous if and only if the function $x \to W_a^x(f)$ is absolutely continuous. An absolutely continuous function is the difference of two increasing and absolutely continuous functions.*[15]

COROLLARY. *An absolutely continuous function on an interval $[a, b]$ has a finite derivative almost everywhere which is summable on $[a, b]$.*

From Th. 6.2.8, we obtain

THEOREM 7.4.3. *If the function f is summable on the interval $[a, b]$, then*

$$x \to \int_a^x f(t)\,\mathrm{d}t$$

is an absolutely continuous function.

By Th. 7.3.3, we have

THEOREM 7.4.4. *If f is an absolutely continuous function on the interval $[a, b]$, then*

$$\int_a^b f'(x)\,\mathrm{d}x = f(b) - f(a). \tag{7.4.2}$$

Remark. The formula (7.4.2) holds (for any pair of numbers a and b in a given interval) only if the function f is absolutely continuous (for if $\int_a^x f'(t)\,\mathrm{d}t = f(x) - f(a)$ then by Th. 7.4.3, $f(x) = f(a) + \int_a^x f'(t)\,\mathrm{d}t$ is absolutely continuous).

A consequence of Th. 7.4.4 is

THEOREM 7.4.5. *Let f be an absolutely continuous function on the interval $[a, b]$. If $f'(x) \geqslant 0$ ($f'(x) \leqslant 0$) almost everywhere on $[a, b]$, then the function f is increasing (decreasing) on $[a, b]$. If $f'(x) = 0$ almost everywhere on $[a, b]$, then the function f is constant on $[a, b]$.*

THEOREM 7.4.6 (Integration by Parts). *If the functions F and G are absolutely continuous on the interval $[a, b]$, then*

$$\int_a^b F(x)G'(x)\,\mathrm{d}x = F(b)G(b) - F(a)G(a) - \int_a^b F'(x)G(x)\,\mathrm{d}x. \tag{7.4.3}$$

Indeed, the function FG is absolutely continuous and

$$(F(x)G(x))' = F'(x)G(x) + F(x)G'(x)$$

a.e. on $[a,b]$. Hence, by Th. 7.4.4 (using the summability of the functions $F'G$ and FG'),

$$F(b)G(b) - F(a)G(a) = \int_a^b (FG)'\,dx = \int_a^b F'G\,dx + \int_a^b FG'\,dx.$$

THEOREM 7.4.7. *If f is an increasing function which is absolutely continuous in the interval $[a,b]$, then*

$$m_1(f(E)) = \int_E f'(x)\,dx \tag{7.4.4}$$

for any measurable set $E \subset [a,b]$.

Proof. Note first that if E, F are disjoint sets contained in $[a,b]$ then $f(E) \cap f(F)$ is at most a countable set. This follows from the fact that this set is contained in the image of the set of rational numbers in $[a,b]$; for if $z \in f(E) \cap f(F)$ then $z = f(x) = f(y)$ where $x \in E$, $y \in F$, $x \neq y$, so that choosing a rational number w in the interval with endpoints x and y we have $z = f(w)$ because f is monotonic. Thus, the left member of (7.4.4), regarded as a function of the set E, is a measure on the algebra of measurable subsets of $[a,b]$. If, therefore, $\{E_i\}_{i=1}^{\infty}$ is a family of disjoint, measurable subsets of $[a,b]$ whose union is E, then since

$$m_1(f(E_i) \cap f(E_j)) = 0 \qquad \text{for } i \neq j,$$

therefore, by Th. 5.1.9.

$$m_1(f(E)) = m_1\left(\bigcup_{i=1}^{\infty} f(E_i)\right) = \sum_{i=1}^{\infty} m_1(f(E_i)).$$

The right-hand member of (7.4.4), regarded as a function of the set E, is also a measure on the algebra of measurable subsets of $[a,b]$ (Th. 6.1.3). By Th. 7.4.4, both measures coincide for intervals and hence also for sets which are open relative to $[a,b]$ (since such sets are unions of at most a countable number of disjoint intervals). By Th. 5.2.5, both measures coincide on $\mathscr{B}([a,b])$, that is, (7.4.4) holds if E is a Borel set contained in $[a,b]$. But any measurable set $E \subset [a,b]$ is, by Th. 5.4.2′, the union of a set J of type F_σ and a set Z of measure zero. Hence $f(E) = f(J) \cup f(Z)$, where $m_1(f(Z)) = 0$, by Th. 7.4.1, and so

$$m_1(f(E)) = m_1(f(J)) = \int_J f'(x)\,dx = \int_E f'(x)\,dx. \quad \blacksquare$$

THEOREM 7.4.8 (Integration by Substitution). *If f is a summable function on the interval $[a,b]$ and if φ is an increasing and absolutely continuous function on the interval $[\alpha,\beta]$, where $a = \varphi(\alpha)$ and $b = \varphi(\beta)$, then the function $t \to f(\varphi(t))\varphi'(t)$*

is summable[16] *on* $[\alpha, \beta]$ *and*

$$\int_a^b f(x)\,dx = \int_\alpha^\beta f(\varphi(t))\varphi'(t)\,dt.^{17} \tag{7.4.5}$$

More generally, for any measurable set $A \subset [\alpha, \beta]$[18]

$$\int_{\varphi(A)} f(x)\,dx = \int_A f(\varphi(t))\varphi'(t)\,dt. \tag{7.4.6}$$

Proof. It suffices to prove (7.4.6) for the case in which $f = \chi_E$, where E is any measurable subset of $[a, b]$. Indeed, if the result is true in this case then it is true for any simple, measurable function on $[a, b]$, for this is a linear combination of characteristic functions. Hence also, by Th. 6.1.5, it is true for any summable, non-negative function on $[a, b]$, for such a function is the limit of an increasing sequence of simple, measurable, non-negative functions (Th. 4.4.10). Hence, finally, it is true for any summable function on $[a, b]$, for such a function is the difference of two summable, non-negative functions on $[a, b]$.

Thus, let $f = \chi_E$, $E \subset [a, b]$. We note that from the validity of the theorem when E has measure zero, it follows that it is true for any set contained in E, for then both functions being integrated are equal to zero almost everywhere. Hence, by Ths 5.4.2–5.4.2′, it suffices to consider the case in which E is a Borel set. But then, by Th. 4.3.4, $\varphi^{-1}(E)$ is measurable and (7.4.6) takes the form

$$m_1(E \cap \varphi(A)) = \int_{A \cap \varphi^{-1}(E)} \varphi'\,dt,$$

which is true by Th. 7.4.7, because

$$\varphi(A \cap \varphi^{-1}(E)) = \varphi(A) \cap E. \quad \blacksquare$$

A function h of bounded variation on $[a, b]$ is called *singular* if $h'(x) = 0$ a.e. on $[a, b]$. By Th. 7.3.4, we have

THEOREM 7.4.9 (Canonical Lebesgue Decomposition). *Every function* $f(x)$ *of bounded variation on the interval* $[a, b]$ *can be expressed as the sum of an absolutely continuous function and a singular function,* e.g.

$$f(x) = \int_a^x f'(x)\,dx + h(x). \tag{7.4.7}$$

Moreover, both terms in such a decomposition are uniquely defined up to an additive constant.[19]

Remark. When f is increasing then both terms in the decomposition are increasing.

We now give an example of a continuous, increasing, singular function which is not constant. Consider the *Cantor Ternary Set*

$$C = [0,1]\backslash\Delta_1^{(1)}\backslash\cdots\backslash\bigcup_{i=1}^{2^{\nu-1}}\Delta_i^{(\nu)}\backslash\ldots,$$

where the open intervals $\Delta_i^{(\nu)}$ are defined as follows. Let $\theta_1^{(0)} = [0,1]$; having defined the closed intervals $\theta_1^{(\nu)},\ldots,\theta_{2^\nu}^{(\nu)}$, we decompose each of them into three intervals of equal length

$$\theta_i^{(\nu)} = \theta_{2i-1}^{(\nu+1)} \cup \Delta_i^{(\nu+1)} \cup \theta_{2i}^{(\nu+1)},$$

where $\Delta_i^{(\nu+1)}$ is the middle interval and is open. Since $|\Delta_i^{(\nu)}| = 1/3^\nu$, therefore

$$m(C) = 1 - \tfrac{1}{3} - \cdots - \frac{2^{\nu-1}}{3^\nu} - \cdots = 0.$$

Let us now define a sequence of functions f by induction. Let $f_0(x) = x$ in $[0,1]$. Having defined the function f_ν (continuous, constant on the intervals $\Delta_1^{(1)},\ldots,\Delta_1^{(\nu)},\ldots,\Delta_{2^\nu-1}^{(\nu)}$ and linear in the intervals $\theta_1^{(\nu)},\ldots,\theta_{2^\nu}^{(\nu)}$), we modify it on the interior of each of the intervals $\theta_i^{(\nu)}$ so that on $\Delta_i^{(\nu+1)}$ it takes a constant value equal to the arithmetic mean of the values of f_ν at the end points of $\theta_i^{(\nu)}$, and so that it is linear on $\theta_{2i-1}^{(\nu+1)}$ and $\theta_{2i}^{(\nu+1)}$. In this way we obtain the function $f_{\nu+1}$. It follows from this construction that the difference of the values of the function f_ν at the endpoints of the intervals $\theta_i^{(\nu)}$ equals $1/2^\nu$. Hence

$$|f_{\nu+1}(x) - f_\nu(x)| < \frac{1}{2^\nu} \quad \text{in } [0,1].$$

It follows that the sequence f_ν is uniformly convergent to a function f. The function f is therefore continuous, increasing and such that $f(0) = 0, f(1) = 1$. But f is constant on the intervals $\Delta_i^{(\nu)}$ and so $f'(x) = 0$ on $[0,1]\backslash C$. Thus f is a singular function.

From Th. 7.3.5, we have

THEOREM 7.4.10. *Let f is a continuous (finite) function on the interval $[a,b]$ and if g is an absolutely continuous function on this interval, then*

$$\int_a^b f(x)\,\mathrm{d}g(x) = \int_a^b f(x)g'(x)\,\mathrm{d}x.^{20} \tag{7.4.8}$$

We now establish *Fubini's theorem on the differentiation of series.*

LEMMA (Riesz). *Let φ_ν be a sequence of functions which are non-negative and increasing in the interval $[a,b]$. Let $\sum_{\nu=1}^\infty \varphi_\nu(b) < \infty$. Then $\varphi'_\nu(x) \to 0$ almost everywhere in $[a,b]$.*

Proof. The function

$$\varphi(x) = \sum_{\nu=1}^\infty \varphi_\nu(x)$$

is increasing and bounded on $[a, b]$. The derivatives $\varphi'(x), \varphi_1'(x), \varphi_2'(x), \ldots$ exist and are bounded on $[a, b]\backslash Z$, where $m_1(Z) = 0$. Because the function

$$\varphi(x) - \sum_{\nu=1}^{n} \varphi_\nu(x) = \sum_{\nu=n+1}^{\infty} \varphi_\nu(x)$$

is increasing, we have, for $x \in [a, b]\backslash Z$ that

$$\left(\varphi(x) - \sum_{\nu=1}^{n} \varphi_\nu(x)\right)' = \varphi'(x) - \sum_{\nu=1}^{n} \varphi_\nu'(x) \geqslant 0,$$

that is $\sum_{\nu=1}^{n} \varphi_\nu'(x) \leqslant \varphi'(x) < \infty$. Therefore $\varphi_\nu'(x) \to 0$. ■

THEOREM 7.4.11 (Fubini). *If f_ν is a sequence of non-negative, increasing functions on $[a, b]$ and $\sum_{\nu=1}^{\infty} f_\nu(b) < \infty$, then*

$$\left(\sum_{\nu=1}^{\infty} f_\nu(x)\right)' = \sum_{\nu=1}^{\infty} f_\nu'(x)$$

almost everywhere on $[a, b]$.

Proof. The functions

$$f(x) = \sum_{\nu=1}^{\infty} f_\nu(x),$$

$$\varphi_n(x) = f(x) - \sum_{\nu=1}^{n} f_\nu(x) = \sum_{\nu=n+1}^{\infty} f_\nu(x) \qquad (n = 1, 2, \ldots)$$

are bounded, non-negative and increasing on $[a, b]$ and therefore have finite derivatives on $[a, b]\backslash Z$, where $m_1(Z) = 0$. Since

$$\varphi_n(x) - \varphi_{n+1}(x) = f_{n+1}(x).$$

therefore, for $x \in [a, b]\backslash Z$ we have

$$\varphi_n'(x) - \varphi_{n+1}'(x) = f_{n+1}'(x) \geqslant 0.$$

Thus the sequence $\varphi_n'(x)$ is decreasing and hence convergent on $[a, b]\backslash Z$. From the sequence $\varphi_n(b)$, which converges to zero, we can choose a subsequence $\varphi_{\alpha_n}(b)$ such that $\varphi_{\alpha_n}(b) < 1/2^n$. Then

$$\sum_{\nu=1}^{\infty} \varphi_{\alpha_\nu}(b) < \infty$$

so that by the lemma $\varphi_{\alpha_\nu}'(x) \to 0$ on $[a, b]\backslash Z_1$, where $m_1(Z_1) = 0$. Hence, for $x \in [a, b]\backslash Z_1\backslash Z$, $\varphi_\nu'(x) \to 0$, that is

$$f'(x) - \sum_{\nu=1}^{n} f_\nu'(x) \to 0. \quad ■$$

We now give an example of a singular function which is continuous and *strictly* increasing.

Let $[u_n, v_n]$ be a sequence of all the intervals with rational endpoints which are contained in $[0, 1]$. By the construction used in the previous example (using a suitable linear transformation and multiplying by $1/2^n$), there exists a function f_n which is singular, continuous and increasing in $[u_n, v_n]$ and is such that $f_n(u_n) = 0, f_n(v_n) = 1/2^n$. Then, the function

$$g_n(x) = \begin{cases} 0 & \text{for } x \leqslant u_n, \\ f_n(x) & \text{in } [u_n, v_n], \\ 1/2^n & \text{for } x \geqslant v_n, \end{cases}$$

is continuous and increasing and $g_n'(x) = 0$ a.e. in $[0, 1]$. But then the function

$$g(x) = \sum_{n=1}^{\infty} g_n(x)$$

is continuous, strictly increasing and, by Th. 7.4.11, $g'(x) = 0$ a.e. in $[0, 1]$. This is therefore a singular function satisfying the required conditions.

Arc Length

Consider a rectifiable curve with equations

$$x = x(t), \quad y = y(t), \quad \alpha \leqslant t \leqslant \beta,$$

where the functions x, y are continuous on $[\alpha, \beta]$. By Th. 1.3.6, x, y are functions of bounded variation. Let

$$c(t, t') = \sqrt{\{[x(t') - x(t)]^2 + [y(t') - y(t)]^2\}}.$$

By the triangle inequality

$$c(t, t'') \leqslant c(t, t') + c(t', t''). \tag{7.4.9}$$

The length L of the curve is the supremum of the polygonal sums $\sum_{i=1}^{k} c(t_{i-1}, t_i)$ for subdivisions $\alpha = t_0 < \cdots < t_k = \beta$. By (7.4.9), a further subdivision can only lead to an increase in the polygonal sum. We show that for any $\varepsilon > 0$ there exists $\delta > 0$ such that

$$L - \varepsilon < \sum_{i=1}^{k} c(t_{i-1}, t_i) \qquad \text{if } \max_i (t_i - t_{i-1}) < \delta. \tag{7.4.10}$$

Indeed, let $\varepsilon > 0$. There exists a subdivision $\alpha = \bar{t}_0 < \cdots < \bar{t}_p = \beta$ such that

$$L - \tfrac{1}{2}\varepsilon < \sum_{j=1}^{p} c(\bar{t}_{j-1}, \bar{t}_j).$$

There exists $\delta > 0$ such that if $|t'' - t'| < \delta$ then $c(t', t'') < \varepsilon/4p$. Let $\alpha = t_0 < \cdots < t_k = \beta$ be a subdivision such that $\max_i(t_i - t_{i-1}) < \delta$. Adding to this subdivision the points $\bar{t}_j$ (distinct from t_i) we obtain a subdivision $\alpha = t_0' < \cdots < t_l' = \beta$ (which is a subdivision of the previous one), for which

$$\sum_{\nu=1}^{l} c(t_{\nu-1}', t_\nu') \leqslant \sum_{i=1}^{k} c(t_{i-1}, t_i) + \tfrac{1}{2}\varepsilon,$$

for the sum of those $c(t'_{\nu-1}, t'_\nu)$ for which $t'_{\nu-1}$ or t'_ν is one of the additional points does not exceed $2p(\varepsilon/4p) = \frac{1}{2}\varepsilon$. Then

$$L - \tfrac{1}{2}\varepsilon < \sum_{j=1}^{p} c(\bar{t}_{j-1}, \bar{t}_j) \leqslant \sum_{\nu=1}^{l} c(t'_{\nu-1}, t'_\nu) \leqslant \sum_{i=1}^{k} c(t_{i-1}, t_i) + \tfrac{1}{2}\varepsilon$$

which gives inequality (7.4.10)

It follows from property (7.4.10) that the polygonal sums tend to the length of the curve if $\max_i (t_i - t_{i-1}) \to 0$. Consequently

$$L(t_1, t_3) = L(t_1, t_2) + L(t_2, t_3) \qquad \text{if } t_1 < t_2 < t_3$$

where $L(t', t'')$ denotes the length of the curve corresponding to the interval $t' \leqslant t \leqslant t''$. If, therefore, we put $s(t) = L(\alpha, t)$, then $s(t'') - s(t') = L(t', t'')$. We have

$$c(t', t'') \leqslant s(t'') - s(t') \qquad (t' < t'') \tag{7.4.11}$$

(the length of a chord does not exceed the length of the curve). From inequalities (1.3.9), we have

$$\left.\begin{matrix} W_{t'}^{t''}(x) \\ W_{t'}^{t''}(y) \end{matrix}\right\} \leqslant s(t'') - s(t') < W_{t'}^{t''}(x) + W_{t'}^{t''}(y).$$

It follows that the absolute continuity of the function s in $[\alpha, \beta]$ is equivalent to the absolute continuity of the functions $t \to W_\alpha^t(x)$ and $t \to W_\alpha^t(y)$, that is, by Th. 7.4.2, to the absolute continuity of the functions x and y in $[\alpha, \beta]$.

THEOREM 7.4.12 (Tonelli). *If the curve $x = x(t)$, $y = y(t)$, $\alpha \leqslant t \leqslant \beta$ is rectifiable, then*

$$s'(t) = \sqrt{\{[x'(t)]^2 + [y'(t)]^2\}} \tag{7.4.12}$$

almost everywhere in $[\alpha, \beta]$. Then

$$L \geqslant \int_\alpha^\beta \sqrt{\{[x'(t)]^2 + [y'(t)]^2\}}\, dt \tag{7.4.13}$$

and equality holds if and only if x and y are absolutely continuous on $[\alpha, \beta]$.[21]

Proof. Let $\delta = \delta_n$ be a number corresponding to $\varepsilon = 1/2^n$ in (7.4.10). For the function

$$\varphi_n(t) = \sup\left\{ s(t) - \sum_{i=1}^{k} c(t_{i-1}, t_i) : \alpha = t_0 < \cdots < t_k = t, \max(t_i - t_{i-1}) < \delta_n \right\} \tag{7.4.14}$$

we have therefore

$$\varphi_n(\beta) \leqslant 1/2^n. \tag{7.4.15}$$

Let $0 < t' - t < \delta_n$. If $\alpha = t_0 < \cdots < t_k = t$ and $\max_i(t_i - t_{i-1}) < \delta_n$, then

$$s(t) - \sum_{i=1}^{k} c(t_{i-1}, t_i) + s(t') - s(t) - c(t, t')$$

$$= s(t') - \left(\sum_{i=1}^{k} c(t_{i-1}, t_i) + c(t, t') \right) \leqslant \varphi_n(t'),$$

whence, taking the supremum on the left-hand side

$$\varphi_n(t) + s(t') - s(t) - c(t, t') \leqslant \varphi_n(t').$$

Hence, using (7.4.11),

$$0 \leqslant s(t') - s(t) - c(t, t') \leqslant \varphi_n(t') - \varphi_n(t) \quad \text{if } 0 < t' - t < \delta_n. \tag{7.4.16}$$

Thus the functions φ_n are increasing and non-negative so that by (7.4.15) and Riesz's lemma, $\varphi_n'(t) \to 0$ a.e. in $[\alpha, \beta]$. Since $s'(t), x'(t), y'(t)$ exist almost everywhere in $[\alpha, \beta]$, then equally

$$\lim_{t' \to t+} \left(\frac{s(t') - s(t)}{t' - t} - \frac{c(t, t')}{t' - t} \right) = s'(t) - \sqrt{\{[x'(t)]^2 + [y'(t)]^2\}}$$

and, by (7.4.16), $0 \leqslant s'(t) - \sqrt{\{[x'(t)]^2 + [y'(t)]^2\}} \leqslant \varphi_n'(t)$. Now, taking the limit as $n \to \infty$ we obtain the equality (7.4.12) a.e. in $[\alpha, \beta]$.

By Th. 7.4.9, we have

$$s(t) = \int_\alpha^t s'(t)\,\mathrm{d}t + h(t)$$

where $h(t)$ is a singular function which is increasing and for which obviously $h(\alpha) = 0$. It now follows that (7.4.12) implies (7.4.13) and that equality occurs if and only if $h(\beta) = 0$, that is if $h(t) = 0$ in $[\alpha, \beta]$ or, equivalently, if $s(t)$ is absolutely continuous in $[\alpha, \beta]$. This last condition is, in turn, equivalent to the condition that x and y be absolutely continuous in $[\alpha, \beta]$. ■

Dini Derivatives

Let f be a finite function and let x_0 be a number in the domain of the function. The limits

$$D^+ f(x_0) = \limsup_{x \to x_0+} \frac{f(x) - f(x_0)}{x - x_0}, \quad D_+ f(x_0) = \liminf_{x \to x_0+} \frac{f(x) - f(x_0)}{x - x_0}$$

$$D^- f(x_0) = \limsup_{x \to x_0-} \frac{f(x) - f(x_0)}{x - x_0}, \quad D_- f(x_0) = \liminf_{x \to x_0-} \frac{f(x) - f(x_0)}{x - x_0}$$

are called the right upper, right lower, left upper and left lower *Dini derivatives* respectively. The right (left) derivatives are always well defined if x_0 is a right (left) accumulation point of the domain of the function. The right-hand derivative

$$f'_+(x_0) = \lim_{x \to x_0+} \frac{f(x) - f(x_0)}{x - x_0}$$

exists if and only if $D_+ f(x_0) = D^+ f(x_0)$. A similar assertion holds also for left-hand derivatives. Obviously $D_+ f \leqslant D^+ f$ and $D_- f \leqslant D^- f$.

A deeper relation between Dini derivatives is given by the

THEOREM (Denjoy–Young–Saks). *$D_+ f = D^- f$ (finite) almost everywhere on the sets $\{D_+ f > -\infty\}$ and $\{D^- f < \infty\}$ (and similarly for the pair $D_- f$ and $D^+ f$).*

Simple consequences of this theorem are:

(1) $m(\{|f'_+| = \infty\}) = m(\{|f'_-| = \infty\}) = 0$,
(2) f is differentiable almost everywhere on the set of points at which both right-hand (or both left-hand) derivatives are finite,
(3) monotone functions are differentiable almost everywhere.

We also have the

THEOREM (Sierpiński–Young). *The sets $\{D^- f < D_+ f\}$ and $\{D^+ f < D_- f\}$ are at most countable.*[22]

In what follows we will study only right-hand derivatives; analogous properties of left-hand derivatives can be obtained by replacing the function f by the function $g(x) = -f(-x)$. For in this case

$$D^- f(x_0) = D^+ g(-x_0) \quad \text{and} \quad D_- f(x_0) = D_+ g(-x_0).$$

From the properties of limits (see (0.2.9) and (0.2.12)) it follows that

$$D^+(-f) = -D_+ f \tag{7.4.17}$$

and

$$\begin{aligned} D^+(f+g)(x_0) &= D^+ f(x_0) + g'_+(x_0), \\ D_+(f+g)(x_0) &= D_+ f(x_0) + g'_+(x_0), \end{aligned} \tag{7.4.18}$$

if $g'_+(x_0)$ exists and if the right-hand side is not of the form $\pm\infty \mp \infty$.

Next we have

$$D^+(f+\gamma) \geqslant D^+ f \quad \text{and} \quad D_+(f+\gamma) \geqslant D_+ f, \tag{7.4.19}$$

if γ is an increasing function. For in this case, if $x > x_0$,

$$[f(x) + \gamma(x)] - [f(x_0) + \gamma(x_0)] \geqslant f(x) - f(x_0),$$

so that after dividing by $x - x_0$ and taking the supremum (infimum) as $x \to x_0+$, we obtain the inequalities (7.4.19).

LEMMA (Zygmund). *If the function f is continuous in the interval $[a, b]$ and*

$D^+f(x) > 0$ in $[a,b]\setminus Z$, where $f(Z)$ does not contain any interior points, then the function f is increasing.

Proof. Suppose that the function f is not increasing. Then, there exist $c, d \in [a,b]$ such that $c < d$ and $f(c) > f(d)$. To obtain a contradiction we show that

$$(f(d), f(c)) \subset f(Z).$$

Thus, let $f(d) < y < f(c)$. The set $\{x : f(x) = y, x \leqslant d\}$ is non-empty and compact and therefore has a maximum x_0. Since $f(x_0) = y$ and $f(d) < y$, therefore $x_0 < d$ and $f(x) < f(x_0)$ in $(x_0, d]$. Hence

$$D^+f(x_0) = \limsup_{x \to x_0+} \frac{f(x) - f(x_0)}{x - x_0} \leqslant 0$$

so that $x_0 \in Z$. Thus $y = f(x_0) \in f(Z)$. ■

THEOREM 7.4.13. *If f is continuous on $[a,b]$ and $D^+f(x) \geqslant 0$ except for at most a countable set in $[a,b]$, then f is increasing on $[a,b]$.*

Proof. Let $g_n(x) = f(x) + x/n$. By (7.4.18), $D^+g_n(x) = D^+f(x) + 1/n > 0$ on $[a,b]\setminus Z$, where Z is at most countable. Since $g_n(Z)$ is also at most countable, therefore, by Zygmund's lemma, g_n is increasing. Thus $f(x) = \lim_{n \to \infty} g_n(x)$ is also increasing. ■

COROLLARY. *Let f be continuous on $[a,b]$. If $D_+f(x) \leqslant 0$ except for at most a countable set in $[a,b]$, then $f(x)$ is decreasing. If $D_+f(x) \leqslant 0 \leqslant D^+f(x)$ except for at most a countable set, then f is constant.*

THEOREM 7.4.14. *If the function f is continuous on $[a,b]$, $D^+f(x) > -\infty$ except for at most a countable set in $[a,b]$ and $D^+f(x) \geqslant 0$ almost everywhere in $[a,b]$, then f is increasing.*

LEMMA. *If $E \subset [a,b]$ and $m_1(E) = 0$, then there exists a finite function γ which is continuous,[23] increasing in $[a,b]$ and such that $\gamma'(x) = \infty$ for every $x \in E$.*

Proof of lemma. By lemma 5.4.3 or Th. 5.4.2, there exists a sequence of open sets G_n such that $E \subset G_n$ and $m_1(G_n) < 1/2^n$. Moreover, we can require that this sequence be decreasing. Then, the function

$$\psi(t) = \sum_{n=1}^{\infty} \chi_{G_n}(t)$$

is summable on $[a,b]$, for

$$\int_a^b \psi(t)\,dt = \sum_{n=1}^{\infty} \int_a^b \chi_{G_n}(t)\,dt \leqslant \sum_{n=1}^{\infty} \frac{1}{2^n} < \infty.$$

Let

$$\gamma(x) = \int_a^x \psi(t)\,dt$$

and let $x_0 \in E$. Let N be any natural number. Since $x_0 \in G_N$ there exists δ such that $(x_0 - \delta, x_0 + \delta) \subset G_N \subset \cdots \subset G_1$. It follows that $\psi(t) \geqslant N$ in the interval $(x_0 - \delta, x_0 + \delta)$. Thus, if $0 < |x - x_0| < \delta$, then

$$\frac{\gamma(x) - \gamma(x_0)}{x - x_0} = \frac{1}{x - x_0}\int_{x_0}^x \psi(t)\,dt \geqslant N.$$

Thus $\gamma'(x_0) = \infty$. ■

Proof of theorem. The set $Z = \{x : D^+ f(x) = -\infty\}$ is at most countable and the set $E = \{x : D^+ f(x) < 0\}$ has measure zero. Let γ be a function corresponding to the set E as in the lemma. The function

$$g_n = f + \frac{1}{n}\gamma$$

satisfies the hypotheses of Th. 7.4.13. Indeed, the function is continuous. Let $x_0 \in [a, b) \backslash Z$, then $D^+ f(x_0) > -\infty$ and if $x_0 \in E$ then $\gamma'(x_0) = \infty$ and so by (7.4.18) $D^+ g_n(x_0) = \infty$. But if $x_0 \in \backslash E$ then, by (7.4.19) $D^+ g_n(x_0) \geqslant D^+ f(x_0) \geqslant 0$. Hence by Th. 7.4.13, the function g_n is increasing and so, therefore, is the function $f = \lim_{n \to \infty} g_n$. ■

THEOREM 7.4.15. (Lebesgue)[(24)] *Let f be a continuous function on the interval $[a, b]$. If*

$$D_+ f \leqslant \lambda \leqslant D^+ f \qquad (\text{or } D_- f \leqslant \lambda \leqslant D^- f) \tag{7.4.20}$$

for some function λ[25] *which is summable on $[a, b]$ and finite except on at most a countable set in $[a, b]$, then the function f is absolutely continuous on $[a, b]$.*

Proof. It clearly suffices to confine the argument to right-hand derivatives. In order to prove the absolute continuity of f it suffices. by Th. 7.4.3, to show that

$$|f(t) - f(t_0)| \leqslant \int_{t_0}^t |\lambda(t)|\,dt \qquad \text{for } a \leqslant t_0 < t \leqslant b.$$

If, therefore, we write

$$\varphi(t) = |f(t) - f(t_0)| \quad \text{and} \quad \psi(t) = \int_{t_0}^t |\lambda(y)|\,dy,$$

it suffices to show that the function $\psi - \varphi$ is increasing in $[t_0, b]$ for $\psi(t_0) - \varphi(t_0) = 0$.

Now, we have the inequality

$$D_+ \varphi(t) \leqslant |\lambda(t)| \qquad \text{if } a \leqslant t < b. \tag{7.4.21}$$

Indeed, taking a sequence $t_\nu \to t^+$ such that

$$\frac{f(t_\nu) - f(t)}{t_\nu - t} \to \lambda(t)^{26}$$

we have $\varphi(t_\nu) - \varphi(t) = |f(t_\nu) - f(t_0)| - |f(t) - f(t_0)| \leqslant |f(t_\nu) - f(t)|$. Now, dividing by $t_\nu - t$ and tending to the limit, we obtain inequality (7.4.21). Since ψ is increasing, therefore by (7.4.19) and (7.4.17), $D^+(\psi - \varphi) \geqslant D^+(-\varphi) = -D_+\varphi$ and so by (7.4.21) we obtain

$$D^+(\psi - \varphi) > -\infty$$

with the exception of at most a countable set in $[a, b]$. If $\psi'(t_t)$ exists and is finite, then by (7.4.18) and (7.4.17) $D^+(\psi - \varphi)(t_t) = \psi'(t_t) - D_+\varphi(t_t)$ and since $\psi'(t) = |\lambda(t)| < \infty$ a.e. in $[a, b]$, therefore, by (7.4.21),

$$D^+(\psi - \varphi) \geqslant 0$$

a.e. in $[a, b]$. Moreover the function $\psi - \varphi$ is continuous. We have therefore shown that the function $\psi - \varphi$ satisfies the hypotheses of Th. 7.4.14 and is therefore increasing in $[a, b]$. ■

By Th. 7.4.15 and the corollary to Th. 7.4.2, we can assert that *in the class of differentiable*[27] *functions on* $[a, b]$ *or, more generally, in the class of functions which are continuous and differentiable except on at most a countable set, absolute continuity is equivalent to the summability of the derivative*[28] *and then*

$$\int_a^b f'(x)\,dx = f(b) - f(a).$$

Non-differentiable Continuous Functions

In the set of all real-valued functions of a real variable let there be defined a 'convergence' s (a relation in the set of pairs $(\{f_n\}, f)$, which we denote by $f_n \underset{s}{\to} f$), satisfying the conditions:

(1) if $f_n \underset{s}{\to} f$ and $f_n \underset{s}{\to} g$ then $f = g$;

(2) if $f_n \underset{s}{\to} f$ then $f_{k_n} \underset{s}{\to} f$;

(3) if $f_n \underset{s}{\to} f$ and $g_n \underset{s}{\to} g$ then $af_n + bg_n \underset{s}{\to} af + bg$;

(4) if f_n are linear (non-homogeneous) and $f_n \underset{s}{\to} f$, then f is linear (non-homogeneous);

(5) $f_n \underset{s}{\to} f$ if f_n, f are continuous and $f_n \to f$ almost uniformly on $\mathbb{R}$, that is, uniformly on every interval $(-c, c)$ after the exclusion of a finite number of terms (depending on c).

For a given function f we define the *s-derivative* $f'_s(a)$ at the point a by the condition

$$\frac{f(a+h_n x)-f(a)}{h_n} \underset{s}{\rightarrow} f_s'(a)x, \quad \text{if only} \quad 0 \neq h_n \rightarrow 0.$$

In particular, if s is almost uniform convergence on $\mathbb{R}$, the s-derivative at a is the derivative in the ordinary sense of a function defined in the neighbourhood of a.

Let p be a function which has a continuous (ordinary) derivative in $\mathbb{R}$ and let p be periodic and non-constant. Also, let $h_n > 0$ be a sequence such that $\sum_1^\infty h_n < \infty$. Then, the function

$$f(x) = \sum_{n=1}^{\infty} h_n p\left(\frac{x}{h_n}\right)$$

is continuous in $\mathbb{R}$. Let ω be a function which is increasing in $[0, \infty)$ and is such that

$$\lim_{x\to 0} \omega(x) = 0 \quad \text{and} \quad \lim_{x\to 0} \frac{\omega(x)}{x} = \infty.$$

We prove the following

THEOREM 7.4.16. *Using a suitable sequence h_n the function f is not s-differentiable at any point of $\mathbb{R}$. Moreover, we can require that ω be the 'modulus of continuity' of f, that is, that $|f(y)-f(x)| \leqslant \omega(|x-y|)$ for $x, y \in \mathbb{R}$.*

Proof. Writing

$$g_n(x) = \sum_{i=1}^{n-1} h_i p\left(\frac{x}{h_i}\right) \quad \text{and} \quad r_n(x) = \sum_{i=n+1}^{\infty} h_i p\left(\frac{x}{h_i}\right)$$

we have

$$g_n(x) = f(x) - h_n p\left(\frac{x}{h_n}\right) - r_n(x),$$

and so for $a \in \mathbb{R}$,

$$\frac{g_n(a+h_n x)-g_n(a)}{h_n} = \frac{f(a+h_n x)-f(a)}{h_n} - \left(p\left(\frac{a}{h_n}+x\right) - p\left(\frac{a}{h_n}\right)\right) - \frac{r_n(a+h_n x)-r_n(a)}{h_n}. \tag{7.4.22}$$

Choose $h_n > 0$ so that

$$\left|\frac{g_n(a+u)-g_n(a)}{u} - g_n'(a)\right| \leqslant \frac{1}{n^2} \qquad \text{for } 0<|u|\leqslant nh_n \text{ and } a\in\mathbb{R}$$

and such that $\sum_{n+1}^\infty h_i < (1/n)h_n$. Then, for arbitrary a

$$\left|\frac{g_n(a+h_n x)-g_n(a)}{h_n} - g_n'(a)x\right| \leqslant \frac{1}{n} \qquad \text{for } |x|\leqslant n$$

and

$$\left|\frac{r_n(a+h_nx)-r_n(a)}{h_n}\right| \leqslant \frac{2M}{n} \qquad \text{for } x \in \mathbb{R}$$

where $M = \sup|p(x)|$. Hence, correspondingly,

$$\frac{g_n(a+h_nx)-g_n(a)}{h_n} - g'_n(a)x \underset{s}{\rightarrow} 0, \quad \frac{r_n(a+h_nx)-r_n(a)}{h_n} \underset{s}{\rightarrow} 0$$

for any a. Suppose that the s-derivative $f'_s(a)$ exists at some point a. By the periodicity of p we have

$$p\left(\frac{a}{h_n}+x\right) - p\left(\frac{a}{h_n}\right) = p(c_n+x) - p(c_n),$$

for some c_n forming a bounded sequence. Choosing a subsequence $c_{k_n} \rightarrow c$ we have

$$p\left(\frac{a}{h_{k_n}}+x\right) - p\left(\frac{a}{h_{k_n}}\right) \underset{s}{\rightarrow} p(c+x) - p(c).$$

Hence, by (7.4.22)

$$g'_{k_n}(a)x \underset{s}{\rightarrow} f'_s(a)x - (p(c+x) - p(c))$$

from which it would follow that p is linear (non-homogeneous) contrary to hypothesis.

To obtain the inequality $|f(x+u) - f(x)| \leqslant \omega(|u|)$ it suffices to choose h_n such that

$$2^n L \leqslant v(h_n T),$$

where

$$L = \sup|p'(x)|, \quad v(x) = \inf_{0<t\leqslant x} \frac{\omega(t)}{t}$$

and T is the period of p.

Since $sv(s) \leqslant \omega(s)$, we have then in the case $|u| \leqslant h_n T$,

$$\left|h_n p\left(\frac{x+u}{h_n}\right) - h_n p\left(\frac{x}{h_n}\right)\right| \leqslant L|u| \leqslant 2^{-n}|u|v(|u|) \leqslant 2^{-n}\omega(|u|)$$

and in the case $|u| \geqslant h_n T$,

$$\left|h_n p\left(\frac{x+u}{h_n}\right) - h_n p\left(\frac{x}{h_n}\right)\right| \leqslant L h_n T \leqslant 2^{-n} h_n T v(h_n T) \leqslant 2^{-n}\omega(h_n T) \leqslant 2^{-n}\omega(|u|). \quad \blacksquare$$

7.5 COUNTABLY ADDITIVE SET FUNCTIONS

Let $\mathscr{K}$ be a class of sets containing the empty set. A function λ defined on the class $\mathscr{K}$ is called *countably additive* if it satisfies the following conditions:

(1°) $\lambda(\varnothing)=0$;
(2°) $\lambda(E)>-\infty$ always, or $\lambda(E)<\infty$ always;
(3°) $\lambda(\bigcup_{i=1}^{\infty}E_i)=\sum_{i=1}^{\infty}\lambda(E_i)$, if $\bigcup_{i=1}^{\infty}E_i\in\mathscr{K}$ and the E_i are disjoint sets in $\mathscr{K}$.[29]

In particular $\lambda(E)=\int_E f\,d\mu$ is a countably additive function (on the class of sets E for which the integral exists).

Conditions (1°) and (3°) together imply finite additivity. Condition (3°) implies the unconditional convergence of the series $\sum_{i=1}^{\infty}\lambda(E_i)$ and hence also absolute convergence:

$$\sum_{i=1}^{\infty}|\lambda(E_i)|<\infty \quad \text{only if} \quad \left|\lambda\left(\bigcup_{i=1}^{\infty}E_i\right)\right|<\infty.$$

A linear combination $\alpha\lambda+\beta\nu$ of countably additive functions λ and ν is also a countably additive function, provided $\alpha\lambda>-\infty$ and $\beta\nu>-\infty$ always or $\alpha\lambda<\infty$ and $\beta\nu<\infty$ always.

In what follows we will study countably additive functions defined on a countably additive algebra $\mathscr{S}$ whose sets will be called measurable. Then, if E is measurable then $|\lambda(E)|<\infty$ implies that $|\lambda(F)|<\infty$ for every measurable $F\subset E$ (for then $\lambda(E)=\lambda(F)+\lambda(E\backslash F)$ and so $|\lambda(F)|<\infty$).

In particular, a measure is a non-negative countably additive function defined on a countably additive algebra.

THEOREM 7.5.1 (Hahn). *If λ is a countably additive function defined on a countably additive algebra $\mathscr{S}$ of a space X, then there exists a decomposition*

$$X=A\cup B,\quad A\cap B=\varnothing$$

(the Hahn decomposition for the function λ) such that if $E\in\mathscr{S}$ then $A\cap E$, $B\cap E\in\mathscr{S}$ and

$$\lambda(A\cap E)\geqslant 0,\quad \lambda(B\cap E)\leqslant 0.$$

Proof. Suppose, for definiteness, that $\lambda>-\infty$ always and denote by $\mathscr{N}$ the class of all measurable sets E such that $\lambda(F)\leqslant 0$ if $F\subset E$. We show that every measurable set E such that $\lambda(E)<0$ contains some set $F\in\mathscr{N}$ such that $\lambda(F)<0$.

Suppose then that the set E does not contain any such set. We define a sequence E_n by induction. We have

$$\varepsilon_1=\sup_{H\subset E}\lambda(H)>0$$

(for otherwise $E\in\mathscr{N}$). Then, there exists a (measurable) set $E_1\subset E$ such that

$$\lambda(E_1)>\min(1,\tfrac{1}{2}\varepsilon_1)>0.$$

If $E_1,\ldots,E_{n-1}$, contained in E, have already been defined, are disjoint and are such that $\lambda(E_1)>0,\ldots,\lambda(E_{n-1})>0$, then

$$\lambda(E\backslash E_1\backslash\cdots\backslash E_{n-1})<\lambda(E\backslash E_1\backslash\cdots\backslash E_{n-1})+\lambda(E_1)+\cdots+\lambda(E_{n-1})=\lambda(E)<0,$$

therefore

$$\varepsilon_n = \sup\{\lambda(H): H \subset E \backslash E_1 \backslash \cdots \backslash E_{n-1}\} > 0$$

(for otherwise $E \backslash E_1 \backslash \cdots \backslash E_{n-1} \in \mathcal{N}$). Thus, there is a measurable set $E_n \subset E \backslash E_1 \backslash \cdots \backslash E_{n-1}$ such that

$$\lambda(E_n) > \min(1, \tfrac{1}{2}\varepsilon_n).$$

Let $F = E \backslash \bigcup_{n=1}^{\infty} E_n$. Since the E_n are disjoint, therefore

$$\lambda(E) = \lambda(F) + \sum_{n=1}^{\infty} \lambda(E_n) \tag{7.5.1}$$

and so, since $|\lambda(E)| < \infty$ we must have $\lambda(E_n) \to 0$ and so $\varepsilon_n \to 0$. If now $H \subset F$, then for every n we have $H \subset E \backslash E_1 \backslash \cdots \backslash E_{n-1}$ and so $\lambda(H) \leqslant \varepsilon_n$. Therefore $\lambda(H) \leqslant 0$. Thus $F \in \mathcal{N}$ and, by (7.5.1) $\lambda(F) < 0$, or we have a contradiction.

A finite or countable union $\bigcup_i E_i$ of sets in $\mathcal{N}$ is also a set in $\mathcal{N}$; for if $F \in \mathcal{S}$ and $F \subset \bigcup_i E_i$ then

$$F = (F \cap E_1) \cup (F \cap (E_2 \backslash E_1)) \cup (F \cap (E_3 \backslash E_1 \backslash E_2)) \cup \cdots,$$

then

$$\lambda(F) = \lambda(F \cap E_1) + \lambda(F \cap (E_2 \backslash E_1)) + \lambda(F \cap (E_3 \backslash E_1 \backslash E_2)) + \cdots \leqslant 0.$$

The class $\mathcal{N}$ is not empty for it contains, for example, the empty set. Thus, there exists a sequence $B_n \in \mathcal{N}$ such that $\lambda(B_n) \to \inf_{\mathcal{N}} \lambda(E)$. Since

$$B = \bigcup_{n=1}^{\infty} B_n \in \mathcal{N},$$

therefore, for every n we have $\lambda(B \backslash B_n) \leqslant 0$ and so $\lambda(B) = \lambda(B_n) + \lambda(B \backslash B_n) \leqslant \lambda(B_n)$. It follows that

$$\lambda(B) = \inf_{\mathcal{N}} \lambda(E). \tag{7.5.2}$$

If E is a measurable set contained in $A = X \backslash B$, then $\lambda(E) \geqslant 0$. For otherwise (from the first part of the proof) there would exist a set $F \in \mathcal{N}$ such that $F \subset E$ and $\lambda(F) < 0$. But then $B \cup F \in \mathcal{N}$ and $\lambda(B \cup F) = \lambda(B) + \lambda(F) < \lambda(B)$[30] contradicting (7.5.2).

Thus if $E \in \mathcal{S}$ then the sets $A \cap E = E \backslash B$, $B \cap E$ are measurable and we have $\lambda(A \cap E) \geqslant 0$, $\lambda(B \cap E) \leqslant 0$. ■

THEOREM 7.5.2 (Jordan Canonical Decomposition). *Every countably additive function λ on a countably additive algebra $\mathcal{S}$ is the difference of two measures*

$$\lambda = \lambda^+ - \lambda^-, \tag{7.5.3}$$

where

$$\lambda^+(E) = \sup_{F \subset E} \lambda(F), \quad \lambda^-(E) = -\inf_{F \subset E} \lambda(F) = \sup_{F \subset E} -\lambda(F) \quad (E, F \in \mathcal{S}). \tag{7.5.4}$$

If $X = A \cup B$ is the Hahn decomposition for λ, then

$$\lambda^+(E) = \lambda(A \cap E), \quad \lambda^-(E) = -\lambda(B \cap E) \qquad \text{for } E \in \mathcal{S}. \tag{7.5.5}$$

Thus λ^+ vanishes on subsets of B and λ^- vanishes on subsets of A.[31]

Proof. Define the functions λ^+ and λ^- by the formulae (7.5.5). We then see that they are measures and that equation (7.5.3) holds. It therefore suffices to establish the results (7.5.4). If $F \subset E$, $(E, F \in \mathscr{S})$, then

$$\lambda(F) = \lambda(A \cap F) + \lambda(B \cap F) \leqslant \lambda(A \cap F) \leqslant \lambda(A \cap F) + \lambda(A \cap E \backslash A \cap F) = \lambda(A \cap E),$$

and since $A \cap E \subset E$ therefore $\sup_{F \subset E} \lambda(F) = \lambda(A \cap E) = \lambda^+(E)$. The second result (7.5.4) has a similar proof. ■

In the case of the function

$$\lambda(E) = \int_E f \, d\mu,$$

where we suppose that f is defined on X and that the integral exists on every measurable set, we can take, as is easily checked,

$$A = \{x : f(x) \geqslant 0\} \quad \text{and} \quad B = \{x : f(x) < 0\}$$

and

$$\lambda^+(E) = \int_E f_+ \, d\mu \quad \text{and} \quad \lambda^-(E) = \int_E f_- \, d\mu.$$

The measures λ^+ and λ^- are called the *upper variation* and the *lower variation* of a countably additive function λ. The measure

$$|\lambda| = \lambda^+ + \lambda^- \tag{7.5.6}$$

is called the *total variation* of the function λ.

Clearly $|\lambda(E)| \leqslant |\lambda|(E)$ and if either term is infinite then so is the other. It follows from (7.5.4) that

$$(\alpha\lambda)^+ = \begin{cases} \alpha\lambda^+, & \text{if } \alpha \geqslant 0, \\ -\alpha\lambda^-, & \text{if } \alpha \leqslant 0, \end{cases} \qquad (\alpha\lambda)^- = \begin{cases} \alpha\lambda^-, & \text{if } \alpha \geqslant 0, \\ -\alpha\lambda^+, & \text{if } \alpha \leqslant 0, \end{cases} \tag{7.5.7}$$

and also

$$(\lambda + \nu)^+ \leqslant \lambda^+ + \nu^+, \quad (\lambda + \nu)^- \leqslant \lambda^- + \nu^-,$$

(if $\lambda, \nu > -\infty$ always, or $\lambda, \nu < \infty$ always). Hence

$$|\alpha\lambda + \beta\nu| \leqslant |\alpha| \, |\lambda| + |\beta| \, |\nu| \tag{7.5.8}$$

(if $\alpha\lambda, \beta\nu > -\infty$ always or $\alpha\lambda, \beta\nu < \infty$ always).

Theorem 7.5.2 implies that many properties of measure, in particular Ths 5.1.5–5.1.6, can be extended to the case of countably additive functions.

Similarly, we can define the *integral relative to a countably additive function* λ on a countably additive algebra $\mathscr{S}$. Let f be a measurable function on a measurable set E. We define

$$\int_E f \, d\lambda = \int_E f \, d\lambda^+ - \int_E f \, d\lambda^-, \tag{7.5.9}$$

if the right-hand side is not of the form $\pm\infty \mp \infty$. If this integral is finite then the function f is called *summable* relative to λ on the set E.

The definition (7.5.9) implies a number of properties analogous to the properties of integral with respect to a measure. In particular, Ths 6.2.6–6.2.7 can be extended automatically.

We now note that if μ_1, μ_2 are measures on $\mathscr{S}$, f is a non-negative measurable function on a measurable set E, $\alpha_1 \geqslant 0$, $\alpha_2 \geqslant 0$, then

$$\int_E f \,\mathrm{d}(\alpha_1\mu_1 + \alpha_2\mu_2) = \alpha_1 \int_E f \,\mathrm{d}\mu_1 + \alpha_2 \int_E f \,\mathrm{d}\mu_2.$$

Indeed, the formula is easily verified in the case in which f is a simple function. It then follows from Th. 4.4.10 and Th. 6.1.5 that it is true for any measurable, non-negative function f.

If $\mu_1 \leqslant \mu_2$ and $f \geqslant 0$ then

$$\int_E f \,\mathrm{d}\mu_1 \leqslant \int_E f \,\mathrm{d}\mu_2.$$

This follows directly from the definition (6.1.1) of the integral.

If the integral (7.5.9) exists then noting that

$$\int_E |f| \mathrm{d}\lambda^+ + \int_E |f| \mathrm{d}\lambda^- = \int_E |f| \mathrm{d}|\lambda|,$$

we have the inequality

$$\left|\int_E f \,\mathrm{d}\lambda\right| \leqslant \int_E |f| \mathrm{d}|\lambda|, \tag{7.5.10}$$

and the summability of f relative to λ is equivalent to the summability of $|f|$ relative to $|\lambda|$ (on E), for it is equivalent to the finiteness of both integrals

$$\int_E |f| \mathrm{d}\lambda^+ \quad \text{and} \quad \int_E |f| \mathrm{d}\lambda^-.$$

THEOREM 7.5.3. *Suppose that the functions λ, ν are countably additive on a countably additive algebra and that $\alpha\lambda$, $\beta\nu > -\infty$ always, or $\alpha\lambda$, $\beta\nu < \infty$ always. If f is summable relative to λ and ν on the set E then it is summable relative to $\alpha\lambda + \beta\nu$ and*

$$\int_E f \,\mathrm{d}(\alpha\lambda + \beta\nu) = \alpha \int_E f \,\mathrm{d}\lambda + \beta \int_E f \,\mathrm{d}\nu.$$

Proof. Using inequality (7.5.8) we have

$$\int_E |f| \mathrm{d}|\alpha\lambda + \beta\nu| \leqslant |\alpha| \int_E |f| \mathrm{d}|\lambda| + |\beta| \int_E |f| \mathrm{d}|\nu| < \infty,$$

so that f is summable relative to $\alpha\lambda + \beta\nu$. Since the integral of the function f is the difference of the integrals of f_+ and f_- it follows that we can confine ourselves to the case in which $f \geqslant 0$. From definition (7.5.9) and the relations

(7.5.7) we have

$$\int_E f\,\mathrm{d}(\alpha\lambda) = \alpha \int_E f\,\mathrm{d}\lambda.$$

We can therefore confine ourselves to the case $\alpha = \beta = 1$. We have

$$(\lambda + \nu)^+ - (\lambda + \nu)^- = \lambda^+ - \lambda^- + \nu^+ - \nu^-$$

(where also $\lambda^+, \nu^+, (\lambda+\nu)^+ < \infty$ or $\lambda^-, \nu^-, (\lambda+\nu)^- < \infty$), and so

$$(\lambda + \nu)^+ + \lambda^- + \nu^- = (\lambda + \nu)^- + \lambda^+ + \nu^+.$$

Hence

$$\int_E f\,\mathrm{d}(\lambda+\nu)^+ + \int_E f\,\mathrm{d}\lambda^- + \int_E f\,\mathrm{d}\nu^- = \int_E f\mathrm{d}(\lambda+\nu)^- + \int_E f\,\mathrm{d}\lambda^+ + \int_E f\,\mathrm{d}\nu^+.$$

And so (since both sides are bounded), $\int_E f\,\mathrm{d}(\lambda+\nu) = \int_E f\,\mathrm{d}\lambda + \int_E f\,\mathrm{d}\nu$ which concludes the proof. ■

Riesz's Theorem

Now let X be a locally compact metric space.[32] We denote by $\mathscr{C}$ the class of all continuous functions which vanish outside a compact set (which depends on the function). It is easily verified that a linear combination of functions in $\mathscr{C}$ is again a function in $\mathscr{C}$. Let λ be a countably additive function which takes finite values on the class of relatively compact Borel sets. If $\varphi \in \mathscr{C}$ then $\varphi(x) = 0$ outside some compact set E; also, λ is a countably additive function on the algebra of Borel sets which are contained in E. Thus the integral $\int_E \varphi\,\mathrm{d}\lambda$ is well defined and does not depend upon the particular choice of the set E. We will denote it simply by $\int \varphi\,\mathrm{d}\lambda$.

The functional $I(\varphi) = \int \varphi\,\mathrm{d}\lambda$ is linear on $\mathscr{C}$:

$$I(\alpha\varphi + \beta\psi) = \alpha I(\varphi) + \beta I(\psi).$$

It is also continuous in the following sense: $I(\varphi_n) \to 0$ if

$$\left.\begin{array}{c}\varphi_n \to 0 \text{ uniformly on } X \text{ and } \varphi_n = 0 \ (n = 1, 2, \ldots) \\ \text{outside some compact set } E.\end{array}\right\} \tag{7.5.11}$$

For in this case

$$|I(\varphi_n)| = \left|\int_E \varphi_n\,\mathrm{d}\lambda\right| \leqslant \int_E |\varphi_n|\mathrm{d}|\lambda| \to 0.$$

Conversely we have

THEOREM 7.5.4 (Riesz). *If $I(\varphi)$ is a (finite) linear functional which is continuous on $\mathscr{C}$ (in the sense defined above), then there exists a unique countably additive function λ which is finite on the class of relatively compact Borel sets and which is such that*

$$I(\varphi) = \int \varphi\,\mathrm{d}\lambda \qquad \textit{for } \varphi \in \mathscr{C}.^{33}$$

Proof. Let $\mathscr{C}_+$ denote the class of all non-negative functions in $\mathscr{C}$. We assert that for $\varphi \in \mathscr{C}_+$ we have $I(\varphi) = I_+(\varphi) - I_-(\varphi)$, where

$$I_+(\varphi) = \sup_{0 \leqslant \chi \leqslant \varphi} I(\chi) \quad \text{and} \quad I_-(\varphi) = \sup_{0 \leqslant \chi \leqslant \varphi} -I(\chi)$$

are non-negative and bounded. Indeed, the set $\mathscr{J} = \{\chi \in \mathscr{C}_+ : 0 \leqslant \chi \leqslant \varphi\}$ is mapped onto itself by the mapping $\chi \to \varphi - \chi$, hence

$$\sup_{\mathscr{J}} I = \sup_{\chi \in \mathscr{J}} I(\varphi - \chi) = I(\varphi) + \sup_{\mathscr{J}} (-I).$$

Next, if $I_+(\varphi) = \infty$, then $k_n = I(\chi_n) \to \infty$ for some $\chi_n \in \mathscr{J}$ so that the sequence $\bar{\chi}_n = \chi_n / k_n$ would satisfy conditions (7.5.11) but $I(\bar{\chi}_n) = 1$, contradicting the continuity of I.

To establish the existence of λ it suffices to show that I_+, I_- are additive (and we may consider only I_+) for then, by Riesz's theorem (§6.4),

$$I_+(\varphi) = \int \varphi \, \mathrm{d}\mu_+, \quad I_-(\varphi) = \int \varphi \, \mathrm{d}\mu_- \qquad \text{on } \mathscr{C}_+,$$

where μ_+, μ_- are measures on $\mathscr{B}(X)$ which are finite on compact sets. Hence

$$I(\varphi) = \int \varphi \, \mathrm{d}\lambda \qquad \text{for } \varphi \in \mathscr{C}_+,$$

where $\lambda = \mu_+ - \mu_-$ on the set of relatively compact Borel sets. Hence also, since $\varphi = \varphi_+ - \varphi_-$, the result is true for $\varphi \in \mathscr{C}$.

So let $\varphi, \psi \in \mathscr{C}_+$. We have

$$I_+(\varphi) + I_+(\psi) = \sup \{I(\xi + \eta) : 0 \leqslant \xi \leqslant \varphi, 0 \leqslant \eta \leqslant \psi\}$$

But the right-hand side is equal to $\sup \{I(\zeta) : 0 \leqslant \zeta \leqslant \varphi + \psi\}$, for $(\xi, \eta) \to \xi + \eta$ maps the set $\{\xi \in \mathscr{C}_+ : 0 \leqslant \xi \leqslant \varphi\} \times \{\eta \in \mathscr{C}_+ : 0 \leqslant \eta \leqslant \psi\}$ onto the set $\{\zeta \in \mathscr{C}_+ : 0 \leqslant \zeta \leqslant \varphi + \psi\}$; indeed, if $0 \leqslant \zeta \leqslant \varphi + \psi$ then $\zeta = \xi + \eta$, where $\xi = \min(\zeta, \varphi) \leqslant \varphi$ and $\eta = \zeta - \xi = \max(0, \zeta - \varphi) \leqslant \psi$. Therefore

$$I_+(\varphi) + I_+(\psi) = I_+(\varphi + \psi).$$

Now let $\bar{\lambda}$ be another function satisfying the conditions of the theorem. Then

$$\int \varphi \, \mathrm{d}\nu = 0 \qquad \text{for } \varphi \in \mathscr{C},$$

where $\nu = \bar{\lambda} - \lambda$. We must show that $\nu = 0$.

Let G be an open set which is relatively compact. Then $\nu = \nu_+ - \nu_-$ on $\mathscr{B}(G)$, where ν_+, ν_- are measures on $\mathscr{B}(G)$. Therefore $\int \varphi \, \mathrm{d}\nu_+ = \int \varphi \, \mathrm{d}\nu_-$ for $\varphi \in \mathscr{C}_+$ which vanish outside compact subsets of the set G and so, by the uniqueness of measure in Riesz's theorem (§6.4) $\nu_+ = \nu_-$ so that $\nu = 0$ on $\mathscr{B}(G)$. Hence $\nu = 0$ for every relatively compact Borel set. ■

Let $\mathscr{C}_0$ be the class of all functions φ which are continuous on X and 'vanish at infinity', that is, are such that $\{|\varphi| \geqslant \varepsilon\}$ is compact for any $\varepsilon > 0$. Clearly $\mathscr{C} \subset \mathscr{C}_0$. If

λ is a countably additive function on $\mathscr{B}(X)$ with bounded variation, then

$$I(\varphi)=\int_X \varphi\,\mathrm{d}\lambda$$

is a functional on $\mathscr{C}_0$ which is linear and continuous relative to uniform convergence in $\mathscr{C}_0$. Conversely we have

THEOREM 7.5.4′ (Riesz). *If I is a (finite) linear functional which is continuous on $\mathscr{C}_0$ (in the sense of uniform convergence), then there exists a unique function λ which is countably additive on $\mathscr{B}(X)$, has a variation $|\lambda|$ which is finite and regular and is such that*

$$I(\varphi)=\int_X \varphi\,\mathrm{d}\lambda \qquad \textit{for } \varphi\in\mathscr{C}_0.$$

The proof does not differ substantially from the proof of Th. 7.5.4 if we note that we only need the above equality for $\varphi\in\mathscr{C}$, since every function in $\mathscr{C}_0$ is the limit of a uniformly convergent sequence of functions in $\mathscr{C}$.[34] We must observe that

$$|I(\varphi)|\leqslant M\sup|\varphi| \qquad \text{in } \mathscr{C}$$

for some constant $M<\infty$. Hence the same inequality holds for I_+ and I_- and (from the construction of measure in Riesz's theorem by (6.4.3)),

$$\mu_+(X)\leqslant M \quad \text{and} \quad \mu_-(X)\leqslant M.$$

For the proof of regularity we refer to Note 32.

Instead of requiring $|\lambda|$ to be regular, we could replace $\mathscr{B}(X)$ by $\mathscr{B}^0(X)$,[35] where we define $\int\varphi\,\mathrm{d}\lambda$ as the integral of the restriction of φ to a σ-compact set, outside which $\varphi=0$. The regularity condition can be omitted if X is σ-compact.[36] Recall the uniqueness of measure in Riesz's theorem (§6.4).

Connection with Interval Functions

We show now that just as the content of an interval determines the Lebesgue measure, so an arbitrary additive and non-negative interval function determines a certain measure, and a function of bounded variation determines a countably additive function.

We have the following

THEOREM 7.5.5. *If F is an interval function (n-dimensional) which is additive, finite and non-negative and is defined for (closed) intervals contained in an open set $\Omega\subset\mathbb{R}^n$, then there exists exactly one measure F^* on $\mathscr{B}(\Omega)$ such that*

$$F^*(\operatorname{Int}R)\leqslant F(R)\leqslant F^*(R) \qquad \textit{for } R\subset\Omega. \tag{7.5.12}$$

LEMMA 7.5.1. *Under the hypotheses of the theorem, for every interval $R\subset\Omega$*

and for any $\varepsilon > 0$ *there exists a system of intervals* $R_1, \ldots, R_k$ *such that* $\delta(R_i) < \varepsilon$ $(i = 1, \ldots, k)$,

$$\operatorname{Int} R = \bigcup_{i=1}^{k} \operatorname{Int} R_i \quad \textit{and} \quad \sum_{i=1}^{k} F(R_i) \leqslant F(R) + \varepsilon.$$

Proof. It suffices to prove that for any $R \subset \Omega$ and $\eta > 0$ there exist R', R'' such that $\operatorname{Int} R = \operatorname{Int} R' \cup \operatorname{Int} R''$, $F(R') + F(R'') \leqslant F(R) + \eta$ and $\delta(R'), \delta(R'') \leqslant \alpha\delta(R)$, where $\alpha = \sqrt{(1 - 5/9n)} < 1$. For then, selecting in this way intervals R_1', R_2' corresponding to R' and R_1'', R_2'' corresponding to R'' etc., we obtain after the kth step 2^k intervals R_i such that

$$\operatorname{Int} R = \bigcup_{i=1}^{2^k} \operatorname{Int} R_i, \quad \sum_{i=1}^{2^k} F(R_i) \leqslant F(R) + (2^k - 1)\eta \quad \text{and} \quad \delta(R_i) < \alpha^k \delta(R)$$

and so it suffices to choose k and $\eta > 0$ so that $\alpha^k \delta(R) < \varepsilon$ and $(2^k - 1)\eta < \varepsilon$.

So let $\eta > 0$ and let $R = [a_1, b_1] \times \cdots \times [a_n, b_n] \subset \Omega$. Let, for definiteness, $b_1 - a_1 = \max_i (b_i - a_i)$. The function

$$\varphi(x) = F([a_1, x] \times \cdots \times [a_n, b_n])$$

is increasing and so it must have a point c of continuity in the interval $(a_1 + (b_1 - a_1)/3, b_1 - (b_1 - a_1)/3)$. Hence, there exists a number d in this interval, with $d > c$, such that

$$\varphi(d) - \varphi(c) < \eta.$$

Putting $R' = [a_1, d] \times \cdots \times [a_n, b_n]$, $R'' = [c, b_1] \times \cdots \times [a_n, b_n]$ we have

$$\operatorname{Int} R = \operatorname{Int} R' \cup \operatorname{Int} R'', \quad \delta(R'), \delta(R'') \leqslant \alpha\delta(R).$$[37]

Also, since $F(R'') = \varphi(b_1) - \varphi(c)$,

$$F(R') + F(R'') = \varphi(d) - \varphi(c) + \varphi(b_1) - \varphi(a_1) \leqslant F(R) + \eta. \quad \blacksquare$$

Proof of theorem. For each $E \subset \Omega$ we let

$$F^*(E) = \inf\left\{\sum_\nu F(R_\nu) : E \subset \cup \operatorname{Int} R_\nu\right\} \tag{7.5.13}$$

(taken over finite or countable unions). From lemma 7.5.1 we easily derive the conclusion that for any $E \subset \Omega$ and $\varepsilon > 0$

$$F^*(E) = \inf\left\{\sum_\nu F(R_\nu) : E \subset \cup \operatorname{Int} R_\nu, R_\nu \cap E \neq \varnothing, \delta(R_\nu) < \varepsilon\right\}.$$

The function F^* defined in this way is an metric outer measure on Ω; the proof of this is the same as the proof of Th. 5.4.1. Thus, by Th. 5.2.3, F^* is a measure on $\mathscr{B}(\Omega)$.

Let $R \subset \Omega$. By (7.5.13) we have

$$F^*(\operatorname{Int} R) \leqslant F(R).$$

If $R \subset \cup \operatorname{Int} R_\nu$, then (by the Borel–Lebesgue theorem) $R \subset \bigcup_{\nu=1}^{k} \operatorname{Int} R_\nu$ for some k, and hence, by Th. 5.3.3, $F(R) \leqslant \sum_{\nu=1}^{k} F(R_\nu)$: therefore

$$F(R) \leqslant F^*(R).$$

Thus the measure F^* satisfies the inequality (7.5.12).

It remains to show that there is only one such measure. Suppose that μ and ν are measures on $\mathscr{B}(\Omega)$ satisfying condition (7.5.12). Let $R \subset \Omega$ and let $R_\nu \subset \Omega$ be a decreasing sequence of intervals such that $R = \lim \operatorname{Int} R_\nu$. Then also $R = \lim R_\nu$.

Since $\mu(\operatorname{Int} R_\nu) \leqslant F(R_\nu) \leqslant \nu(R_\nu)$, therefore[38] $\mu(R) \leqslant \nu(R)$; similarly $\nu(R) \leqslant \mu(R)$ so that $\nu(R) = \mu(R)$. Thus, both measures are equal on closed intervals contained in Ω. Since every interval open on the right is the limit of an increasing sequence of closed intervals, therefore the measures are equal for intervals open on the right and so, by lemma 5.4.1, for open sets contained in Ω. Hence, by Th. 5.2.5, $\mu = \nu$ on $\mathscr{B}(\Omega)$. ■

The measure F^* defined by condition (7.5.12) is called the *measure associated with the* (*additive, non-negative*) *interval function* F. In particular, Lebesgue measure (restricted to $\mathscr{B}_n$) is associated with the content function of intervals. Obviously, every measure μ on $\mathscr{B}(\Omega)$ which is finite on (closed) intervals is associated with some additive non-negative interval function, e.g. with the function

$$\mu^{\#}(R) = \mu(\hat{R}),$$

where $\hat{R} = [a_1, b_1) \times \cdots \times [a_n, b_n)$, when $R = [a_1, b_1] \times \cdots \times [a_n, b_n]$. From condition (7.5.12) it follows that

$$(\alpha F + \beta G)^* = \alpha F^* + \beta G^* \qquad \text{if } \alpha, \beta \geqslant 0.$$

Let R be a (closed) interval contained in Ω. If P_ν, R_ν are monotone[39] sequences of intervals such that

$$P_\nu \subset \operatorname{Int} R, \quad R \subset \operatorname{Int} R_\nu, \quad \operatorname{Int} P_\nu \to \operatorname{Int} R, \quad R_\nu \to R, \tag{7.5.14}$$

then

$$F^*(\operatorname{Int} R) = \lim_{\nu \to \infty} F(P_\nu) \quad \text{and} \quad F^*(R) = \lim_{\nu \to \infty} F(R_\nu). \tag{7.5.15}$$

This follows from the inequalities

$$F^*(\operatorname{Int} P_\nu) \leqslant F(P_\nu) \leqslant F^*(P_\nu) \leqslant F^*(\operatorname{Int} R)$$

and

$$F^*(R) \leqslant F^*(\operatorname{Int} R_\nu) \leqslant F(R_\nu) \leqslant F^*(R_\nu).$$

We say that the (closed) interval $R \subset \Omega$ is an *interval of continuity for* F, if

$$F^*(R \backslash \operatorname{Int} R) = 0,$$

that is, from (7.5.12), if

$$F^*(\operatorname{Int} R) = F(R) = F^*(R).$$

By (7.5.15) this condition holds if and only if $F(R_\nu)-F(P_\nu)\to 0$ for some (and hence for any) pair of sequences P_ν, R_ν satisfying conditions (7.5.14).

The hyperplane Π with equation $x_i=c$ is called a *hyperplane of continuity* for F, if

$$F^*(\Pi\cap\Omega)=0.$$

Otherwise Π is called a *hyperplane of discontinuity*.

There exists at most a countable number of hyperplanes of discontinuity. Indeed, Ω is the union of some sequence of closed intervals R_ν. If therefore Π is a hyperplane of discontinuity then $F^*(\Pi\cap R_\nu)>0$ for some ν. If now $\Pi_1,\ldots,\Pi_k$ are distinct hyperplanes with equations $x_i=c_1,\ldots,x_i=c_k$ then $F^*(\Pi_1\cap R_\nu)+\cdots+F^*(\Pi_k\cap R_\nu)\leqslant F^*(R_\nu)$ so that the set of hyperplanes Π with equations $x_i=c$ such that $F^*(\Pi\cap R_\nu)\geqslant 1/j$ is finite. The union of these sets ($j=1,2,\ldots$; $\nu=1,2,\ldots$; $i=1,2,\ldots,n$) is precisely the set of all hyperplanes of discontinuity.

It follows from the above definition that when all faces of an interval R are located on hyperplanes of continuity, then R is an interval of continuity.

We now consider the case of set functions taking positive or negative values. A set of intervals $\mathscr{K}$ is called *dense* in an open set Ω if every closed interval $R\subset\Omega$ is the limit of a decreasing sequence of intervals in $\mathscr{K}$.[40] In particular, the set of intervals of continuity of an interval function (additive, non-negative, defined for intervals contained in Ω) is dense in Ω. For, noting that the set of hyperplanes of discontinuity is at most countable, any interval $[a_1,b_1]\times\cdots\times[a_n,b_n]$ is the limit of a decreasing sequence of intervals $[c_1^{(\nu)},d_1^{(\nu)}]\times\cdots\times[c_n^{(\nu)},d_n^{(\nu)}]$ such that the hyperplanes $x_1=c_1^{(\nu)}$, $x_1=d_1^{(\nu)},\ldots,x_n=c_n^{(\nu)}$, $x_n=d_n^{(\nu)}$ are hyperplanes of continuity.

LEMMA 7.5.2. *If λ, ν are countably additive functions on $\mathscr{B}(\Omega)$ (where Ω is an open set in $\mathbb{R}^n$) which are equal and finite on a set of intervals $\mathscr{K}$ which are dense in Ω, then $\lambda=\nu$ on $\mathscr{B}(\Omega)$.*

Proof. First suppose that λ and ν are measures. Since every closed interval is the limit of a decreasing sequence of intervals in $\mathscr{K}$, therefore $\lambda=\nu$ for closed intervals contained in Ω. It follows, as in the second half of the proof of Th. 7.5.5, that $\lambda=\nu$ on $\mathscr{B}(\Omega)$. In the general case $\lambda=\lambda^+-\lambda^-$, $\nu=\nu^+-\nu^-$ and $\lambda=\nu$ on $\mathscr{K}$. Therefore

$$\lambda^+ + \nu^- = \nu^+ + \lambda^- \quad \text{on } \mathscr{K}\text{, and so on } \mathscr{B}(\Omega).$$

It follows that we have

$$\lambda=\lambda^+-\lambda^-=\nu^+-\nu^-=\nu \quad \text{on } \mathscr{B}(\Omega)$$

(if, for example $\lambda^+(E)=\infty$ then $\lambda^-(E)<\infty$, $\nu^+(E)=\infty$, and so $\lambda^-,\nu^-<\infty$ always). ■

Now suppose that F is an additive function of finite variation on the interval P.[41] By Th. 7.2.1,

$$F=F_1-F_2$$

for some finitely additive, non-negative F_i where also $F_1^* - F_2^*$ does not depend on the particular choice of the pair F_1, F_2 (because for another such pair G_1, G_2 we have $F_1 + G_2 = G_1 + F_2$ and so $F_1^* + G_2^* = G_1^* + F_2^*$). We therefore have a uniquely defined function

$$F^* = F_1^* - F_2^*,$$

corresponding to F, which is countably additive and finite on $\mathscr{B}(\operatorname{Int} P)$.[42] We call F^* *the countably additive function corresponding to the function* F (additive and of finite variation). Clearly, every function λ which is countably additive and finite on $\mathscr{B}(\operatorname{Int} P)$ corresponds to some interval function of bounded variation on P, for example, to the function

$$\lambda^{\#}(R) = \lambda(\hat{R} \cap \operatorname{Int} P),$$

where $\hat{R} = [a_1, b_1) \times \cdots \times [a_n, b_n)$, when $R = [a_1, b_1] \times \cdots \times [a_n, b_n]$. Also

$$(\alpha F + \beta G)^* = \alpha F^* + \beta G^*.$$

If R is a (closed) interval contained in $\operatorname{Int} P$ and if P_ν, R_ν are sequences satisfying conditions (7.5.14), then

$$F^*(\operatorname{Int} R) = \lim_{\nu \to \infty} F(P_\nu) \quad \text{and} \quad F^*(R) = \lim_{\nu \to \infty} F(R_\nu). \tag{7.5.16}$$

The closed interval $R \subset \operatorname{Int} P$ is called an *interval of continuity* for F if it is an interval of continuity for F^+ and for F^- or, equivalently (since $|F|^* = (F^+)^* + (F^-)^*$) if it is an interval of continuity for the variation $|F|$. Thus R is an interval of continuity if and only if

$$|F|(R_\nu) - |F|(P_\nu) \to 0$$

for some (and hence for any) pair of sequences P_ν, R_ν satisfying conditions (7.5.14). If R is an interval of continuity for F, then

$$F^*(\operatorname{Int} R) = F(R) = F^*(R).$$

THEOREM 7.5.6. *For a function F which is absolutely continuous, every interval ($R \subset \operatorname{Int} P$) is an interval of continuity.*

Proof. Let P_ν, R_ν be sequences satisfying conditions (7.5.14); it is required to prove that $|F|(R_\nu) - |F|(P_\nu) \to 0$. Take $\varepsilon > 0$. Let $\delta > 0$ be a number corresponding to ε in the definition of absolute continuity of $|F|$ (see Th. 7.3.1). We have

$$|R_\nu| - |P_\nu| < \delta \qquad \text{for } \nu \geqslant N$$

for some N. Fix $\nu \geqslant N$ and let $R_\nu = P_\nu \cup Q_1 \cup \cdots \cup Q_k$, where Q_i are non-overlapping intervals which also do not overlap P_ν.[43] Then $|Q_1| + \cdots + |Q_k| < \delta$, and so

$$|F|(R_\nu) - |F|(P_\nu) = |F|(Q_1) + \cdots + |F|(Q_k) < \varepsilon.$$

Then $|F|(R_\nu) - |F|(P_\nu) \to 0$. ∎

Since the set of intervals of continuity is dense in $\operatorname{Int} P$, therefore by Lemma 7.5.2, we have

THEOREM 7.5.7. *The function F^* associated with F is the unique countably additive function on $\mathscr{B}(\operatorname{Int} P)$ such that $F^*(R) = F(R)$ for intervals of continuity.*

Remark. It follows that F^* depends only on the values taken by F on intervals of continuity. A countably additive function may be associated with two different interval functions, provided only that they agree on their common intervals of continuity. Hence their difference is a singular function (for it vanishes on a dense set of intervals and so its derivative is zero wherever it exists).

Now let g be a function of bounded variation on $[a, b]$. Let g^* denote the countably additive function associated with g (that is, associated with the interval function determined by g). If $a < \alpha < \beta < b$, then by (7.5.16)

$$g^*((\alpha, \beta)) = g(\beta - 0) - g(\alpha + 0) \tag{7.5.17}$$

and this formula remains true if $a \leqslant \alpha < \beta \leqslant b$ for we can proceed to the limit. If $a < c < b$, then

$$g^*((c)) = g(c + 0) - g(c - 0), \tag{7.5.18}$$

for $g^*((c)) = \lim g^*((\alpha_\nu, \beta_\nu))$, where $\alpha_\nu \to c-$, $\beta_\nu \to c+$, and it suffices to make use of (7.5.17). It follows from this that the interval $[\alpha, \beta] \subset (a, b)$ is an interval of continuity for g if and only if the variation $x \to W_a^x(g)$ is continuous at α and β, that is (Th. 1.3.4 with the remark) if and only if g is continuous at α and at β.

Lebesgue–Stieltjes Integral

Let F be an additive function of bounded variation on the interval P. If f is a (Baire) function which is F^*-summable on a Borel set $E \subset \operatorname{Int} P$, then

$$(\mathrm{L}) \int_E f \, \mathrm{d}F = \int_E f \, \mathrm{d}F^*$$

is called the *Lebesgue–Stieltjes integral* of the function f relative to the function F on the set E and we say that the function f is *summable relative to the function* F on the set E. In particular, if g is a function of a single variable of bounded variation, then

$$(\mathrm{L}) \int_E f \, \mathrm{d}g = \int_E f \, \mathrm{d}g^*,$$

where g^* is the countably additive set function associated with g.

THEOREM 7.5.8. *If for a bounded Baire function f the Riemann–Stieltjes integral $\int_P f \, \mathrm{d}F$ exists, then for every interval of continuity R of the function F*

we have

$$\text{(L)} \int_R f \, \mathrm{d}F = \int_R f \, \mathrm{d}F.^{44} \tag{7.5.19}$$

The Lebesgue–Stieltjes integral $\lambda(E) = (\mathrm{L})\int_E f \, \mathrm{d}F$, regarded as a countably additive set function defined on $\mathscr{B}(\operatorname{Int} P)$ is the function associated with the Riemann–Stieltjes integral

$$\Phi(R) = \int_R f \, \mathrm{d}F,$$

regarded as an interval function.

Sketch Proof. Just as for the case of functions of a single variable (Th. 1.5.3) it is shown that the integrals $\int_P f \, \mathrm{d}F^+, \int_P f \, \mathrm{d}F^-$ exist and that their difference is equal to $\int_P f \, \mathrm{d}F$. Therefore (by Th. 7.5.3) it suffices to consider only the case $F \geqslant 0$.

In this case the proof of (7.5.19) is analogous to the proof of Th. 6.2.13. It is only necessary to choose a sequence of subdivisions $R = \bar{P}_{1\nu} \cup \cdots \cup \bar{P}_{k_\nu \nu}$ so that the $\bar{P}_{k\nu}$ are intervals of continuity; thus for $R = [c_1, d_1] \times \cdots \times [c_n, d_n]$ we select $c_i = \xi_{i0}^{(\nu)} < \cdots < \xi_{i\nu}^{(\nu)} = d_i$ in such a way that for every $\nu = 1, 2, \ldots, i = 1, \ldots, n$, $j = 1, \ldots, \nu - 1$ the hyperplane $x_i = \xi_{ij}^{(\nu)}$ is a hyperplane of continuity and then for the intervals $P_{k\nu}$ we take the intervals $[\xi_{1,j_1-1}^{(\nu)}, \xi_{1,j_1}^{(\nu)}] \times \cdots \times [\xi_{n,j_n-1}^{(\nu)}, \xi_{n,j_n}^{(\nu)}]$. Then, just as for Riemann integrals, the sums S_ν and s_ν tend to $\int_R f \, \mathrm{d}F$. Moreover, $\varphi_\nu(x) \to m(x)$ and $\Phi_\nu(x) \to M(x)$ on $R \setminus \bigcup_{\nu=1}^{\infty} B_\nu$, where $F^*(B_\nu) = 0$, for B_ν is the union of the boundaries of the intervals $P_{1\nu}, \ldots, P_{k_\nu \nu}$.

Since the set of intervals, all of whose faces lie on hyperplanes which are simultaneously continuity hyperplanes for F and for Φ, is dense in $\operatorname{Int} P$, therefore, by lemma 7.5.2 and (7.5.19), $\lambda = \Phi^*$ on $\mathscr{B}(\operatorname{Int} P)$. ∎

It is not difficult to verify that if f is a function of one variable which is integrable on the interval $[a, b]$ relative to a function g of bounded variation, then

$$\int_a^b f \, \mathrm{d}g = [g(a+0) - g(a)]f(a) + (\mathrm{L})\int_{(a,b)} f \, \mathrm{d}g + [g(b) - g(b-0)]f(b). \tag{7.5.20}$$

THEOREM 7.5.9 (Integration by Parts). *If f and g are functions of bounded variation on the interval $[a, b]$, then*

$$\begin{aligned} &(\mathrm{L})\int_{(a,b)} f(x+0) \, \mathrm{d}g(x) + (\mathrm{L})\int_{(a,b)} g(x-0) \, \mathrm{d}f(x) \\ &\quad = f(b-0)g(b-0) - f(a+0)g(a+0), \\ &(\mathrm{L})\int_{(a,b)} f(x-0) \, \mathrm{d}g(x) + (\mathrm{L})\int_{(a,b)} g(x+0) \, \mathrm{d}f(x) \\ &\quad = f(b-0)g(b-0) - f(a+0)g(a+0). \end{aligned} \tag{7.5.21}$$

If in addition

$$f(x)=\tfrac{1}{2}\{f(x-0)+f(x+0)\} \quad \text{and} \quad g(x)=\tfrac{1}{2}\{g(x-0)+g(x+0)\} \qquad (7.5.22)$$

at every common point of discontinuity of the functions f and g, then

$$(\mathrm{L})\int_{(a,b)} f\,\mathrm{d}g+(\mathrm{L})\int_{(a,b)} g\,\mathrm{d}f=f(b-0)g(b-0)-f(a+0)g(a+0). \qquad (7.5.23)$$

Proof. It suffices to prove the first of the equations (7.5.21), for the second follows from the first by interchanging the rôles of f and g. Also, (7.5.23) can be obtained by adding equations (7.5.21) and dividing by 2 (where, at each point x which is a point of discontinuity for only one of the functions f and g, we change the value of $f(x)$ or $g(x)$ in such a way that relations (7.5.22) hold, which by (7.5.18) does not affect the value of the integrals). Let χ be the characteristic function of the set $\{(x, y): a<y<x<b\}$. By Fubini's theorem (Th. 6.3.4 applied here specifically to the lower and upper variations of the associated functions f^* and g^*) we have

$$(\mathrm{L})\int_{(a,b)}\left[(\mathrm{L})\int_{(a,b)}\chi(x, y)\,\mathrm{d}f(x)\right]\mathrm{d}g(y)=(\mathrm{L})\int_{(a,b)}\left[(\mathrm{L})\int_{(a,b)}\chi(x, y)\,\mathrm{d}g(y)\right]\mathrm{d}f(x).$$

By (7.5.17) the integrals in the square brackets are equal respectively to $f(b-0)-f(y+0)$ and $g(x-0)-g(a+0)$. Therefore

$$(\mathrm{L})\int_{(a,b)}[f(b-0)-f(y+0)]\,\mathrm{d}g(y)=(\mathrm{L})\int_{(a,b)}[g(x-0)-g(a+0)]\,\mathrm{d}f(x).$$

Hence, using (7.5.17), we obtain the first of the formulae (7.5.21). ■

7.6 THE RADON–NIKODYM THEOREM

Let λ, ν be countably additive functions on a countably additive algebra $\mathscr{S}$ in a space X. We say that the function λ is *absolutely continuous relative to the function* ν, if the following condition is satisfied:

$$\lambda(E)=0 \quad \text{if} \quad |\nu|(E)=0. \qquad (7.6.1)$$

It follows from the definition that a linear combination of functions which are absolutely continuous relative to ν is an absolutely continuous function relative to ν.

If λ is absolutely continuous relative to ν then λ^+, λ^- and $|\lambda|$ are also absolutely continuous relative to ν. For if $E\in\mathscr{S}$ and $|\nu|(E)=0$, then for any $F\subset E$ $(F\in\mathscr{S})$ we have $|\nu|(F)=0$, and so $\lambda(F)=0$. Hence, by (7.5.4) and (7.5.6)

$$\lambda^+(E)=0, \quad \lambda^-(E)=0 \quad \text{and} \quad |\lambda|(E)=0.$$

If λ is finite[45] and absolutely continuous relative to ν, then

$$\lambda(E)\to 0 \quad \text{if} \quad |\nu|(E)\to 0. \qquad (7.6.2)$$

For suppose that λ does not have property (7.6.2). Then, there exists $\varepsilon > 0$ and a sequence E_n such that

$$|\nu|(E_n) < \frac{1}{2^n} \quad \text{and} \quad |\lambda|(E_n) \geqslant |\lambda(E_n)| \geqslant \varepsilon.$$

The sequence $F_n = \bigcup_{i=n}^{\infty} E_i$ is decreasing. Let $F = \lim F_n$. Then we have

$$|\nu|(F) \leqslant |\nu|(F_n) \leqslant \frac{1}{2^{n-1}} \quad \text{and} \quad |\lambda|(F_n) \geqslant \varepsilon.$$

Hence in the limit, since $|\lambda|$ is finite, we obtain

$$|\nu|(F) = 0 \quad \text{and} \quad |\lambda|(F) \geqslant \varepsilon.$$

But then $|\lambda|$ would not be absolutely continuous relative to ν and the same would hold for λ.

Clearly, condition (7.6.2) itself implies that λ is absolutely continuous reative to ν.

We say that the countably additive function λ on the space X is *σ-finite* if

$$X = \bigcup_{1}^{\infty} E_i \qquad \text{where } |\lambda(E_i)| < \infty.$$

Now let μ be a measure on a countably additive algebra $\mathscr{S}$ in a space X. If f is a measurable function which is integrable on every measurable set, then

$$\lambda(E) = \int_E f \, d\mu$$

is a countably additive function on $\mathscr{S}$ which is absolutely continuous relative to μ.

Conversely we have

THEOREM 7.6.1 (Radon–Nikodym). *If λ is a countably additive, σ-finite function on $\mathscr{S}$ which is absolutely continuous relative to a σ-finite measure μ on $\mathscr{S}$, then there exists a function f, measurable and integrable on X relative to the measure μ, such that*

$$\lambda(E) = \int_E f \, d\mu \qquad \textit{for } E \in \mathscr{S}. \tag{7.6.3}$$

LEMMA 7.6.1. *If μ and λ are finite measures on $\mathscr{S}$ such that λ is absolutely continuous relative to μ and $\lambda(X) > 0$, then there exists a measurable and non-negative function f such that*

$$\int_X f \, d\mu > 0 \quad \textit{and} \quad \int_E f \, d\mu \leqslant \lambda(E) \qquad \textit{for } E \in \mathscr{S}.$$

Proof. Let $X = A_n \cup B_n$ be the Hahn decomposition for the function $\lambda - (1/n)\mu$

(see Th. 7.5.1). Then

$$\frac{1}{n}\mu(E) \leqslant \lambda(E) \qquad \text{if } E \subset A_n, E \in \mathscr{S}, \tag{7.6.4}$$

and $(1/n)\mu \geqslant \lambda$ on measurable subsets of the set B_n. Let $B = \bigcap_{n=1}^{\infty} B_n$, then we have $(1/n)\mu(B) \geqslant \lambda(B)$ for all n so that $\lambda(B) = 0$. Therefore $\lambda(X \setminus B) > 0$. But $X \setminus B = \bigcup_{n=1}^{\infty} A_n$ and so $\lambda(A_p) > 0$ for some p, and so, by absolute continuity, $\mu(A_p) > 0$. If we now take $f = (1/p)\chi_{A_p}$, then $\int_X f \, d\mu = (1/p)\mu(A_p) > 0$ and by (7.6.4) we have

$$\int_E f \, d\mu = \frac{1}{p}\mu(E \cap A_p) \leqslant \lambda(E \cap A_p) \leqslant \lambda(E)$$

for $E \in \mathscr{S}$. ■

Proof of theorem. It suffices to consider the case in which λ, μ are finite measures. For, suppose that the theorem has been proved in this case. By hypothesis

$$X = \bigcup_{i=1}^{\infty} E_i \qquad \text{where } \mu(E_i), |\lambda(E_i)| < \infty,$$

and we can assume that the E_i are disjoint. Then there exist functions φ_i, ψ_i measurable and non-negative on E_i such that

$$\lambda^+(A) = \int_A \varphi_i \, d\mu, \quad \lambda^-(A) = \int_A \psi_i \, d\mu \qquad \text{for } A \subset E_i, A \in \mathscr{S}.$$

Define $\varphi(x) = \varphi_i(x)$ on E_i $(i = 1, 2, \ldots)$ and $\psi(x) = \psi_i(x)$ on E_i $(i = 1, 2, \ldots)$, then we have

$$\lambda^+(E) = \sum_{i=1}^{\infty} \lambda^+(E \cap E_i) = \sum_{i=1}^{\infty} \int_{E \cap E_i} \varphi_i \, d\mu = \int_E \varphi \, d\mu$$

and similarly $\lambda^-(E) = \int_E \psi \, d\mu$ for all $E \subset \mathscr{S}$. Hence

$$\lambda(E) = \int_E \varphi \, d\mu - \int_E \psi \, d\mu = \int_E f \, d\mu, \qquad \text{where } f = \varphi - \psi.$$[46]

So let λ, μ be finite measures on $\mathscr{S}$. We denote by $\mathscr{K}$ the class of all functions h which are measurable and non-negative on X and which satisfy the inequality

$$\int_E h \, d\mu \leqslant \lambda(E) \tag{7.6.5}$$

for all $E \in \mathscr{S}$. If f_n is an increasing sequence of functions in $\mathscr{K}$, then $\lim f_n \in \mathscr{K}$. If $f, g \in \mathscr{K}$ then $\max(f, g) \in \mathscr{K}$, for (7.6.5) holds for $E \subset \{f \leqslant g\}$ and for $E \subset \{f > g\}$ and hence for the disjoint union of such sets. Let

$$\alpha = \sup_{h \in \mathscr{K}} \int_X h \, d\mu. \tag{7.6.6}$$

There exists a sequence $f_\nu \in \mathscr{K}$ such that $\int_X f_\nu \, d\mu \to \alpha$ and we may assume that this sequence is increasing, by replacing it by the sequence $\max(f_1, \ldots, f_n)$ if necessary. Then, writing $f = \lim_{n\to\infty} f_n$, we have $f \in \mathscr{K}$ and

$$\int_X f \, d\mu = \alpha. \tag{7.6.7}$$

It suffices to prove that

$$\int_E f \, d\mu = \lambda(E) \qquad \text{for } E \in \mathscr{S}.$$

Suppose the contrary; then there would exist a set $E_0 \in \mathscr{S}$ such that $\int_{E_0} f \, d\mu < \lambda(E_0)$. Then, for the function

$$\lambda_1(E) = \lambda(E) - \int_E f \, d\mu,$$

which is an absolutely continuous measure on $\mathscr{S}$ relative to μ, we would have $\lambda_1(X) \geqslant \lambda_1(E_0) > 0$. So, by lemma 7.6.1, there would exist a measurable, non-negative function f_1 such that

$$\int_X f_1 \, d\mu > 0 \quad \text{and} \quad \int_E f_1 \, d\mu \leqslant \lambda_1(E)$$

for $E \in \mathscr{S}$. But then, by (7.6.7), we would have

$$\int_X (f + f_1) \, d\mu > \alpha \quad \text{and} \quad \int_E (f + f_1) \, d\mu \leqslant \int_E f \, d\mu + \lambda_1(E) = \lambda(E)$$

and so $f + f_1 \in \mathscr{K}$, contradicting (7.6.6). ■

In Th. 7.6.1, the function f is defined almost everywhere relative to μ. We have

LEMMA 7.6.2. *If μ is a σ-finite measure on $\mathscr{S}$, the functions f and g are integrable on X relative to μ and*

$$\int_E f \, d\mu \leqslant \int_E g \, d\mu \qquad \textit{for } E \in \mathscr{S},$$

then $f \leqslant g$ almost everywhere (μ) on X. In particular, the equality of the integrals for $E \in \mathscr{S}$ implies that the functions are equal almost everywhere.

Indeed, the set $\{g < f\}$ is a countable union of sets of the form

$$A = E \cap \{g < c < d < f\}$$

where $\mu(E) < \infty$ and c and d are constants.[47] But then $\mu(A) = 0$ for otherwise we would have $\int_A g \, d\mu \leqslant c\mu(A) < d\mu(A) \leqslant \int_A f \, d\mu$ contrary to hypothesis.

For a given countably additive function λ and σ-finite measure μ, the function f defined a.e. (μ) by condition (7.6.3) is called the *Radon–Nikodym derivative* of the function λ relative to the measure μ and we denote it by $d\lambda/d\mu$.

It follows from lemma 7.6.2 that

$$\frac{d\lambda}{d\mu} \leqslant \frac{d\nu}{d\mu} \qquad \text{a.e. } (\mu), \text{ if } \lambda \leqslant \nu$$

and that

$$\frac{d}{d\mu}(\alpha\lambda + \beta\nu) = \alpha\frac{d\lambda}{d\mu} + \beta\frac{d\nu}{d\mu} \qquad \text{a.e. } (\mu),$$

(provided $\alpha\lambda, \beta\nu > -\infty$ always or $\alpha\lambda, \beta\nu < \infty$ always).

THEOREM 7.6.2 (Change of Measure). *Under the hypotheses of the Radon–Nikodym theorem, if f is summable relative to λ on E, then $f\,d\lambda/d\mu$ is summable relative to μ on E and*

$$\int_E f\,d\lambda = \int_E f\frac{d\lambda}{d\mu}\,d\mu.^{48} \tag{7.6.8}$$

Proof. It suffices to prove the theorem for $f \geqslant 0$ and $\lambda \geqslant 0$ for then

$$\int_E f\,d\lambda = \int_E f_+\,d\lambda^+ - \int_E f_-\,d\lambda^+ - \int_E f_+\,d\lambda^- + \int_E f_-\,d\lambda^-$$

$$= \int_E f_+\frac{d\lambda^+}{d\mu}\,d\mu - \int_E f_-\frac{d\lambda^+}{d\mu}\,d\mu - \int_E f_+\frac{d\lambda^-}{d\mu}\,d\mu + \int_E f_-\frac{d\lambda^-}{d\mu}\,d\mu$$

$$= \int_E f\frac{d\lambda^+}{d\mu}\,d\mu - \int_E f\frac{d\lambda^-}{d\mu}\,d\mu = \int_E f\frac{d\lambda}{d\mu}\,d\mu.$$

So let $\lambda \geqslant 0$; then $d\lambda/d\mu \geqslant 0$. If f is the characteristic function of the set, then (7.6.8) is equivalent to (7.6.3). Hence (7.6.8) is true for any simple, measurable and non-negative function, since it is a linear combination of characteristic functions. Hence, by Th. 6.1.5, it is true for any measurable, non-negative function f. ■

Let λ, ν be countably additive functions on $\mathscr{S}$. We say that the functions λ, ν are *mutually singular* (or that λ is *singular relative to* ν) if there exists a decomposition $X = A \cup B$, $A \cap B = \varnothing$ such that if $E \in \mathscr{S}$ then $A \cap E$, $B \cap E \in \mathscr{S}$ and $|\lambda|(A \cap E) = |\nu|(B \cap E) = 0$. It follows from the definition that a linear combination of functions which are singular relative to ν is also singular relative to ν. If λ is singular relative to ν then so are λ^+, λ^- and $|\lambda|$. If λ is both singular and absolutely continuous relative to ν, then $\lambda = 0$. We have the following

THEOREM 7.6.3 (Lebesgue Canonical Decomposition). *Let μ be a countably additive, σ-finite function on $\mathscr{S}$. Every function ν which is countably additive and σ-finite on $\mathscr{S}$ is uniquely representable in the form $\nu = \rho + \sigma$, where ρ is absolutely continuous and σ is singular relative to μ.*

Proof. Uniqueness follows from the fact that if $\nu = \bar{\rho} + \bar{\sigma}$ is another such decomposition, then for any $E \in \mathscr{S}$ of finite ν-measure, the function

$$\rho_{S_E} - \bar{\rho}_{S_E} = \bar{\sigma}_{S_E} - \sigma_{S_E}$$

is both absolutely continuous and singular relative to μ_{S_E} and hence is zero.

In the proof of existence it suffices to consider the case in which μ, ν are measures. Since ν is absolutely continuous relative to $\nu + \mu$, therefore

$$\nu(E) = \int_E f \, \mathrm{d}(\mu + \nu) \qquad \text{for } E \in \mathscr{S},$$

where $f = \mathrm{d}\nu/\mathrm{d}(\mu + \nu)$, and since $0 \leqslant \nu \leqslant \mu + \nu$, we can assume that

$$0 \leqslant f \leqslant 1 \qquad \text{on } X.$$

Let $A = \{x : f(x) < 1\}$, $B = \{x : f(x) = 1\}$. If $E \subset A$ and $\mu(E) = 0$ then $\nu(E) = 0$, for this holds, if $\nu(E) < \infty$, by

$$\int_E (1 - f) \, \mathrm{d}\nu = 0.$$

If $E \subset B$, then $\mu(E) = 0$, for since $\nu(E) = \mu(E) + \nu(E)$, this holds if $\nu(E) < \infty$. So, defining $\rho(E) = \nu(A \cap E)$ and $\sigma(E) = \nu(B \cap E)$ we have $\nu = \rho + \sigma$, where ρ is absolutely continuous and σ is singular relative to μ. ■

Now let us turn to the case of a set function on the space $\mathbb{R}^n$. Let Ω be an open set in $\mathbb{R}^n$.

THEOREM 7.6.4. *If λ is a countably additive function on $\mathscr{B}(\Omega)$ which is absolutely continuous relative to m_n and is finite on closed intervals contained in Ω then $F(R) = \lambda(R)$ is an additive interval function which is absolutely continuous on every closed interval contained in Ω*[49] *and*

$$DF = \frac{\mathrm{d}\lambda}{\mathrm{d}m_n} \qquad \textit{almost everywhere in } \Omega. \tag{7.6.9}$$

Proof. If $R_1, R_2 \subset \Omega$ are adjacent closed intervals, then since $m_n(R_1 \cap R_2) = 0$ we have $\lambda(R_1 \cap R_2) = 0$, so that

$$F(R_1 \cup R_2) = \lambda(R_1 \cup R_2) = \lambda(R_1) + \lambda(R_2) = F(R_1) + F(R_2)$$

showing that F is an additive function. It follows from (7.6.2) that F is absolutely continuous on every closed interval contained in Ω. Finally, by Th. 7.6.1,

$$F(R) = \lambda(R) = \int_R \frac{\mathrm{d}\lambda}{\mathrm{d}m_n} \, \mathrm{d}x,$$

so that, from Th. 7.1.7, we obtain the relation (7.6.9). ■

THEOREM 7.6.5. *If F is an additive interval function which is absolutely continuous on an interval P, then the associated function F^* is a countably additive* *function on $\mathscr{B}(\operatorname{Int} P)$ which is absolutely continuous relative to m_n. Also*

$$DF = \frac{dF^*}{dm_n} \quad \text{almost everywhere in } P \tag{7.6.10}$$

and

$$F^*(E) = \int_E DF\, dx \quad \text{for } E \in \mathscr{B}(\operatorname{Int} P). \tag{7.6.11}$$

Proof. By Th. 7.5.6, every interval $R \subset \operatorname{Int} P$ is an interval of continuity for F so that

$$F(R) = F^*(R). \tag{7.6.12}$$

In proving the absolute continuity of F^* we may assume that F is non-negative (since $F^* = (F^+)^* - (F^-)^*$). Suppose that $E \in \mathscr{B}(\operatorname{Int} P)$ and $m_n(E) = 0$. Choose $\varepsilon > 0$. Let $\delta > 0$ correspond to ε in the definition of absolute continuity of F and let $E \subset \bigcup_{i=1}^{\infty} P_i$, where P_i are closed intervals contained in $\operatorname{Int} P$ such that $\sum_{i=1}^{\infty} |P_i| < \delta$. Fix n and let $\bigcup_{i=1}^{n} P_i = \bigcup_{i=1}^{k} R_i$, where R_i are non-overlapping intervals. But $\sum |R_i| \leqslant |P_1| + \cdots + |P_n| < \delta$ so that, by (7.6.12) we have $F^*(\bigcup_{i=1}^{n} P_i) \leqslant \sum F^*(R_i) < \varepsilon$. Hence in the limit we obtain $F^*(\bigcup_{i=1}^{\infty} P_i) \leqslant \varepsilon$. Thus $F^*(E) = 0$. This shows that F^* is absolutely continuous relative to m_n. Therefore, by (7.6.12) and Th. 7.6.4 (taking $\lambda = F^*$) we have the relation (7.6.10) and, by the definition of the Radon–Nikodym derivative, the relation (7.6.11). ∎

Let F be an additive interval function of bounded variation on an interval P. From Ths 7.6.4–7.6.5, it follows that F is absolutely continuous if and only if F^* is absolutely continuous relative to m_n. We have similarly

THEOREM 7.6.6. *The function F is singular if and only if F^* is singular relative to m_n.*

Proof. Suppose first that F^* is singular relative to m_n; then so is $\lambda = |F^*|$ so that

$$\lambda(E) = m_n(P \backslash E) = 0 \tag{7.6.13}$$

for some $E \in \mathscr{B}(\operatorname{Int} P)$. To show that $DF = 0$ a.e. on P, noting that $|F(R)| \leqslant \lambda^{\#}(R)$ on common intervals of continuity of F and $\lambda^{\#}$, it suffices to prove that $D\lambda^{\#} = 0$ a.e. on E. But if this were not so we would have $D\lambda^{\#} > \varepsilon$ on some closed $C \subset E$ with positive m_n- measure, for some $\varepsilon > 0$. Therefore

$$\lambda(C) \geqslant \varepsilon m_n(C),$$

because for some decreasing sequence of open sets $G_n \subset P$ we have

$$C = \cap G_n \quad \text{and} \quad \lambda(G_n) \geqslant \varepsilon m_n(C);$$

for, taking the decomposition $G_n = \cup Q_i$, given by lemma 5.4.1, we will have

$$\lambda(Q_i) \geqslant \varepsilon m_n(Q_i \cap C)$$

by lemma 7.1.2. Hence $\lambda(E) > 0$, which contradicts (7.6.13).

Suppose, conversely, that F is singular. By Th. 7.6.3, noting the first part of the proof, we have

$$F^* = G^* + H^*,$$

where G^*, G are absolutely continuous and H^*, H are singular. Then, recalling the remark following Th. 7.5.7, $F = G + H + T$ where T is singular. Hence, by the uniqueness of the Lebesgue decomposition, we must have $G = 0$. Hence $F^* = H^*$ is singular. ■

Thus, if $F = G + H$ (where F, G, H are functions of bounded variation on P), then

$$F^* = G^* + H^*$$

is a Lebesgue decomposition if and only if

$$F = G + H$$

is a Lebesgue decomposition, in which case, by Th. 7.6.5, we have

$$G^*(E) = \int_E DF\,\mathrm{d}x \qquad \text{for } E \in \mathscr{B}\,(\mathrm{Int}\,P).$$

A direct corollary of Th. 7.6.2 and Th. 7.6.5 is given by the following generalization of Th. 7.3.5:

THEOREM 7.6.7. *If F is an additive interval function which is absolutely continuous on an interval P and if the function f is summable relative to F on the set $E \in \mathscr{B}$ (Int P), then fDF is summable on E and*

$$(\mathrm{L}) \int_E f\,\mathrm{d}F = \int_E f\,DF\,\mathrm{d}x. \tag{7.6.14}$$

Change of Variable in a Lebesgue Integral

We now prove a theorem on changing the variables in a Lebesgue integral.

THEOREM 7.6.8. *Let $u = (u_1, \ldots, u_n) \to \varphi(u) = (\varphi_1(u), \ldots, \varphi_n(u))$ be a homeomorphism of an open set $\Omega \subset \mathbb{R}^n$ onto an open set $G \subset \mathbb{R}^n$. Suppose that the functions $\varphi_1, \ldots, \varphi_n$ satisfy a Lipschitz condition on Ω.[50] If the set $E \subset \Omega$ is measurable then the set $\varphi(E) \subset G$ is measurable and if the function f is summable on $\varphi(E)$ then the function $u \to f(\varphi(u))|\Delta(u)|$ is summable on E and*

$$\int_{\varphi(E)} f(x)\,\mathrm{d}x = \int_E f(\varphi(u))|\Delta(u)|\,\mathrm{d}u, \tag{7.6.15}$$

where

$$\Delta(u) = \frac{\partial(\varphi_1, \ldots, \varphi_n)}{\partial(u_1, \ldots, u_n)}$$

is the Jacobian of the mapping $u \to \varphi(u)$, defined almost everywhere (*by* Th. 7.1.8) *in* Ω. *In particular* (*taking* $f(x)=1$)

$$m_n(\varphi(E)) = \int_E |\Delta(u)|\, du.$$

Proof. The measurability of $\varphi(E)$ follows from lemma 5.4.4. It suffices to consider the case $E=\Omega$ (in the general case we can extend f to G by putting $f=0$ on $G\backslash\varphi(E)$). The function

$$\lambda(E) = m_n(\varphi(E))$$

is easily shown to be a measure on $\mathscr{B}(\Omega)$ and (lemma 5.4.4) it is absolutely continuous relative to m_n. Given a function f defined on G we let $g = f\circ\varphi$. If f is $\mathscr{B}(G)$-measurable, then, by Th. 4.5.3, g is $\mathscr{B}(\Omega)$-measurable. If, in addition, f is summable, then g is summable relative to λ and

$$\int_G f \mathrm{d}x = \int_\Omega g\, \mathrm{d}\lambda;$$

indeed, this holds if f is the characteristic function of a set in $\mathscr{B}(G)$ (Th. 4.3.4) and hence successively, assuming $\mathscr{B}(G)$-measurability, if f is simple and non-negative, if f is summable and non-negative (Th. 6.1.5) and finally, if f is summable (from the definition of the integral). Thus, by Th. 7.6.2, if the function f is $\mathscr{B}(G)$-measurable and summable then the function $g\,\mathrm{d}\lambda/dm_n$ is summable and

$$\int_G f \mathrm{d}x = \int_\Omega g \frac{\mathrm{d}\lambda}{dm_n} du. \tag{7.6.16}$$

If $f=0$ a.e. in G, then $g\,\mathrm{d}\lambda/dm_n = 0$ a.e. in Ω; for then there exists a set $Z\in\mathscr{B}(G)$ such that $m_n(Z)=0$ and $f=0$ on $G\backslash Z$ (lemma 5.4.3 or Th. 5.4.2), hence $g=0$ on $\Omega\backslash Z_1$, where $Z_1 = \varphi^{-1}(Z)\in\mathscr{B}(\Omega)$ and also $\mathrm{d}\lambda/dm_n = 0$ a.e. on Z_1, by Th. 6.1.4, since

$$\int_{Z_1} \frac{\mathrm{d}\lambda}{dm_n} \mathrm{d}x = \lambda(Z_1) = m_n(Z) = 0.$$

Thus (Th. 5.5.4), if f is summable on G, then $g\,\mathrm{d}\lambda/dm_n$ is summable on Ω and relation (7.6.16) holds. But, by Th. 7.6.4, we also have the equality

$$\frac{\mathrm{d}\lambda}{dm_n}(u) = \lim_{\delta(Q)\to 0,\, u\in Q} \frac{m_n(\varphi(Q))}{|Q|}$$

a.e. on Ω. Now, by Th. 7.1.8 and lemma 5.4.6, the right-hand side of this equation equals $|\Delta(u)|$ a.e. on Ω, so that

$$\int_G f(x)\,\mathrm{d}x = \int_\Omega g(u)|\Delta(u)|\,du. \quad \blacksquare$$

Remark. If we assume that the mapping φ is injective on the open set Ω

(continuity follows from the Lipschitz condition) then by the theorem on the invariance of domain,[51] $G=\varphi(\Omega)$ is an open set and φ is a homeomorphism.

NOTES

1. The assumption of boundedness is not essential.
2. This proof was given by Z. Opial.
3. They are properly defined if there exists at least one sequence Q_ν such that $F(Q_\nu)$ is defined, $x\in Q_\nu$ and $\delta(Q_\nu)\to 0$.
4. And if there exists a sequence as in Note 3 which is common to both F and G.
5. In the n-dimensional case, given a function f of n variables, there is an associated additive function of n-dimensional intervals defined by

$$F([a_1^{(1)},a_1^{(2)}]\times\cdots\times[a_n^{(1)},a_n^{(2)}])=\sum(-1)^{\sum_{i=1}^n \varepsilon_i} f(a_1^{(\varepsilon_1)},\ldots,a_n^{(\varepsilon_n)}).$$

   Conversely, every additive function is of this form. If f is of class $\mathscr{C}^n$, then $DF(x)=\partial^n f/\partial x_1\cdots\partial x_n(x)$.
6. In accordance with the definition in §5.3, we require that this function be finite.
7. Approximate continuity suffices assuming boundedness, but this is not sufficient in the case of summability.
8. A direct proof of this theorem is given in Appendix 1.
9. The assumption of continuity is not essential.
10. A proof of Stepanov's theorem is given in Appendix 3.
11. In the case of functions of a single variable, this definition coincides with that of the integral $\int_a^b f(x)\,\mathrm{d}g(x)$, where $P=[a,b]$ and g is the function, of bounded variation, with which the function G is associated.
12. These formulae imply that for interval functions F associated with functions f of one variable, the Jordan decomposition corresponds to the Jordan decomposition of f given by Th. 1.4.1: F^+ and F^- are associated respectively with φ and ψ.
    If $F=F'-F''$, where F', F'' are additive and non-negative then $F^+\leqslant F'$ and $F^-\leqslant F''$ (this follows from (7.2.4)).
13. It can be shown that for a function H of bounded variation on an interval P to be singular it is necessary and sufficient that for any $\varepsilon>0$ there exists a system of non-overlapping intervals $P_1,\ldots,P_k$ contained in P such that $\sum_i |H|(P_i)<\varepsilon$ and $m_n(P\setminus\bigcup_i P_i)<\varepsilon$.
14. More generally $\varphi(f,g)$ is absolutely continuous if f, g are absolutely continuous and φ satisfies a Lipschitz condition (locally). In particular, it holds if φ is of class $\mathscr{C}^1$.
15. For example the functions φ, ψ in a Jordan decomposition (Th. 1.4.1).
16. The function $f(\varphi(t))$ itself may not even be measurable.
17. If the function φ is absolutely continuous on $[\alpha,\beta]$ then (7.4.5) holds assuming that f is measurable and bounded on $[a,b]$. In the case where f is continuous, it follows immediately from the fact that the function $F\circ\varphi$, where $F(x)=\int_a^x f(y)\,\mathrm{d}y$, is absolutely continuous and $F'(x)=f(x)$ on $[a,b]$.
18. Then, by Th. 7.4.1, $\varphi(A)$ is measurable.
19. Note that a function of one variable is determined to within an additive constant by its associated interval function.
20. It can be shown that (7.4.8), and (7.3.3) also, holds if f is only integrable relative to g (absolutely continuous). In fact, in the case where g is increasing the method used in the proof of Th. 6.2.13 is applicable, while the general case can be reduced to the previous case by using the canonical Jordan decomposition.

21. This certainly holds if, for example, t is the natural (arc-length) parameter, for then $s(t)=t$ is absolutely continuous.
22. Proofs of both theorems are given in Appendix 4.
23. Even absolutely continuous.
24. This version is due to de la Vallée Poussin.
25. This might, for example, be one of the Dini derivatives. Among the λ satisfying (7.4.20) the one with the smallest absolute value is $\lambda_+ f=(D_+f)_+-(D^+f)_-$ (respectively, $\lambda_- f=(D_-f)_+-(D^-f)_-$).
26. The existence of such a sequence is guaranteed by the following

    LEMMA. *If* $\varphi:(0,c)\to\mathbb{R}$ *is continuous, then the set of its limit points as* $t\to 0+$ *is the whole interval* $[\liminf \varphi(t), \limsup \varphi(t)]$.

27. That is, possessing a finite derivative.
28. Even more generally (see Note 25), in the class of continuous functions f such that $\{|f'_+|=\infty\}$ (respectively $\{|f'_-|=\infty\}$) is at most countable, absolute continuity is equivalent to the summability of $\lambda_+ f$ (respectively $\lambda_- f$).
29. If $\mathscr{K}$ is a finitely additive algebra then condition (2°) is a consequence of the others, where in equation (3°) we assume that the right-hand side is well defined.
30. Note that $|\lambda(B)|<\infty$.
31. $\lambda^+(E)<\infty$, if $\lambda(E)<\infty$, $\lambda^-(E)<\infty$, if $\lambda(E)>-\infty$ (by (7.5.5)). If $\lambda=\lambda'-\lambda''$, where λ', λ'' are measures, then $\lambda^+\leqslant\lambda'$ and $\lambda^-\leqslant\lambda''$ (which follows, for example, from (7.5.4)).
32. It is not essential for the space to be metric. In Th. 7.5.4 for a locally compact, Hausdorff topological space we impose on λ the following regularity condition: $|\lambda_{\mathscr{B}(G)}|$ is regular for every open, relatively compact set G. This is equivalent (see Notes 6.17, 6.20, 6.22) to the condition that every open, relatively compact set is the union of a σ-compact set and a set of measure zero (i.e. such that $\lambda=0$ on its subsets). The proof remains unchanged, where we make use of the fact that the sum of finite regular measures is regular and that the regularity of the finite measure μ implies the regularity of the measure $\nu\leqslant\mu$ (which is an easy consequence of the equivalence in Note 6.17). Instead of imposing a regularity condition on λ we could replace $\mathscr{B}(X)$ by $\mathscr{B}^\sigma(X)$ (see Note 6.22)).
33. See Note 32.
34. For, let $\varphi\in C_0$. By Tietze's theorem (see e.g. Kuratowski, *Topology*, I, p. 127) there exist functions ε_n continuous on X such that $0\leqslant\varphi_n\leqslant 1$ in X, $\varepsilon_n=0$ on $\{|\varphi|\leqslant 1/(n+1)\}$ and $\varepsilon_n=1$ on $\{|\varphi|\geqslant 1/n\}$. Then $\varphi\varepsilon_n\in C$ and $\varphi\varepsilon_n\to\varphi$ uniformly on X.
35. Or by $\mathscr{B}^\sigma(X)$ in the case of a topological space (see Note 6.20).
36. In the case of a topological space only the regularity condition may be simplified (see Note 6.20).
37. For $\frac{4}{9}(b_1-a_1)^2+\sum_2^n(b_i-a_i)^2=\sum_1^n(b_i-a_i)^2-\frac{5}{9}(b_1-a_1)^2\leqslant(1-(5/9n))\sum_1^n(b_i-a_i)^2$.
38. From condition (7.5.12) it follows that these measures are finite on closed intervals.
39. It is easily shown that the assumption of monotonicity is not essential.
40. This condition is equivalent to the condition that the set of points $(a_1,b_1,\ldots,a_n,b_n)$ in the space $\mathbb{R}^{2n}$ for which $[a_1,b_1]\times\cdots\times[a_n,b_n]\in\mathscr{K}$, is dense in the open set of points $(a_1,b_1,\ldots,a_n,b_n)$ of $\mathbb{R}^{2n}$ for which $[a_1,b_1]\times\cdots\times[a_n,b_n]\subset\Omega$.
41. We consider this case for the sake of simplicity; for if, as before, we consider functions of bounded variation defined on closed subintervals of Ω, then, without the assumption of the finiteness of variation of F on Ω (defined analogously), the natural domain of F would not be a countably additive algebra.

42. $F_i^*(E) \leqslant F_i(P)$.
43. For the Q_i we can take, for example, the intervals in a normal system (see §5.3) for pairs P_ν, R_ν contained in R_ν and not overlapping P_ν.
44. As in the single-variable case (§1.5–1.6) the Riemann–Stieltjes integral then exists on every interval $R \subset P$ and is an additive function of bounded variation.
45. σ-finiteness (see subsequent definition) is not sufficient.
46. Indeed, one of the integrals is finite, e.g. $\int_E \psi \, d\mu < \infty$. Since $f_- = \max(0, \psi - \varphi) \leqslant \psi$, therefore $\int_E f_- \, d\mu < \infty$, hence $\int_E f \, d\mu$ exists. From the equality $f_+ - f_- = \varphi - \psi$ it follows that $f_+ + \psi = \varphi + f_-$ (at points where f_- and ψ are finite, that is, a.e. in E), hence

$$\int_E f_+ \, d\mu + \int_E \psi \, d\mu = \int_E \varphi \, d\mu + \int_E f_- \, d\mu,$$

so that $\int_E f \, d\mu = \int_E \varphi \, d\mu - \int_E \psi \, d\mu$. See also Note 6.6.
47. Take $E = E_i$, where $X = \bigcup_1^\infty E_i$, $\mu(E_i) < \infty$ and c, d rational.
48. If, in addition, λ is a measure and ρ is a countably additive, σ-finite function on $\mathscr{S}$ which is absolutely continuous relative to λ, then ρ is also absolutely continuous relative to μ and taking $f = d\rho/d\lambda$ we obtain, as a corollary of Th. 7.6.2, the formula

$$\frac{d\rho}{d\mu} = \frac{d\rho}{d\lambda} \cdot \frac{d\lambda}{d\mu}.$$

49. By Th. 7.5.7, λ is therefore the function associated with F.
50. It suffices to assume that these functions satisfy a Lipschitz condition in a neighbourhood of each point of Ω (with a constant which may depend on the neighbourhood).
51. See K. Kuratowski, *Topology*, II, §53, IV, 10.

APPENDIX 1

ANOTHER PROOF OF LEBESGUE'S THEOREM ON THE DIFFERENTIABILITY OF THE INTEGRAL

(Th. 7.1.7)[1]

It suffices to consider the case $f \geqslant 0$.

Let x be both a point of approximate continuity of f and a point at which DF exists. Then, there exists a measurable set $E \subset P$ for which x is a density point and also a point of continuity of f_E. If Q is a cube of sufficiently small diameter, containing x, we therefore have $m_n(E \cap Q) > 0$ and

$$\frac{F(Q)}{|Q|} \geqslant \frac{m_n(E \cap Q)}{|Q|} \cdot \frac{1}{m_n(E \cap Q)} \int_{E \cap Q} f \, \mathrm{d}x \geqslant \frac{m_n(E \cap Q)}{|Q|} \inf_{E \cap Q} f,$$

so that in the limit as $\delta(Q) \to 0$ we obtain $DF(x) \geqslant f(x)$.

Hence, by Ths 7.1.3–7.1.4,

$$DF - f \geqslant 0 \qquad \text{a.e. in } P.$$

On the other hand, by Th. 7.2.3, $\int_P DF \, \mathrm{d}x \leqslant F(P)$, or

$$\int_P (DF - f) \, \mathrm{d}x \leqslant 0,$$

so that, by Th. 7.1.4, $DF - f = 0$, or $DF = f$ a.e. in P. ■

NOTE

1. This follows an idea in the book: S. Hartman and J. Mikusiński, *Theory of Lebesgue Measure and Integration* (Polish).

APPENDIX 2

ANOTHER PROOF OF FUBINI'S THEOREM FOR LEBESGUE INTEGRALS

(Ths 6.3.7–6.3.8)[1]

In §6.3, we established the inclusion $\mathscr{L}_p \times \mathscr{L}_q \subset \mathscr{L}_{p+q}$. In particular, this implies that:

If E is $(\mathscr{L}_p)$-measurable and F is $(\mathscr{L}_q)$-measurable, then $E \times F$ is $(\mathscr{L}_{p+q})$-measurable.

We now give a direct proof of this assertion. Now, if $m_p(E) = 0$ and Q is an interval in $\mathbb{R}^q$, then

$$m_{p+q}(E \times Q) = 0$$

(because, for a covering $E \times Q \subset \cup(P_\nu \times Q)$, where $E \subset \cup P_\nu$, the sum of the contents $\sum_\nu |P_\nu \times Q| = (\sum_\nu |P_\nu|)|Q|$ can be made arbitrarily small). Thus, if $m_p(E) = 0$, then

$$m_{p+q}(E \times \mathbb{R}^q) = 0$$

(because $\mathbb{R}^q = \bigcup_\nu Q_\nu$ so that $E \times \mathbb{R}^q = \bigcup_\nu (E \times Q_\nu)$, where Q_ν are intervals). If now E is $(\mathscr{L}_p)$-measurable then, by Th. 5.4.2′, $E = \bigcup_\nu A_\nu \cup Z$ where the A_ν are closed and $m_p(Z) = 0$, therefore

$$E \times \mathbb{R}^q = \bigcup_\nu (A_\nu \times \mathbb{R}^q) \cup (Z \times \mathbb{R}^q).$$

It follows by Th. 5.4.2′, that $E \times \mathbb{R}^q$ is $(\mathscr{L}_{p+q})$-measurable. Similarly, if F is $(\mathscr{L}_q)$-measurable then $\mathbb{R}^p \times F$ is $(\mathscr{L}_{p+q})$-measurable. Now since $E \times F = (E \times \mathbb{R}^q) \cap (\mathbb{R}^p \times F)$ the proof is complete.

PROOF OF FUBINI'S THEOREM FOR MEASURABLE, NON-NEGATIVE FUNCTIONS
(Th. 6.3.7)[2]

It suffices to consider the case

$$E = \mathbb{R}^p \quad \text{and} \quad F = \mathbb{R}^q,$$

for in the general case we can extend f to $\mathbb{R}^{p+q}$ by putting $f = 0$ on $\backslash(E \times F)$.

Let the function f be defined on $\mathbb{R}^{p+q}$ and for $y \in \mathbb{R}^q$ denote by f^y the function defined on $\mathbb{R}^p$ by the formula $f^y(x) = f(x, y)$. We say that f has property (F) when it satisfies the following four conditions:

(1°) f is measurable and non-negative,
(2°) f^y is measurable for almost all y,
(3°) $g(y) = \int f^y \, dx$ (defined for almost all y) is measurable,
(4°) $\int f \, d(x, y) = \int (\int f^y \, dx) \, dy$.

It is therefore sufficient to prove that every measurable, non-negative function has property (F).

We verify that:

(1) if $f_1, \ldots, f_k$ have property (F) and $\alpha_1, \ldots, \alpha_k \geqslant 0$, then $\sum_1^k \alpha_j f_j$ has property (F);
(2) if f, g have property (F), $f \leqslant g$ and $\int g \, d(x, y) < \infty$, then $g - f$ has property (F);
(3) If g has property (F), $g = 0$ almost everywhere and $0 \leqslant f \leqslant g$, then f has property (F);
(4) if f_ν have property (F) and $f_\nu \leqslant f_{\nu+1}$, $\nu = 1, 2, \ldots$, then $\lim_{\nu \to \infty} f_\nu$ has property (F).

We say that the set $E \subset \mathbb{R}^{p+q}$ has property (J) if χ_E has property (F). We check that any interval has property (J). It follows from (4) that the limit of any increasing sequence of sets with property (J) has property (J). From (1) and (4) it follows that the union of at most a countable number of disjoint sets with property (J) has property (J). It follows from (2) that if E, F have property (J), $E \subset F$ and F is bounded, then $F \backslash E$ has property (J). Finally, (3) implies that any subset of a set of measure zero with property (J) has property (J).

We then obtain successively the following consequences: every open set in $\mathbb{R}^{p+q}$ has property (J) (lemma 5.4.1); every bounded closed set has property (J), since it equals $G_2 \backslash G_1$, where $G_1 \subset G_2$ are open and bounded; every bounded set of type F_σ has property (J); every bounded set of type G_δ has property (J), for it equals $G \backslash S$, where $S \subset G$, G is open and bounded and S is of type F_σ; every bounded set of measure zero has property (J), by lemma 5.4.3; every bounded measurable set has property (J), by Th. 5.4.2′; every measurable set has property (J).

Consequently, the characteristic function of any measurable set in $\mathbb{R}^{p+q}$ has property (F). Hence, by (1), any simple, measurable, non-negative function has property (F). Hence, finally, by (4) and by Th. 4.4.10, any measurable, non-negative function has property (F). ■

PROOF OF FUBINI'S THEOREM FOR SUMMABLE FUNCTIONS (Th. 6.3.8)

This is covered by the proof of Th. 6.3.5.

COROLLARY 1. *If E is $(\mathscr{L}_p)$-measurable and F is $(\mathscr{L}_q)$-measurable, then*

$$m_{p+q}(E \times F) = m_p(E) m_q(F).$$

It suffices to apply Fubini's theorem to the function $\chi_{E\times F}$.

COROLLARY 2. *If f is summable, or measurable and non-negative, on a measurable set $E \subset \mathbb{R}^p$, and if g has the same property on a measurable set $F \subset \mathbb{R}^q$, then $(x, y) \rightarrow f(x)g(y)$ is measurable and*

$$\int_{E\times F} f(x)g(y)\,\mathrm{d}(x, y) = \left(\int_E f\,\mathrm{d}x\right)\left(\int_F g\,\mathrm{d}y\right).$$

The functions $(x, y) \rightarrow f(x)$ and $(x, y) \rightarrow g(y)$ are $(\mathscr{L}_{p+q})$-measurable and hence so is their product.

NOTES

1. This follows an idea in the book; S. Saks, *Theory of the Integral.*
2. Some of the details of the proof are left to the reader to supply.

APPENDIX 3

STEPANOV'S THEOREM ON DIFFERENTIABILITY ALMOST EVERYWHERE

(Th. 7.1.9)

LEMMA 1. *If a function f defined and finite on a subset A of a metric space X satisfies a Lipschitz condition with constant M,* i.e. $|f(x')-f(x)| \leqslant M\rho(x,x')$ *for $x, x' \in A$, then it has an extension to X which satisfies a Lipschitz condition with the same constant M.*

Indeed, it suffices to take the function $\bar{f}(x) = \inf_{c\in A}(f(c) + M\rho(x,c))$ for $x \in X$.[1] For it is clear that $\bar{f} = f$ on A. If $x, x' \in X$ then, since for any $c \in A$ we have $f(c) + M\rho(x,c) \leqslant f(c) + M\rho(x',c) + M\rho(x,x')$, therefore, taking the infimum, $\bar{f}(x) \leqslant \bar{f}(x') + M\rho(x,x')$. It follows immediately that the function $\bar{f}$ is finite on X and satisfies a Lipschitz condition with constant M. ■

LEMMA 2. *If a is a density point of the set $A \subset \mathbb{R}^n$, then*

$$\lim_{x\to a} \frac{\rho(x,A)}{|x-a|} = 0.$$

Indeed, let $0 < \varepsilon < \frac{1}{4}$. Given $c \in \mathbb{R}^n$ and $r > 0$ we denote by $Q_r(c)$ the cube $\{x : |x_i - c_i| \leqslant \frac{1}{2}r, i = 1, \ldots, n\}$. Thus, there exists $\delta > 0$ such that $m^*(A \cap Q_r(a))/r^n > 1 - \varepsilon^n$ if $0 < r < \delta$. Now suppose that $|x-a| < \delta$. Putting $r = |x-a|$ we then have $Q_{\varepsilon r}(x) \cap A \neq \varnothing$ for otherwise $A \cap Q_r(a) \subset Q_r(a) \backslash Q_{\varepsilon r}(x)$, and since $Q_{\varepsilon r}(x) \subset Q_r(a)$ we would have $m^*(A \cap Q_r(a))/r^n \leqslant (r^n - (\varepsilon r)^n)/r^n = 1 - \varepsilon^n$. Hence $\rho(x,A) \leqslant \frac{1}{2}\sqrt{(n)}\varepsilon r$ so that $\rho(x,A)/|x-a| \leqslant \frac{1}{2}\sqrt{(n)}\varepsilon$. ■

THEOREM (Stepanov). *A function f defined on an open set $G \subset \mathbb{R}^n$ and satisfying, at every point a of the set $E \subset G$, the condition*

$$\limsup_{x\to a} \frac{|f(x) - f(a)|}{|x-a|} < \infty,$$

has a differential at almost every point of E.

Proof. It follows by hypothesis that $E = \bigcup_{k=1}^{\infty} E_k$, where $E_k = \{c \in E : |f(x) - f(c)| \leqslant k|x - c|$, if $x \in G$ and $|x - c| < 1/k\}$. Furthermore, E_k is an at most countable union of sets E_{kr} of diameter $< 1/2k$, and we have $E = \bigcup_{k,r} E_{kr}$. It therefore suffices to prove that f has a differential at almost every point of the set $A = E_{kr}$. The function f_A satisfies a Lipschitz condition with constant k and so (lemma 1) it has an extension g to the set G which satisfies a Lipschitz condition with constant k and which, by Rademacher's theorem, has a differential a.e. in G. Let $\varphi = f - g$ (so that $\varphi = 0$ on A). By Lebesgue's theorem on density points (Th. 7.1.2) it suffices to prove that if $a \in A$ is a density point of the set A, then φ has a differential at a. Thus, let $x \in G$ and let $|x - a| < 1/2k$. For any $\delta > 0$ there exists $z \in A$ such that $|x - z| \leqslant \rho(x, A) + \delta$ so that, noting that $z \in E_k$ and $|x - z| < 1/k$, we have $|\varphi(x) - \varphi(a)| = |\varphi(x) - \varphi(z)| \leqslant |f(x) - f(z)| + |g(x) - g(z)| \leqslant 2k|x - z| \leqslant 2k(\rho(x, A) + \delta)$. Hence $|\varphi(x) - \varphi(a)| \leqslant 2k\rho(x, A)$. Hence, by lemma 2, we obtain

$$\lim_{x \to a} \frac{\varphi(x) - \varphi(a)}{|x - a|} = 0. \quad \blacksquare$$

NOTE

1. We exclude the trivial case where $A = \varnothing$.

APPENDIX 4

PROOFS OF THE SIERPIŃSKI–YOUNG AND DENJOY–YOUNG–SAKS THEOREMS[1]

The Sierpiński–Young theorem for the set $\{D^-f<D_+f\}$ follows from the fact that this set is the union, over rational q, of the sets

$$\{D^-f<q<D_+f\}=\{D^-(f-q_x)<0<D_+(f-q_x)\}.$$

Moreover, each of these sets is at most countable since, we observe, the set of points at which $f-q_x$ attains a local strict minimum is at most countable. The set $\{D^+f<D_-f\}$ is dealt with similarly.

All four cases of the Denjoy–Young–Saks theorem can be obtained from the first case: in the set $\{D_+f>-\infty\}$ we simply replace $f(t)$ by $-f(t)$, $f(-t)$ and by $-f(-t)$ respectively. Moreover, it suffices to prove the result:

$$-\infty<D^-f=D_+f<\infty \text{ a.e. on the set } \{D_+f>0\}, \qquad (*)$$

for then, replacing f by $f+n_x$ we obtain (*) on the set $\{D_+f>-n\}$, $n=1,2,\ldots$.

Let E be the domain of f. It suffices to prove (*) on each non-empty set of the form

$$C=\{t\in E:t<r, f(t)<f(u) \text{ for } t<u<r \text{ (in } E)\},$$

for their union, taken over rational r, contains $\{D_+f>0\}$. Let $c=\inf C$. The function

$$g(x)=\sup_{C\cap(c,x]} f$$

is increasing in (c,r), $f\geqslant g$ in $E\cap(c,r)$ and $f=g$ in C. Hence, since g' exists and is finite a.e. in (c,r) and since D^-f, D_+f are well defined except for at most a countable number of points of E, we have $D^-f\leqslant g'\leqslant D_+f$ a.e. in C. Thus, by the Sierpiński–Young theorem, we obtain (*) in C.

NOTE

1. See §7.4; the proofs follow ideas taken from the books: S. Saks, *Theory of the Integral*, and F. Riesz and B.Sz.Nagy, *Functional Analysis*.

EXERCISES

The following collection of exercises, compiled by M. Kosiek, W. Mlak and Z. Opial, covers a part of the material contained in the book. The reader can find other topics for exercises in the assertions made without proof in some sections of the text.

CHAPTER 1

1 Calculate the variation of the function f on the interval $[a, b]$:

(a) $f(x) = x^4 + x^3 - 3x^2 - x + 2, \qquad a = -3,\ b = 3.$

(b) $f(x) = \begin{cases} \sin(1/x) & x \neq 0, \\ 0 & x = 0. \end{cases} \qquad a = 0,\ b = 1,$

(c) $f(x) = \begin{cases} \exp(-1/x)\sin(\pi/x) & x \neq 0, \\ 0 & x = 0. \end{cases} \qquad a = 0,\ b = 1,$

2 Express the functions given in Exercise 1 as a difference of two increasing functions.

3 Calculate the Riemann–Stieltjes integrals of the following functions:

(a) $\displaystyle\int_0^1 (x^2 - 2x + 5)\mathrm{d}(\mathrm{e}^x),$

(b) $\displaystyle\int_{-\pi}^0 (x^4 - 5)\mathrm{d}(\sin x),$

(c) $\displaystyle\int_{-1}^5 (x^2 + 3x - 2)\mathrm{d}[x]$, where $[x]$ is the largest integer which is less than or equal to x.

4 Show that the function defined on the Cantor set in the example following Th. 7.4.9 has bounded variation.

5 Show that $\int x\,\mathrm{d}h(x) = 0$, where h is the function specified in Exercise 4. *Hint*: make use of Th. 1.6.2.

6 The function f is strictly increasing in the interval $[a, b]$. Show that the inverse function to f is continuous on $f([a, b])$.

7 The function f is invertible and has Darboux's property in the interval $[a,b]$. Show that the function f is continuous on $[a,b]$.

8 The functions f and g are continuous on $[a,b]$ and g is strictly increasing in $[a,b]$. Let

$$F(x)=\int_a^x f(s)\,\mathrm{d}g(s).$$

Show that

$$\frac{\mathrm{d}F}{\mathrm{d}g}=\lim_{h\to 0}\frac{F(x+h)-F(x)}{g(x+h)-g(x)}=f(x)$$

for $x\in[a,b]$.

9 The functions f and g are continuous on $[a,b]$ and g is strictly increasing in $[a,b]$. Suppose that for every $x\in[a,b]$ the limit

$$\lim_{h\to 0}\frac{f(x+h)-f(x)}{g(x+h)-g(x)}$$

exists and is denoted by $\mathrm{d}f(x)/\mathrm{d}g(x)$. Suppose that the function $\mathrm{d}f(x)/\mathrm{d}g(x)$ is continuous with respect to x. Show that

$$f(b)-f(a)=\int_a^b \frac{\mathrm{d}f(s)}{\mathrm{d}g(s)}\,\mathrm{d}g(s).$$

10 A sequence of functions f_n, continuous on $[a,b]$, converges uniformly on $[a,b]$ to the function f. The functions g_n are of bounded variation and there exists a constant M such that $W_a^b(g_n)\leqslant M$. Show that

$$\int_a^b f_n(x)\,\mathrm{d}g_n(x)\to\int_a^b f(x)\,\mathrm{d}g(x),$$

where $g_n(x)\to g(x)$ at each point $x\in[a,b]$.

11 The function g is of bounded variation and is continuous on $[a,b]$. Suppose that for any continuous function f we have the equality

$$\int_a^b f(x)\,\mathrm{d}g(x)=0.$$

Show that g is constant on $[a,b]$.

12 Let $f(x)\geqslant m>0$ for $x\in[a,b]$ and let $W_a^b(f)<+\infty$. Show that there exist two non-decreasing functions g and h such that

$$f(x)=\frac{g(x)}{h(x)}\qquad\text{for } x\in[a,b].$$

13 Let L be a convex curve. Take the origin of a system of polar coordinates (θ,ρ) within this curve. Let the curve L have equation $\rho=\rho(\theta)$ in this coordinate system. Show that the function $\rho(\theta)$ is of bounded variation.

14 Prove *Blaschke's theorem*: from any infinite set of convex curves which are uniformly bounded (i.e. located within a fixed circle), it is possible to select a sequence which converges to a convex curve.

15 Let f be a continuous function on the interval $[a, b]$. Let $a = x_0^{(n)} < \cdots < x_{k_n}^{(n)} = b$ be any sequence of subdivisions of the interval $[a, b]$ with the property that

$$\lim_{n \to \infty} \max_{1 \leqslant i \leqslant k_n} (x_i^{(n)} - x_{i-1}^{(n)}) = 0.$$

Show that

$$\lim_{n \to \infty} \sum_{i=1}^{k_n} |f(x_i^{(n)}) - f(x_{i-1}^{(n)})| = W_a^b f.$$

16 Show that if at the point $x = c$ one of the functions f, g is continuous while the other is bounded in some neighbourhood of this point, then the existence of the integrals

$$\int_a^c f(x)\,\mathrm{d}g(x), \quad \int_c^b f(x)\,\mathrm{d}g(x)$$

implies the existence of the integral

$$\int_a^b f(x)\,\mathrm{d}g(x).$$

17 The function f is continuous and g is of bounded variation in the interval $[a, b]$. Show that the integral

$$\int_a^x f(x)\,\mathrm{d}g(x)$$

is continuous at each point of continuity of the function g.

18 Let a curve have equations $\theta = \theta(t) > 0$, $\rho = \rho(t)$ $(a \leqslant t \leqslant b)$ in polar co-ordinates. Show that a sufficient condition for the curve to be rectifiable is that the functions $\theta(t)$, $\rho(t)$ are of bounded variation on $[a, b]$.

19 Let the function f be continuous and let its variation $W_0^\infty(f) = \lim_{x \to \infty} W_0^x(f)$ be finite on the interval $(0, +\infty)$. Let $h > 0$. We denote by f_h the function which is linear on each of the intervals $[nh, (n+1)h]$, $n = 0, 1, 2, \ldots$, and such that

$$f_h(nh) = f(nh).$$

Show that

$$\int_0^\infty |f_h(x) - f(x)|\,\mathrm{d}x < +\infty.$$

20 Let f be a function of bounded variation on the interval $(0, +\infty)$ such that $\lim_{x \to \infty} f(x) = 0$ and let g be a function which is integrable over any finite interval and such that for any $x \geqslant 0$

$$\left| \int_0^x g(x)\,\mathrm{d}x \right| \leqslant M.$$

Show that the limit

$$\lim_{x \to \infty} \int_0^x f(x)g(x)\,dx$$

exists. (This is the analogue of the well-known Abel theorem in the theory of infinite series.)

CHAPTER 2

21 Find the first five terms in a sequence of Tonelli polynomials converging to the functions

(a) $\cos \pi x$ in the interval $[\frac{1}{3}, \frac{2}{3}]$,
(b) $\sin x$ in the interval $[0, 3]$,
(c) $\exp(\frac{1}{4}(x + y))$ on the set $[0, 2] \times [-1, 1]$.

22 Find the sequence of Bernstein polynomials $B_n(x)$ for the functions:

(a) $f(x) = (x - \frac{1}{2})^{1/3}$ in the interval $[0, 1]$,
(b) $x\,e^{-x}$ in the interval $[0, 1]$.

Calculate n for which $|B_n(x) - f(x)| \leqslant 10^{-3}$. *Hint*: find and estimate $\omega(\delta)$.

23 Use Stone's theorem to show that the family of trigonometric polynomials

$$f(t) = \sum_{n=-k}^{k} a_n e^{int}, \quad t \in [0, 2\pi], \quad a_n\text{-complex},$$

approximate all continuous functions on the interval $[0, 2\pi]$ which are complex valued and such that $f(0) = f(2\pi)$.

CHAPTER 3

24 Choose a, b in such a way that the function

$$f(x) = \begin{cases} x \sin \dfrac{1}{x} & x > 0, \\ a & x = 0 \\ \dfrac{x^2 - 1}{x + 1} & x < 0, x \neq -1, \\ b & x = -1, \end{cases}$$

is continuous.

25 For which x is the function

$$f(x) = \begin{cases} e^x & x\text{-rational}, \\ e^{-x} & x\text{-irrational}, \end{cases}$$

continuous?

26 Find the set of points of continuity for the functions:

(a) $$f(x)=\begin{cases}\dfrac{xy}{x^2+y^2}, & x^2+y^2\neq 0,\\ 0, & x^2+y^2=0.\end{cases}$$

(b) $$f(x)=\begin{cases}\dfrac{\ln(e^x+y)}{x^2+y^2}, & x^2+y^2\neq 0,\\ 0, & x^2+y^2=0.\end{cases}$$

27 Determine whether the sequence of functions $f_n(x)=\cos^n x$ is uniformly convergent:

(a) in the interval $[-\pi,\pi]$,
(b) in the interval $[\alpha,\beta]$, where $0<\alpha<\beta<\pi$.

Find at which point of the interval $[-\pi,\pi]$ is the convergence continuous.
28 Show that the family of linear functions $f(x)=ax+b$, where $m\leqslant a\leqslant M$, is equicontinuous at every point.
29 Show that the family of polynomials $f(x)=a_0+a_1x+\cdots+a_nx^n$ of degree $\leqslant n$, whose coefficients satisfy the inequalities $m\leqslant a_i\leqslant M$ $(i=0,\ldots,n)$ for some constants m, M is equicontinuous on any bounded interval.
30 Show that the family of functions

$$f_n(x)=\int_0^1\frac{x-(s-\frac{1}{2})^n}{s+1}\,ds, \qquad n=0,1,\ldots$$

is equicontinuous on the interval $[0,1]$. (This is a special case of Exercise 40.)
31 Show that the function $f(x)=x[x]$ (where $[x]$ is defined in Exercise 3) is upper semicontinuous in the interval $(0,+\infty)$ and lower semicontinuous in the interval $(-\infty,0)$.
32 According to Baire's theorem the function $f(x)=[x]$ is the limit of a decreasing sequence of continuous functions. Find such a sequence.
33 For the function

$$f(x)=\begin{cases}\sin\dfrac{1}{x} & x\neq 0,\\ 0 & x=0,\end{cases}$$

find $M(0)$, $m(0)$, $\omega(0)$.
34 Using the theorems of §3.5, show that the function

$$f(x)=\begin{cases}x & x\in[0,1]\backslash C,\\ 0 & x\in C,\end{cases}$$

where C is the Cantor set, is a function of Baire's first class (see the example following Th. 7.4.9).
35 Find a sequence of continuous functions which converges to the function defined in Exercise 34.

36 There is given a family R of functions defined on a metric space X. Constants K and $0 < \alpha \leqslant 1$ exist such that for all $f \in R$ the inequality

$$|f(x') - f(x)| \leqslant K[\rho(x', x)]^{\alpha}$$

holds. Show that the family R is equicontinuous on X.

37 Show that the functions f which have continuous derivatives on the interval $[a, b]$ and satisfy the condition

$$\int_a^b |f'(x)|^2 \, dx \leqslant K$$

for some $K > 0$, form an equicontinuous family of functions on $[a, b]$.

38 Show that the set of all polynomials of degree at most n, which are uniformly bounded on $[a, b]$ form an equicontinuous family.

39 Let $f(x, y)$ be a function defined on the rectangle $a \leqslant x \leqslant b$, $c \leqslant y \leqslant d$ which is continuous relative to x for any fixed y and is continuous relative to y for any fixed x. Show that a necessary and sufficient condition for the function $f(x, y)$ to be continuous relative to both variables (x, y) is that the functions $x \to f(x, y)$, $c \leqslant y \leqslant d$, should form an equicontinuous family.

40 The function $K(x, y, z)$ is continuous for $0 \leqslant x \leqslant 1$, $0 \leqslant y \leqslant 1$ and for any z. The functions φ in a family R are continuous on $[0, 1]$ and are uniformly bounded. Show that the functions of the form

$$\psi(x) = \int_0^1 K(x, s, \varphi(s)) \, ds, \qquad \text{where } \varphi \in R,$$

form an equicontinuous family on the interval $[0, 1]$.

41 The functions φ in a family R are equicontinuous on $[a, b]$ and are uniformly bounded. The function h is continuous. Show that the functions $h \circ \varphi$ $(\varphi \in R)$ form an equicontinuous family on $[a, b]$.

42 There is given a family of continuous functions which are uniformly bounded on the interval $[a, b]$. Every function of this family is piecewise linear on $[a, b]$. For any function of this family, the length of any interval of linearity is not less than a fixed number $\delta > 0$. Show that the functions of this family are equicontinuous on $[a, b]$.

43 Suppose that the family $\mathscr{R}$ of functions is equicontinuous on a continuum E. Show that the family is compact if it is bounded at at least one point of the set E (more generally, if $\sup_{\mathscr{R}} \{\inf_E |f(x)|\} < \infty$).

44 Suppose that for a convergent sequence of k-times differentiable functions f_ν on an interval $[a, b]$ the sequence of kth-derivatives is equicontinuous on $[a, b]$. Show that each of the sequences

$$\{f_\nu(x)\}, \{f'_\nu(x)\}, \ldots, \{f_\nu^{(k)}(x)\}$$

is uniformly convergent on $[a, b]$.

45 Show that for a function f, defined on the interval (a, b), to be lower semi-continuous, it is necessary and sufficient that the planar set $\{(x, y) : y < f(x)\}$ be open.

46 Let G be an open set in the plane. Show that the function $f(x) = \sup_{(x,y)\in G} y$ is lower semicontinuous (on the projection of the set G on the x-axis).

CHAPTERS 4 AND 5

47 Prove Th. 4.1.7.
48 Show that the intersection of an arbitrary number of monotone classes is a monotone class.
49 Show that if $\mathscr{S}$ is a countably additive algebra in the Cartesian product $X_1 \times X_2$, then for any $x_1 \in X_1$ the family $\{(A)_{x_1} : A \in \mathscr{S}\}$ is a countably additive algebra in X_2.
50 Show that the function

$$\chi(x) = \begin{cases} 0 & x\text{-irrational}, \\ 1 & x\text{-rational}, \end{cases}$$

is measurable and find an increasing sequence of simple functions which converges to $\chi(x)$.
51 Show that the function $\chi(x)$ (Exercise 50) is of Baire's second class.
52 On the basis of the theorems of Chapter 3, show that the function $\chi(x)$ of Exercise 50 is not of Baire's first class.
53 Modify the construction of the Cantor set (as defined following Th. 7.4.9) in such a way that its measure is equal to a given number ε, where $0 < \varepsilon < 1$. Show that every such set is nowhere dense.
54 Let M be the set of rational numbers in the interval $[0, 1] = X$. Let the points of M be enumerated in some way in a sequence $M = \{x_1, x_2, x_3, \ldots\}$. For any $\varepsilon > 0$ and $i = 1, 2, 3, \ldots$ let $F_i(\varepsilon)$ denote the open interval of length $\varepsilon/2^i$ centred at the point x_i. Let

$$F(\varepsilon) = \bigcup_{i=1}^{\infty} F_i(\varepsilon), \quad F = \bigcap_{n=1}^{\infty} F\left(\frac{1}{n}\right).$$

Establish the following properties:

(a) there exists $\varepsilon > 0$ and $x \in X$ such that $x \notin F(\varepsilon)$,
(b) $F(\varepsilon)$ is an open set and $\mu(F(\varepsilon)) \leqslant \varepsilon$,
(c) the set $X \backslash F(\varepsilon)$ is nowhere dense,
(d) the set $X \backslash F$ is of the first category, and hence F is uncountable, since F is a complete metric space,
(e) the set F has zero measure.

55 Show that if A is a linear, measurable set of positive measure then there exist distinct points x and y in A whose distance from each other is a rational number.
56 Show that every linear set of positive measure contains a non-measurable subset.
57 In the interval $[0, 1]$ construct a measurable set E such that for every interval

$\Delta \subset [0,1]$ we have

$$\mu(\Delta \cap E) > 0, \quad \mu(\Delta \backslash E) > 0.$$

58 The set E is $\mathscr{L}$-measurable. Show that

$$E = \bigcup_{n=1}^{\infty} P_n \cup Q$$

where the P_n are perfect sets and the set Q is of $\mathscr{L}$-measure zero.
59 Show that a necessary and sufficient condition for a continuous function on $[a,b]$ to satisfy Luzin's (N)-condition on $[a,b]$ is that for every measurable set $E \subset [a,b]$ the set $f(E)$ is measurable.
60 The set E is dense in the interval $(-\infty, +\infty)$. Show that the function f is measurable if and only if for every $a \in E$ the set

$$\{x : f(x) \geqslant a\}$$

is measurable.
61 Show that a function of bounded variation on the interval $[a,b]$ in $(\mathscr{B})$-measurable.
62 The complex-valued function $f(t) = \varphi(t) + i\psi(t)$ of a real variable t is called measurable if φ and ψ are $\mathscr{L}$-measurable. Show that f is measurable if and only if for every open set M in the complex plane the set $f^{-1}(M)$ is $\mathscr{L}$-measurable.
63 If for any fixed n, $f_k^{(n)} \xrightarrow{\mu} f^{(n)}$ as $k \to \infty$ and $f^{(n)} \xrightarrow{\mu} f$ as $n \to \infty$, then one can select a sequence from the set $\{f_k^{(n)}\}$ which converges to f in measure.
64 If, on a set E of finite, positive measure, the sequence $\{f_n\}$ of almost everywhere finite functions satisfies the condition

$$\lim_{n,m \to \infty} |f_n(x) - f_m(x)| = 0$$

almost everywhere, then there exists a positive constant c and a measurable set $F \subset E$ of positive measure, such that $|f_n(x)| \leqslant c$ for all $x \in F$ and all n.
65 For functions which are finite almost everywhere on the interval $[a,b]$ we introduce the definition:

$$\textit{essential maximum of } f(x) = \inf[a : m(\{x : f(x) \geqslant a\}) = 0].$$

Show that for a sequence $\{f_n\}$ of essentially bounded functions (i.e. such that ess.max.$|f_n(x)| < +\infty$) to be uniformly convergent almost everywhere on $[a,b]$ to a function f it is necessary and sufficient that

$$\lim_{n \to \infty} \text{ess.max.} |f_n(x) - f(x)| = 0.$$

66 Let $\mathscr{X}$ denote the set of all measurable functions on $[0,1]$. For $x, y \in \mathscr{X}$ let

$$\rho(x,y) = \inf\{\varepsilon : m(\{t : |x(t) - y(t)| > \varepsilon\}) \leqslant \varepsilon\}.$$

Show that the function ρ defined in this way is a metric and that convergence relative to this metric is equivalent to convergence in measure. Show that the metric space defined above is complete.

67 The uniformly bounded sequence of functions f_n converges in measure to f. The function g is continuous. Show that the sequence $g \circ f_n$ converges in measure to $g \circ f$.
68 The sequence f_n converges in measure to f. The function g satisfies a Lipschitz condition. Show that the sequence $g \circ f_n$ converges in measure to $g \circ f$.
69 The sequence f_n converges uniformly on $[a, b]$ to the function f. Each of the functions f_n is continuous almost everywhere on $[a, b]$. Show that the limit function f is also continuous almost everywhere on $[a, b]$.
70 The set E has finite measure. The sequences of functions f_n, g_n converge in measure respectively to f and g. Establish the following properties.

(a) if α and β are real numbers, then the sequence $\alpha f_n + \beta g_n$ converges in measure to $\alpha f + \beta g$,

(b) if $f(x) = 0$ a.e., then $f_n^2(x) \xrightarrow{\mu} 0$,

(c) $f_n g \xrightarrow{\mu} fg$,

(d) $f_n^2 \xrightarrow{\mu} f^2$,

(e) $f_n g_n \xrightarrow{\mu} fg$.

71 On the algebra of all subsets of the set N of all positive integers we introduce a measure: $\mu(E) =$ the number of elements in the set E. Show that for μ-measurable functions convergence in measure is equivalent to uniform convergence.

CHAPTER 6

72 What is the value of $\int s(x) \, d\mu(x)$, if s is a simple function?
73 Show directly from definition (1.1.1) that $\int_0^1 x \, dx = \frac{1}{2}$.
74 Check the summability of the functions

(a) $f(x) = \dfrac{1}{x}$,

(b) $f(x) = \begin{cases} \sin(1/x) & x \neq 0, \\ 0 & x = 0, \end{cases}$

relative to Lebesgue measure in the interval $[0, 1]$.
75 Use Fubini's theorem (Th. 6.3.4) to evaluate the integrals

(a) $$\iint_\Omega (x + 2y) \, dx \, dy,$$

where Ω is the area bounded by the curves $y = -x^2$, $y = x - 2$.

(b) $$\iint_\Omega 2xy \, dx \, dy$$

where Ω is the triangle with vertices (0, 0), (0, − 1), (− 1, 2).

(c) $\iint_{\Omega} (2x + y^2)\,dx\,dy$

where Ω is the parallelogram with vertices (− 1, 0), (0, 1), (1, 0), (2, 1).

76 Use Fubini's theorem to evaluate the integral

$$\iiint_V x^3 y z^2\,dx\,dy\,dz,$$

where V is the volume bounded by the surfaces with equations $x = yz$, $z = y$, $y = 1$, $x = 0$.

77 Using Fubini's theorem, show that the volume of the solid figure obtained by rotating the curve with equation $y = f(x)$, $a \leqslant x \leqslant b$ about the axis $0x$ is given by

$$\pi \int_a^b [f(x)]^2\,dx.$$

78 Define a measure on the set of subsets of N (the positive integrals) as in Exercise 71. The real function f defined on N is summable on N if and only if the series

$$\sum_{n=1}^{\infty} f(n)$$

is absolutely convergent.

79 Show that if the function f is summable on E and if $E_n = \{x : |f(x)| \geqslant n\}$, then $n.\ \mu(E_n) \to 0$.

80 The function f equals zero on Cantor's Ternary Set. On each adjacent interval of length 3^{-p} we define $f(x) = p$. Evaluate the integral

$$\int_0^1 f(x)\,dx.$$

81 Show that if $f_n(x) \geqslant 0$ and $\int_E f_n(x)\,dx \to 0$, then f_n converges to 0 in measure, but not necessarily almost everywhere (give an example).

82 Show that the condition

$$\int_E \frac{|f_n|}{1 + |f_n|}\,dx \to 0$$

is equivalent to the convergence of f_n to zero in measure.

83 Let $\{E_n\}$ be a sequence of measurable sets and let m be any natural number. Show that if G denotes the set of all those points which belong to at least m sets in the sequence $\{E_n\}$, then G is a measurable set and

$$m \cdot \mu(G) \leqslant \sum_{n=1}^{\infty} \mu(E_n).$$

84 On the subsets of the natural numbers N define a measure as in Exercise 71.
(a) On N define a sequence of functions

$$f_n(k) = \begin{cases} 1/n & \text{for } 1 \leqslant k \leqslant n, \\ 0 & \text{for } k > n. \end{cases}$$

Investigate the convergence of the sequence f_n and the sequence of integrals $\int f_n \, d\mu$.
(b) On N define a sequence of functions

$$f_n(k) = \begin{cases} 1/k & \text{for } 1 \leqslant k \leqslant n, \\ 0 & \text{for } k > n. \end{cases}$$

Show that the sequence f_n is uniformly convergent. Investigate the convergence of the sequence

$$\int f_n \, d\mu.$$

85 A function f measurable on a set E is called sth-*power summable*, if the function $x \to |f(x)|^s$ is summable on E. Let $\mu(E) < \infty$. Show that if $0 < p < q$ and f is qth-power summable then it is pth-power summable.
86 Let f_n be a sequence of quadratically summable functions on $[0, 1]$ which converges in Lebesgue measure to the function $F(x)$. Assume that

$$\int_0^1 |f_n(x)|^2 \, dx \leqslant K.$$

Show that for any quadratically summable function f we have

$$\int_0^1 f_n(x) f(x) \, dx \to \int_0^1 F(x) f(x) \, dx.$$

87 Show that the sum of two pth-power summable functions is pth-power summable ($p > 0$).
88 The function f is summable on E. The function h is continuous and bounded on $(-\infty, +\infty)$. Show that the integral

$$\int_E h[f(x)] \, d\mu$$

exists and is finite if $\mu(E) < \infty$.
89 The function $x \to \exp|f(x)|$ is summable on $[0, 1]$. Is f also summable on $[0, 1]$?
90 The function w is continuous on $[-M, M]$. The sequence of functions f_n, uniformly bounded by M, converges in measure to f. Show that

$$\int_0^1 w(f_n(x)) \, dx \to \int_0^1 w(f(x)) \, dx.$$

91 The functions f_n, measurable on $[0, 1]$, converge to f in measure. Show that

$$\int_0^1 \sin(f_n(x))\,dx \to \int_0^1 \sin(f(x))\,dx.$$

92 The function g satisfies a Lipschitz condition. The sequence f_n of $(\mathscr{L})$-measurable functions on $[a, b]$ converges to f in measure. The function g is summable on $[a, b]$ and $|f_n(x)| \leqslant g(x)$. Show that

$$\int_a^b g(f_n(x))\,dx \to \int_a^b g(f(x))\,dx.$$

93 Show that for any $(\mathscr{L})$-measurable function f on $[a, b]$:

$$\lim_{p\to\infty} \left\{ \int_a^b |f(x)|^p\,dx \right\}^{1/p} = \text{ess.max.}\,|f(x)|.$$

94 Suppose that $\sum_{n=1}^{\infty} \alpha_n^{1-\delta} < +\infty$, $\alpha_n \geqslant 0$, $\delta > 0$. Let $\{x_n\}$ be any sequence of points in the interval $[0, 1]$. Show that the series

$$\sum_{n=1}^{\infty} \frac{\alpha_n}{(x - x_n)}$$

is absolutely convergent almost everywhere in $[0, 1]$.
95 Give an example which shows that the statement in Exercise 94 is no longer true under the hypothesis

$$\sum_{n=1}^{\infty} \alpha_n < \infty.$$

96 Let f be a summable function on $[0, 1]$. Let φ denote the characteristic function of the union of intervals $\bigcup_{n=0}^{\infty} [2n, 2n+1]$. Show that

$$\lim_{k\to\infty} \int_0^1 f(x)\varphi(kx)\,dx = \tfrac{1}{2}\int_0^1 f(x)\,dx.$$

State an analogous theorem in which the function φ is continuous and periodic.

CHAPTER 7

97 Investigate whether the functions in Exercise 1 are absolutely continuous or not.
98 Determine whether the following functions are uniformly or absolutely continuous:

(a) $\cos x$ on the interval $(-\infty, +\infty)$,

(b) $\dfrac{1}{x}\cos(x^2)$ on the interval $[1, +\infty)$.

99 Decompose the following functions into a sum of an absolutely continuous function and a singular function:

(a) $f(x)=[x]$,
(b) $f(x)=x[x]$,
(c) $f(x)=x\sin\left(\frac{\pi}{2}[x]\right)$

(see Exercise 3 for the definition of $[x]$).

100 Find the interval functions associated with the functions defined in Exercise 98. Find their variation and their upper and lower variations on finite intervals.

101 On the set $[0,1]$ define an interval function in the following way:

$$F\left(\left\{\frac{m}{2^n}\right\}\right)=\frac{1}{2^{2n-1}} \text{ for any irreducible fraction } m/2^n,$$

$$F\left([0,1]\setminus\left\{\frac{m}{2^n}; n,m=1,2,\ldots\right\}\right)=0.$$

Show that (a) F is singular, (b) F is a measure.
Calculate $\bar{D}F(x)$, $\underline{D}F(x)$ if

(a) $x\in\{m/2^n; n,m=1,2,\ldots\}$,
(b) $x\notin\{m/2^n; n,m=1,2,\ldots\}$.

102 Construct on $[0,1]^3$ an analogous interval function to that constructed in Exercise 101.

103 Calculate the Dini derivatives, at $x_0=0$, of the functions:

(a) $f(x)=\begin{cases} x\sin\dfrac{1}{x} & x\neq 0,\\ 0 & x=0;\end{cases}$

(b) $f(x)=\sin x$;
(c) $f(x)=|x|$.

104 Calculate the Dini derivatives of the singular function defined using the Cantor set C, following Th. 7.4.9:

(a) for $x\in C$,
(b) for $x\notin C$.

105 Calculate the derivatives of the interval functions

(a) $F(\Delta)=\iint_\Delta \sqrt{(|xy|)}\,\mathrm{d}x\,\mathrm{d}y$ at the points $(0,0)$, $(-1,1)$;

(b) $F(\Delta)=\iint_\Delta (x^2+y^2)\sin\dfrac{1}{x^2+y^2}\,\mathrm{d}x\,\mathrm{d}y$ at $(0,0)$.

106 Find $\bar{D}F(x_0,-2x_0)$ and $\underline{D}F(x_0,-2x_0)$ for the interval function F given by

$$F(\Delta)=\iint_\Delta f(x,y)\,\mathrm{d}x\,\mathrm{d}y,$$

where

$$f(x,y)=\begin{cases}\dfrac{\sin(2x+y)}{|2x+y|} & y\neq -2x,\\ 0 & y=-2x.\end{cases}$$

107 The functions f and g are continuous on $[a,b]$ and g is strictly increasing in $[a,b]$. Suppose that the limit

$$\lim_{h\to 0}\frac{f(x+h)-f(x)}{g(x+h)-g(x)}=\frac{\mathrm{d}f(x)}{\mathrm{d}g(x)}$$

exists except for at most a countable number of points x in the interval $[a,b]$ and suppose that

$$\left|\frac{\mathrm{d}f(x)}{\mathrm{d}g(x)}\right|\leqslant M.$$

Show that $|f(x+h)-f(x)|\leqslant M|g(x+h)-g(x)|$.

108 The function f is continuous on $[a,b]$. Let $a<b$ and $f(a)<f(b)$. Let E denote the set of all $x\in[a,b]$ for which $D_+f(x)>0$. Show that the Lebesgue outer measure of the set $f(E)$ is greater than zero (T. Ważewski).

109 Show that a continuous and strictly increasing function which satisfies Luzin's condition is absolutely continuous.

110 Show that the function $\rho(\theta)$ occurring in Exercise 13 is absolutely continuous.

111 Show that for any function f summable on $[0,1]$ and for any regular (i.e. such that $\lim_{n\to\infty}\{\max_i|a_i^{(n)}-a_{i-1}^{(n)}|\}=0$) sequence of subdivisions $0=a_0^{(n)}\leqslant\cdots\leqslant a_{k_n}^{(n)}=1$ of the interval $[0,1]$

$$\lim_{n\to\infty}\sum_{i=1}^{k_n}\left|\int_{a_{i-1}^{(n)}}^{a_i^{(n)}}f(x)\,\mathrm{d}x\right|=\int_0^1|f(x)|\,\mathrm{d}x.$$

Show that for an absolutely continuous function f the variation W_a^bf is equal to the integral $\int_a^b|f'(x)|\,\mathrm{d}x$.

112 Show directly that for an absolutely continuous function f the equality

$$[W_a^xf]'=|f'(x)|$$

holds almost everywhere.

113 Let f be a continuous function on the interval $[a,b]$. Show that

$$\lim_{h\to 0}\int_a^{b-h}\left|\frac{f(x+h)-f(x)}{h}\right|\mathrm{d}x=W_a^bf\qquad(h>0).$$

114 Show that for an absolutely continuous function f on $[a,b]$

$$\lim_{h\to 0}\int_a^{b-h}\left|\frac{f(x+h)-f(x)}{h}-f'(x)\right|\mathrm{d}x=0\qquad(h>0).$$

115 Show that for a function f which is of bounded variation and continuous on $[a, b]$

$$\lim_{h \to 0} \int_a^{b-h} \left| \frac{f(x+h) - f(x)}{h} - f'(x) \right| \mathrm{d}x = W_a^b g,$$

where

$$g(x) = f(x) - \int_a^x f'(x) \, \mathrm{d}x.$$

BIBLIOGRAPHY

English language translations of books in the original bibliography are given if available.

Aumann, G., *Reele Funktionen*, Springer, Berlin, 1954.
Bourbaki, N., *Les structures fondamentales de l'analyse*, Paris 1940–1970.
Federer, H., *Geometric Measure Theory*, Springer, Berlin, 1969.
Hahn, H. and Rosenthal, A., *Set Functions*, Albuquerque, 1948.
Halmos, P., *Measure Theory*, Van Nostrand, New York, 1950.
Hartman, S. and Mikusiński, J., *The Theory of Lebesgue Measure and Integration*, Pergamon, Oxford, 1961.
Haupt, O. and Aumann, G., *Differential- und Integralrechnung*, I, II, III, Berlin, 1938.
Kuratowski, K, *Topology*, I, II, Academic Press, New York, 1966.
Maurin, K., *Analysis*, I, II (Polish), Warsaw, 1971.
Natanson, I. P., *Theory of Functions of a Real Variable* (Russian) Moscow, 1950.
Riesz, F. and Nagy, B.Sz., *Functional Analysis*, Ungar, New York, 1955.
Rudin, W., *Real and Complex Analysis*, New York, 1966.
Saks, S., *Theory of the Integral*, Warsaw, 1937.
Segal, I. E. and Kunze, R. A., *Integrals and Operators*, McGraw-Hill, New York, 1968.
Sierpiński, W., *Analytically Representable Functions* (Polish), Warsaw, 1925.
Sikorski, R., *Real Functions*, I, II, (Polish), Warsaw, 1958–9.
de la Vallée Poussin, Ch. J., *Intégrales de Lebesgue. Fonctions d'ensemble. Classes de Baire*, Paris, 1916.

Some additional recent English language publications:

de Barra, G., *Introduction to Measure Theory*, Van Nostrand, New York, 1955.
Cohn, D. L., *Measure Theory*, Birkhäuser, Boston, 1980.
Folland, G. B., *Real Analysis*, Wiley, New York, 1984.
Lang, S., *Real Analysis* (2nd edn), Addison-Wesley, 1983.
Smith, K. T., *Primer of Modern Analysis*, Springer, New York, 1983.

SUBJECT INDEX

Absolute continuity of integral, 126
Absolutely continuous
  function, 163, 192
  interval function, 160
Absolutely monotone function, 35
Additive interval function, 93
Adjoining intervals, 93
Algebra
  cartesian product of, 68
  completion of, 84
  countably additive, 65
  finitely additive, 66
  generated by sets, 67
  of Borel sets, 69
  restriction of, 67
Almost everywhere, 83
Almost uniform convergence, 112
Approximately continuous function, 148
Approximation
  uniform, 39
Arc-length, 17, 170
Ascoli's Theorem, 50

Baire
  function, 78
  Theorem, 54, 61
Banach–Vitali Theorem, 123
Bernstein
  polynomial, 33, 38
  Theorem, 35
Blaschke's Theorem, 212
Bolzano's Theorem, 46
Borel sets, 69
Bounded variation, 14
  of interval function, 156

Canonical Lebesgue decomposition, 162, 167
Cantor ternary set, 168
Caratheodory's condition, 86, 91
Cartesian product
  of algebras, 68
  of measures, 133
Change
  of integration function, 28
  of measure, 196
Characteristic function of a set, 76
Cluster value, 4
Complete measure, 83
Completion of
  a measure, 84
  an algebra, 84
Content of an interval, 93
Continuous
  convergence, 48
  function, 45
Convergence
  almost uniform, 112
  continuous, 48
  uniform, 47
Countably additive
  algebra, 65
  function, 189
  ideal, 91
  set function, 178
Covering
  Vitali, 145
Curve
  rectifiable, 18

Decomposition
  canonical Jordan, 18, 158, 180
  canonical Lebesgue, 162, 167, 196
  Hahn, 179
Decreasing function, 10
Denjoy–Young–Saks Theorem, 173, 210

Dense set of intervals, 188
Density point of a set, 147
Derivative
  Dini, 172
  of interval function, 148
  Radon–Nikodym, 195
  s-, 176
Diagonal principle of choice, 8
Differential, 105
Dini
  derivatives, 172
  Theorem, 47

Egorov's Theorem, 112
Envelope
  lower, 49
  upper, 49
Equicontinuous family, 49
Essential maximum, 218

Family
  equicontinuous, 49
  uniformly bounded, 49
Fatou's Lemma, 127
Finitely additive
  algebra, 66
  class of sets, 71
Finitely multiplicative
  class of sets, 71
Fréchet's Theorem, 110
Fubini's Theorem, 134, 135, 137, 138, 205, 206
Function
  absolutely continuous, 163, 192
  absolutely monotone, 35
  approximately continuous, 148
  Baire, 78
  characteristic, 76
  continuous, 45
  countably additive, 189
  decreasing, 10
  of Baire's first class, 57, 58
  increasing, 10
  Lebesgue measurable, 109
  lower-semicontinuous, 51
  measurable, 74
  measurable relative to an algebra, 73
  monotone, 10
  mutually singular, 196
  pointwise discontinuous, 60
  Riemann–Stieltjes integrable, 20
  ($\mathscr{S}$)-measurable, 73
  sth-power summable, 221
  saltus, 11, 19
  simple, 76
  singular, 162, 167, 169
  summable, 124, 181, 190
  upper-semicontinuous, 51
  μ-measurable, 108
  σ-finite, 193

Hahn
  decomposition, 179
  Theorem, 179
Hausdorff α-dimensional measure, 89
Helly
  first theorem, 12, 20
  second theorem, 29
Hyperplane
  of continuity, 188
  of discontinuity, 188

Ideal
  countably additive, 91
Increasing function, 10
Inner measure, 101
Integral
  Lebesgue, 117, 124
  lower Darboux, 130
  relative to a
    countably additive function, 181
    measure, 117, 124
  Riemann–Stieltjes, 20, 157
  upper Darboux, 130
Integration
  by parts, 28, 165, 191
  by substitution, 166
Interval(s)
  adjoining, 93
  content, 93
  dense set of, 188
  of continuity, 187, 189
  n-dimensional, 92
  non-overlapping, 92
  normal system of, 94
Interval function, 149
  absolutely continuous, 160
  additive, 93
  bounded variation, 156
  derivative of, 148
  lower derivative of, 148
  variation of, 156
Invariant measure, 104

Jordan
  canonical decomposition, 18, 158, 180

Lebesgue
  canonical decomposition, 196
  integral, 117, 124
  lower sum, 122
  measurable function, 109
  measurable set, 98
  measure, 98
  point of a function, 151
  upper sum, 122
Lebesgue–Hausdorff Theorem, 79
Lebesgue–Stieltjes integral, 190
Lebesgue's Theorem, 112, 120, 127, 128, 150, 159, 175
  on density points, 147
  on differentiation of integrals, 152, 204
Lipschitz condition, 14, 16, 49
Lower
  Darboux integral, 130
  derivative of interval function, 148
  envelope, 49
  Lebesgue sum, 122, 130
  variation, 158, 181
Lower-semicontinuous function, 51
Luzin
  condition, 103
  Theorem, 109

Maximum at a point, 55
Mean-value theorem, 27, 117, 125
Measure, 81
  cartesian product, 133
  change of, 196
  complete, 83
  completion of, 84
  Hausdorff α-dimensional, 89
  inner, 101
  invariant, 104
  Lebesgue, 98
  metric outer, 88
  outer, 85, 91
  outer Lebesgue, 96
  regular, 142
Measurable function
  on a set, 74
  relative to an algebra, 73
Metric outer measure, 88
Minimum at a point, 55
Modulus of continuity, 33, 177
Monotone
  class of sets, 67
  function, 10
μ-convergent sequence, 113
μ-equivalent sets, 83
μ-measurable
  function, 108
  set, 81
μ^*-measurable set, 86
Mutually singular functions, 196

n-cube, 115
n-dimensional interval, 92
Negative part of a function, 76
Non-overlapping intervals, 92
Normal
  subdivision, 93
  system of intervals, 94

Oscillation
  at a point, 56
  of a function on a set, 4
Outer
  Lebesgue measure, 96
  measure, 85, 91

Pointwise
  discontinuous function, 60
Polynomial
  Bernstein, 33, 38
  Tonelli, 31, 32
Positive part
  of a function, 76
pth-order difference, 35

Rademacher's Theorem, 152
Radon–Nikodym
  derivative, 195
  Theorem, 193
Rectifiable curve, 18
Regular measure, 142
Restriction
  of an algebra of sets, 67
Riemann–Stieltjes
  integrable function, 20
  integral, 20, 157
Riesz
  Lemma, 168
  Theorem, 114, 139, 183, 185

s-derivative, 176
($\mathscr{S}$)-measurable function, 73
sth-power summable function, 221
Saltus function, 11, 19
Scattered family of cubes, 145
Section of a set, 69
Separation of points, 42
Sequence
  μ-convergent, 113

Set(s)
Borel, 69
Cantor ternary, 168
characteristic function of, 76
density point of, 147
finitely additive class of, 71
finitely multiplicative class of, 71
Lebesgue measurable, 98
monotone class of, 67
μ-equivalent, 83
μ-measurable, 81
μ^*-measurable, 86
section of, 69
Set function
countably additive, 178
Sierpiński–Young Theorem, 173, 210
σ-compact space, 103
σ-finite
function, 193
space, 132
Simple function, 76
Singular function, 162, 167
strictly increasing, 169
Space
σ-compact, 103
σ-finite, 132
Stepanov's Theorem, 156, 208
Stone's Theorem, 42
Stone–Weierstrass Theorem, 42
Sub-division
normal, 93
sub-, 21
Sum
Lebesgue lower, 122, 130
Lebesgue upper, 122, 130
Summable function, 124, 181
relative to an interval function, 190

Tonelli
polynomial, 31, 32
Theorem, 171
Total variation
of a set function, 181

Uniform
approximation, 39
convergence, 47
Uniformly bounded family, 49
Upper
Darboux integral, 130
derivative of interval function, 148
envelope, 49
Lebesgue sum, 130.
Upper-semicontinuous function, 51
Upper variation
of a set function, 181

Variation, 14
bounded, 14
lower, 158, 181
of a function, 14
of an interval function, 156
total, 181
upper, 158, 181
Vitali
covering, 145
covering theorem, 145
Theorem, 110, 128

Weierstrass' Theorem, 31, 47, 53

Zygmund's Lemma, 173

QA 331 .L7913 1988 AAX-2452
ojasiewicz, Stanis aw.
An introduction to the
theory of real functions